AF615919

Desulfurization of Iron and Steel and Sulfide Shape Control

William G. Wilson

and

Alex McLean

This monograph was written as part of the technical development activities of Niobium Products Company Limited, a wholly owned Subsidiary of Companhia Brasileira de Metalurgia e Mineracao, Araxa, MG, Brasil. William G. Wilson is Manager of High Strength Steel Development at NPC and Dr. Alex McLean is Professor of Metallurgy and Materials Science at the University of Toronto, Canada.

P. O. Box 411
Warrendale, PA 15086

Printed in U.S.A.
Library of Congress Card Cat. No. 80-69613
ISBN No. 0-89520-151-8

TABLE OF CONTENTS

ABSTRACT

DESULFURIZATION OF IRON AND STEEL

AND SULFIDE SHAPE CONTROL

There is an increasing need for steels with sufficient ductility that they can be used when extracting energy from hostile environments such as the Arctic and from off-shore wells. Reduction of the sulfur content of these steels, either in the iron prior to its being charged into the steel-making furnaces, or in the steel itself is most helpful in achieving the necessary ductility. Additional improvements in ductility can be achieved by globularizing the remaining inclusions.

Even though there may not be a requirement for steels for hostile environments, the economics of blast furnace operation favor their producing high sulfur irons which will require desulfurization to meet ordinary sulfur specifications.

The objective of this manuscript is to provide a single reference that describes the underlying fundamentals of desulfurization, and uses these fundamentals to explain the methods of desulfurization presently used in both iron and steel.

In addition, the methods for globularizing the remaining sulfides are described, and the necessity of preventing reoxidation of these elements used for globularization during teeming is shown.

DESULFURIZATION OF IRON AND STEEL

AND SULFIDE SHAPE CONTROL

CHAPTER 1

INTRODUCTION

As a result of the world-wide energy shortage, production of oil and natural gas from locations with hostile environments is now quite common. Typical examples of these hostile environments are the production of oil from deep water in the open sea, and natural gas from arctic regions. The steels necessary for production of oil and gas from these new sources have to be stronger and tougher. Since 1970, there has been an increasing demand for steels with lower sulfur specifications, even for applications which are less demanding than those encountered when extracting energy from hostile environments.[1] The scope of these new demands for low sulfur steels in Europe is shown in Figure 1. These new steels, with carbon contents less than 0.10 pct., rely on the precipitation during controlled rolling of the carbides, nitrides and carbo-nitrides of micro-alloying additions such as columbium, vanadium and titanium, in order to increase the strength and refine the grain size. These new micro-alloyed, high-strength steels are characterized by their ability to withstand multi-axial stresses applied at high loading rates. This property will be referred to throughout the remainder of this monograph as either "ductility" or "toughness". Sulfur removal, as well as globularization of the remaining sulfides, is necessary to obtain maximum toughness, particularly in the transverse direction. The mechanisms for providing a strong ductile matrix for these steels have been presented in detail elsewhere.[2]

Throughout this text, the measure of toughness that will be used whenever possible to illustrate these improvements will be Charpy V Notch (CVN) energy values on specimens whose orientation is transverse to the direction of rolling. Improvements in transverse CVN values are ordinarily associated with improvements in ductility in the through-thickness direction,

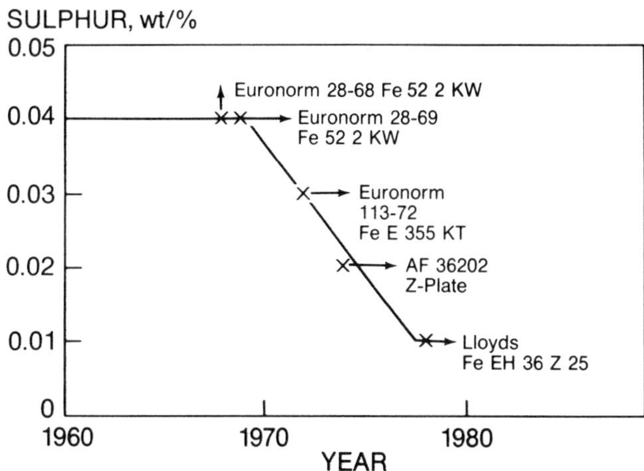

FIGURE 1 Sulfur specification for extra toughness requirements in steels made in Europe

(z direction), and improvement in other values determined with other methods of ductility measurement such as the crack opening displacement test (COD). To document these improvements in ductility in all test directions with all testing procedures is outside the scope of this document.

Curves showing CVN energy values at various temperatures, Figure 2, can be used to illustrate the benefits to be obtained from desulfurization and inclusion shape control. The impact energy-temperature curve for 0.11 pct. carbon steel has a shape typical of the low carbon micro-alloyed steels used for arctic linepipe.[3] The vertical portion of the curve represents the transition from ductile to brittle fracture and is commonly referred to as the transition temperature. The temperature at which this transition takes place at any

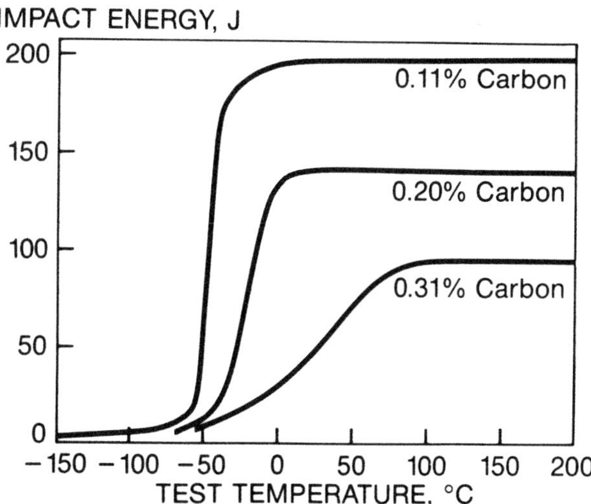

FIGURE 2 The effect of carbon content on impact transition temperature curves

particular carbon level, is controlled primarily by the grain size of the steel. Lower transition temperatures are obtained by controlled rolling in such a way as to achieve the finest possible ferrite grain size. The cleanness of the steel, (number and shape of inclusions), has little or no effect on the transition temperature. The upper horizontal line on the curve at an impact energy of 200 Joules, represents the upper shelf energy. At any carbon content, this value is mainly controlled by the number and shape of the inclusions in the steel. Since the highest stresses in linepipe are in the circumferential direction, the most likely mode of failure is a longitudinal split. It has been determined that in order to stop one of these longitudinal splits, high shelf energy is necessary. The failure of linepipe by this mode is discussed in more detail in appended paper #1.[4] There are many examples of the importance of CVN shelf energy, but the policy of this monograph will be to use only one example to illustrate each of the ideas presented in this report.

In an integrated steel plant where facilities for desulfurization of the iron have been installed to help meet these requirements for low sulfur steels, the increased efficiency of the blast furnace operation may more than compensate for the cost of desulfurization. The economic advantages of running a blast furnace with a low lime charge, to produce high sulfur hot metal, are substantial. Dofasco has achieved a 13% increase in hot metal production by reducing the stone rate from 8.15% to 5.15%. At the same time, coke usage declined by 9%, corresponding to 76 pounds per ton (38Kg/mt) of hot metal.

It has been estimated that one of the major methods for reducing energy requirements in steel plant operations is this coke reduction in the blast furnace. The details of Dofasco's modified blast furnace practice and desulfurization process are described in appended paper #2.[5] There can be no doubt that an increasing number of blast furnace operations will be conducted in this manner in the future.

The production of steels for service in hostile environments, with the lowest possible sulfur content and complete sulfide shape control of the remaining sulfides will involve significant increases in cost due to amortization of capital equipment and the use of expensive raw materials. It is unlikely that these costs will be fully compensated by improvements in blast furnace operation alone. However, it should be noted that the surface obtained on low sulfur steels is superior to that obtained on high sulfur steels, so that less surface preparation is necessary on billets, blooms and slabs prior to rolling to finished size. The magnitude of this improvement is shown in Figure 3.[6] Reduction in surface preparation by these amounts can further compensate for much of the cost of desulfurization. It is possible that steels for less demanding applications can be produced with higher levels of sulfur and less than perfect sulfide shape control. In this case, the steelmaker has a wider range of lower cost options available for desulfurization and shape control treatment.

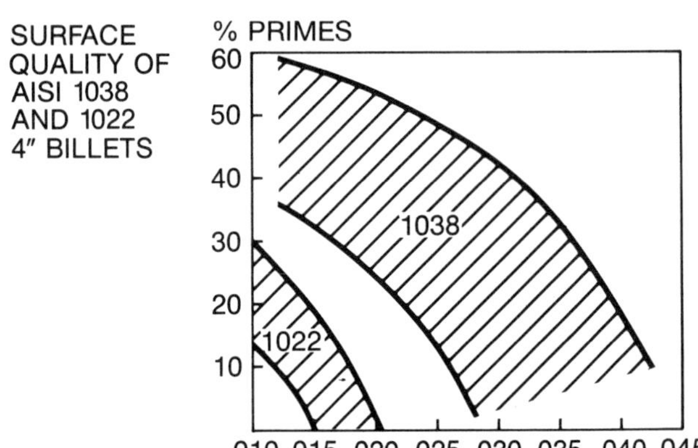

FIGURE 3 The effect of sulfur on surface quality

The subject of desulfurization is complex. There is no simple solution to the problem which applies to every steel plant. This monograph will present the fundamental principles of desulfurization which are applicable to all methods for sulfur removal in both iron and steel, and use these principles to explain the results that are obtained by those methods which are in most common use. Where possible, a comparison will be made of the reactants required to achieve a common level of desulfurization with each method. In addition, the benefits of globularization of the remaining sulfides will be described and the methods for obtaining this globularization will be presented.

It is not within the scope of this monograph to attempt to recommend any specific combination of desulfurizing processes or method for sulfide shape control to the user. Rather, that decision must be based on the individual plant requirements for steels requiring lower sulfur contents and improved ductility. Furthermore, each steel plant must consider the route which they choose to follow, not only on the basis of the results they require, but also on the basis of availability of capital, space to install facilities, and availability of raw materials.

CHAPTER 2

FUNDAMENTAL CONSIDERATIONS

2.1 General Aspects

With a very few exceptions, sulfur is considered undesirable in steel, and there is an increasing demand for steels with lower sulfur levels. As mentioned in Chapter 1, problems associated with sulfur are due mainly to the harmful effect of sulfide inclusions. There are three main methods by which sulfur may be removed from blast furnace iron or molten steel:

1) By reaction with metallic additions such as magnesium or rare earth elements, which in combination with sulfur form very stable sulfides.

2) By reaction with compound additives such as calcium-carbide, CaC_2 or soda-ash, Na_2CO_3.

3) By reaction with fluid, highly basic slags of low iron oxide content. The significance of these characteristic slag properties can be illustrated by consideration of the following equation which represents the removal of sulfur from the metal phase by reaction with lime in the slag phase, to form calcium sulfide in the slag and oxygen in the metal:

$$(CaO) + \underline{S} = (CaS) + \underline{O}$$

This reaction shows how desulfurization of the metal will be favored both by slags containing high concentrations of lime, and metal of a low oxygen content.

For consistent results and highest desulfurization efficiency, certain criteria must be fulfilled:

1) The desulfurizing reagent must be intimately mixed with the sulfur-containing metal. This implies the use of powdered reagents, special methods of addition, and the possible utilization of an appropriate stirring technique.

2) The oxygen potential of the slag/metal system should be as low as possible. This is readily accomplished in the case of blast furnace hot metal which is essentially saturated with respect to carbon. For sulfur removal from molten steel, this generally implies the use of strong deoxidizers, elimination of high FeO slag from the system, and it may necessitate the use of ladles lined with higher stability oxides containing a minimum of silica.

3) Opportunity must be given for removal of the reaction products from the system. For example, when hot metal from the blast furnace is externally desulfurized, it is essential that only desulfurized metal be transferred to the steelmaking vessel, with a minimum carry-over of sulfur-containing slag. Otherwise, sulfur reversion to the metal will occur under the oxidizing conditions generated during steelmaking. This particular aspect also underlines the importance of the use of fluid basic slags which have a high affinity for sulfur and will act as sinks or reservoirs for the desulfurization reaction products.

In light of the above comments, the following factors are noteworthy, in that they have exerted a major influence on the development of a desulfurizing philosophy during the last half of the seventies which is particularly appropriate for world conditions in the 80's, from the standpoints of materials supply, environmental constraints, and energy considerations.

1) While the reducing conditions which exist in the blast furnace are favorable for desulfurization of iron, there are several inherent disadvantages of this system:

   a) The temperatures within the slag/metal system are relatively low. This limits the amount of lime which can be taken into solution, and thus restricts the degree of basicity which can be obtained within the slag phase, as reflected by the CaO/SiO_2 ratio.

   b) Relatively little mixing is achieved between the slag and metal phase. For this reason, hot metal from the blast furnace generally contains sulfur in excess of that which would be in equilibrium with the slag phase.

2) Sulfur removal from hot metal outside of the blast furnace is favored both by the reducing conditions inherent in the treatment of carbon saturated iron, and also by the possibility of vigorously mixing the various phases together,

so that the time required for effective desulfurization is minimized.

3) Although high basicity fluid slags which favor desulfurization can be generated within the primary steel-making vessel, the oxygen content of these slags is high, and this decreases their ability to remove sulfur. The ratio of sulfur in the slag to sulfur in the metal may only be about 10.

4) In many cases, removal of sulfur from molten steel is best accomplished outside the primary steel-making furnace, in separate reaction vessels, which, depending on their design and function may be referred to as secondary refining vessels, ladle furnaces, or simply ladles.

Based on the above concepts, it has now been well established that sulfur removal is most effectively accomplished by external desulfurization of hot metal. For the most demanding applications requiring the lowest sulfur levels, for example, less than 0.005%, further sulfur removal is achieved most readily outside the primary steel-making vessel. In this context, it is worth noting, that vessels used in the past simply for transferring metal from one location to another, have now in many cases, become chemical reactors. Under these circumstances, it will be apparent that when sulfur is to be removed from iron or steel, outside the primary furnaces, major consideration must be given to the various reaction vessels with respect to their design, their refractory lining, and their ability to contain metal during vigorous agitation deliberately created to promote rapid reaction between the various phases. Finally, it will be appreciated that the importance of minimizing carry-over of reaction products from one vessel to another cannot be overestimated. The various aspects outlined here will be discussed in more detail in the sections which follow.

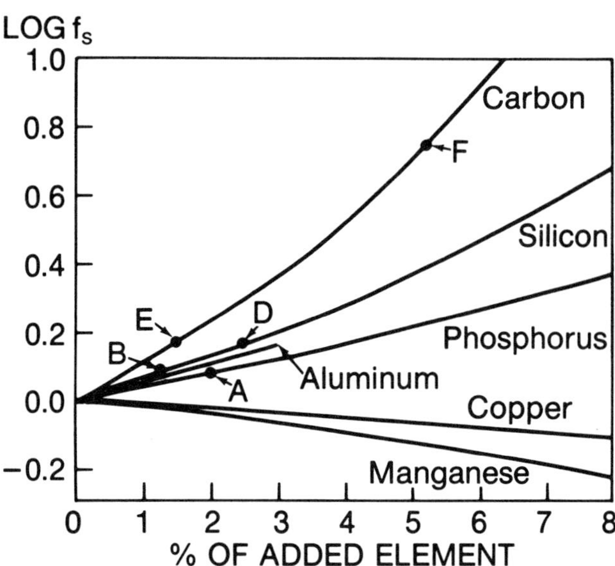

FIGURE 4 Activity coefficient of sulfur as effected by alloying elements in iron

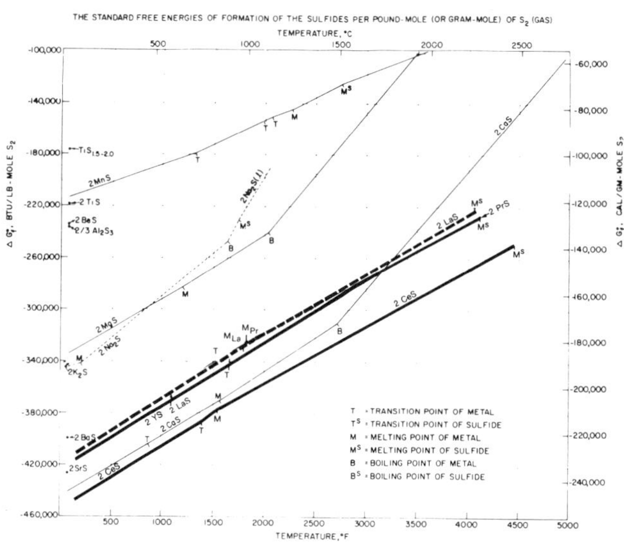

FIGURE 5 Standard free energies of formation of the sulfides

2.2 Sulfur Behavior in Iron Alloys

Figure 4 shows the effect of several elements on the activity coefficient of sulfur in molten iron. Both carbon and silicon increase the activity coefficient of sulfur. For example, in the case of hot metal with 4.2% carbon and 1.5% silicon, the activity coefficient of sulfur is about 5.[7] This means that when computing a solubility product or a distribution coefficient, the activity of sulfur is five times greater than the actual sulfur content. In other words, the sulfur is five times more active than the concentration value would indicate. For this reason, sulfur removal from hot metal is much easier than sulfur removal from steel in which both the carbon and silicon are at very low concentrations. In the case of stainless steels containing high concentrations of chromium, the activity coefficient of sulfur can be as low as 0.1. With other parameters remaining constant, this would imply that it is ten times more difficult to remove sulfur from high chromium stainless steels than it is from low carbon mild steels.

Table I Some of the properties of sulfide forming elements and their sulfides.

Metal	Boiling Point [°F/(°C)]	Solubility Liquid Iron	Sulfide Formed	Free Energy of Formation (BTU/pound-Mole of Sulfur)	Density	Melting Point [°F/(°C)]
Calcium	2624 F (1450 C)	< 0.005%	CaS	−440	2.18	Decomposes
Strontium	2516 F (1380 C)	None	SrS	−425	3.20	3600 F (2480 C)
Barium	2984 F (1640 C)	None	BaS	−400	4.25	2192 F (1200 C)
Magnesium	2025 F (1107 C)	0.040%	MgS	−340	2.85	3600 F (2480 C)
Aluminum	4465 F (2462 C)	No Limit	Al_2S_3	−235	2.02	2012 F (1100 C)
Cerium	5400 F (2982 C)	No Limit	Ce_2O_2S	−750	6.00	3542 F (1950 C)
			Ce_2S_3	−580	5.19	3434 F (1890 C)
Lanthanum	5400 F (2982 C)	No Limit	La_2O_2S	−750	5.87	3524 F (1950 C)
			La_2S_3	−580	4.99	3902 F (2150 C)

2.3 Sulfur Removal With Metals

The free energies of formation of the metal sulfides (Figure 5) can be used to predict those metals which may be useful for removing sulfur from iron and steel.[8] The general reaction for this process may be written as follows:

$$x\,\underline{Me} + y\,\underline{S} \rightarrow Me_x S_y$$

The only metal commonly used for desulfurization of iron is magnesium. However, because of its low boiling point (1105 C), it has been used to only a limited extent for steel desulfurization. The use of magnesium for the production of spheroidal graphite iron dates back to the 1950's. Techniques which had been developed for adding magnesium to foundry irons, for example, magnesium impregnated coke, were subsequently applied to the desulfurization of blast furnace iron, when low sulfur steels were required in greater quantities. Specific details of magnesium desulfurizing techniques will be given in Chapter 3.

The free energy of formation of the rare earth sulfides, together with their high boiling points, and reasonable solubility in liquid iron, would make them desirable for iron desulfurization. The rare earths can indeed desulfurize iron to low levels, however, their atomic weight is almost five times greater than the atomic weight of magnesium. This means that it requires five times as many pounds of rare earths to remove one pound of sulfur from iron. In addition, the price of rare earths has always been about three to four times greater than that of magnesium. Thus, when compared with magnesium for iron desulfurization, rare earths may well be a technical victor, however, they are an economic failure.

The free energies of formation of the sulfides of sodium, potassium, calcium, strontium, and barium would indicate that these metals should desulfurize iron. However, all of these metals have little or no solubility in molten iron. A summary of the properties of some of these metals is presented in Table I. While no reference has been found on the use of calcium metal for desulfurizing iron, it is used in the form of a calcium-silicon alloy, in combination with high sulfur capacity slags, to desulfurize steel. The exact method of sulfur removal under these conditions has not yet been firmly established. For those metals which are soluble in molten iron, there are sufficient experimental data available to enable one to compute the various solubility products. These may be expressed as follows:

$$K_{Mg} = \%\,Mg \times \%\,S = 1.5 \times 10^{-4} \quad (1316\text{-}1427\ C)$$

$$K_{Ce} = \%\,Ce \times \%\,S = 1.5 \times 10^{-3} \quad (1600\ C)$$

$$= 8.9 \times 10^{-4} \quad (1550\ C)$$

$$= 4.7 \times 10^{-4} \quad (1500\ C)$$

$$= 1.3 \times 10^{-5} \quad (1315\ C)$$

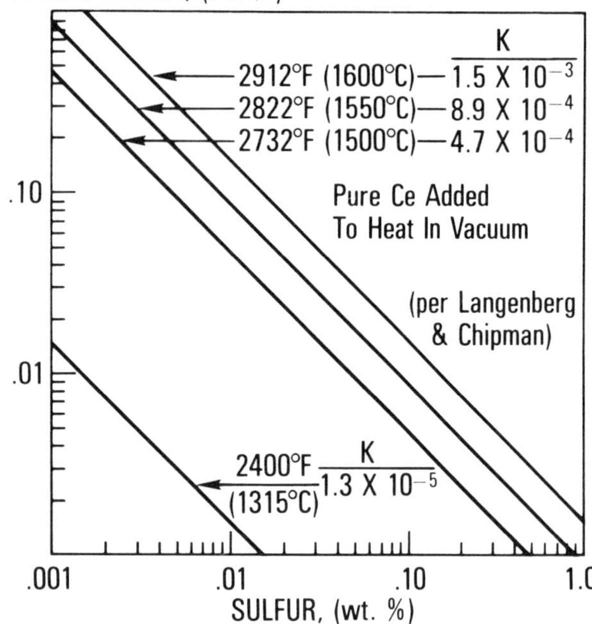

FIGURE 6 Solubility product for cerium sulfide at various temperatures

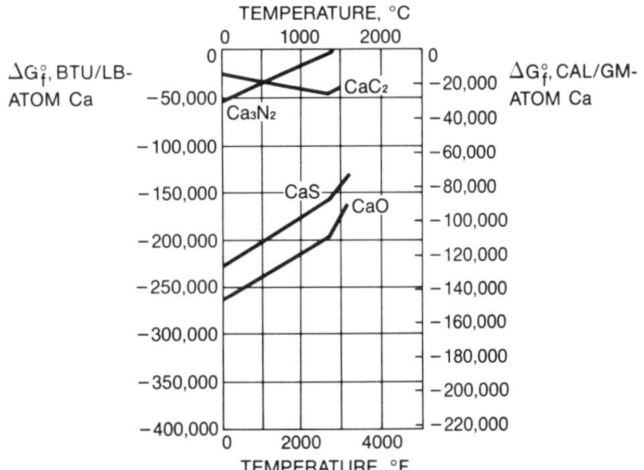

FIGURE 7 Standard free energies of formation of some calcium compounds as a function of temperature

The solubility products for cerium are presented graphically in Figure 6.[9] Any combination of cerium and sulfur above the solubility product line will result in the precipitation of cerium-sulfide, and the precipitation will continue until the cerium and sulfur contents are represented by a point on the solubility product line which corresponds to the temperature at which the reaction occurs. As the temperature decreases, the affinity of cerium for sulfur increases. Thus, during solidification of steel, rare earth elements will continue to react with sulfur and precipitate rare earth sulfides.

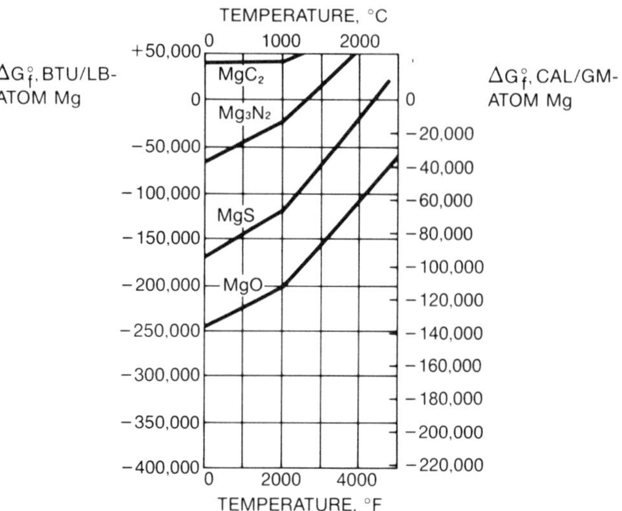

FIGURE 8 Standard free energies of formation of some magnesium compounds as a function of temperature

2.4 Sulfur Removal with Compounds

Figure 7 shows the effect of temperature on the relative stability of the nitride, carbide, sulfide and oxide compounds of calcium. Using existing thermodynamic data, the free energies of formation of these compounds have been computed on the basis of the free energy of formation per gram-atom and also per pound-atom of calcium. Data of this type are helpful in predicting what will happen when a compound such as calcium carbide CaC_2, has the opportunity to react in an environment that contains sulfur. The free energy of formation of calcium sulfide (CaS) at liquid iron temperature is about 65Kcal/gram-atom of Ca more negative than the free energy of CaC_2. Stated another way, CaS is more stable than CaC_2, and when calcium is present in a system containing both carbon and sulfur, calcium sulfide is the compound that will form preferentially. The information contained in Figure 7 can also be used to describe the situation which is likely to exist when calcium oxide is used to desulfurize molten iron. Since the free energy of formation of CaO is more negative than the free energy of formation of CaS, the desulfurizating reaction will only proceed provided the oxygen released during the reaction is continuously removed from solution by reaction with carbon or another deoxidizing element.

Figure 8, which is similar in construction to Figure 7, gives the standard free energies of formation of the equivalent magnesium compounds as a function of temperature. Since the scales on both Figures 7 and 8 are the same, it can be seen that the relative difference in stability between the oxides and sulfides, is greater in the case of magnesium. For a given oxygen potential, such as that corresponding to carbon saturation in molten iron, this would imply that sulfur removal with magnesia is significantly less than would be the case when adding lime under the same conditions. The thermodynamic data and practical experience would indicate that MgO has little ability to desulfurize iron.

Many of the compounds used for desulfurizing iron are not liquid at normal blast furnace iron temperatures. Under these conditions, the reactions then take place between sulfur dissolved in liquid iron and a solid additive. There is no doubt that desulfurizing reactions will proceed more rapidly if both of the reacting phases were in the liquid state. One method of obtaining at least a portion of the compound used for desulfurization in the liquid form is to add fluxing agents with the desulfurizer. Specific examples of the use of flux additives will be given in the next section.

2.5 Sulfur Removal With Slags

High sulfur capacity slags are primarily a method for producing low melting point solutions, the major component of which, from the standpoint of sulfur removal, is lime. These solutions contain a minimum amount of other oxides such as FeO and MnO which would react with the metals and metalloids which are used to combine with oxygen released during the formation of CaS from CaO, as indicated by the reaction given in section 2.1. Two of the more stable oxides generally used to flux lime are alumina or silica, or combinations of both. One further constituent found in many of these slags is calcium fluoride, CaF_2, which is primarily a diluent for the other slagmaking components and a method of lowering their melting points. The fluidity of any slag is directly proportional to the amount of superheat above the melting point. It will be shown later, that desulfurization is better with slags that are at several hundred degrees C above the melting point, where their fluidity is reasonably high.

About 1920, R. Perrin, a French engineer from Ugine, developed a process for the desulfurization of steel in the ladle. A synthetic lime-alumina slag was melted in a submerged arc furnace and tapped into a ladle. Molten steel from a separate furnace was then poured onto the slag. Desulfurization was rapid and efficient.[10] This process is still practiced in France, Italy, Poland and the Soviet Union. However, it has never been widely accepted in North America.

The lime-alumina system modified by the presence of fluorspar, is shown as the $CaO-Al_2O_3-CaF_2$ ternary diagram in Figure 9. At about 40% CaO, 40% CaF_2, 20% Al_2O_3, there is an area in this diagram where the melting point of this slag is 1200 C. Other compositions of slags near this area that are slightly higher or lower in Al_2O_3, have melting points which are still below 1350 C. The sulfur capacity of the slags with these low melting points at 1500 C are shown in Figure 10.[11] The small numbers beside the lines on this diagram indicate the sulfur capacities of the various slags. Slags

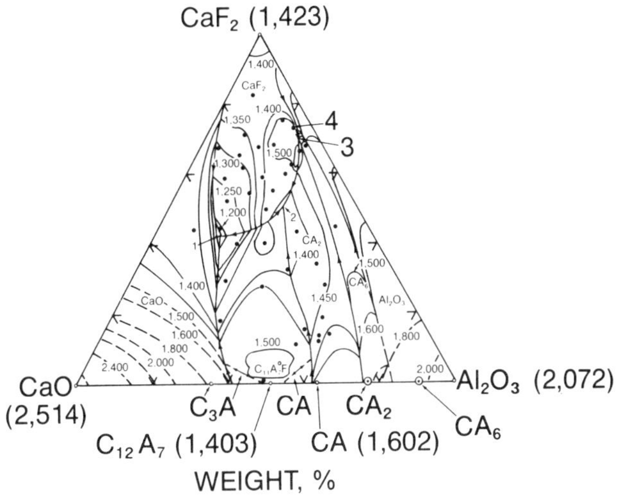

FIGURE 9 $CaO-Al_2O_3-CaF_2$ phase diagram

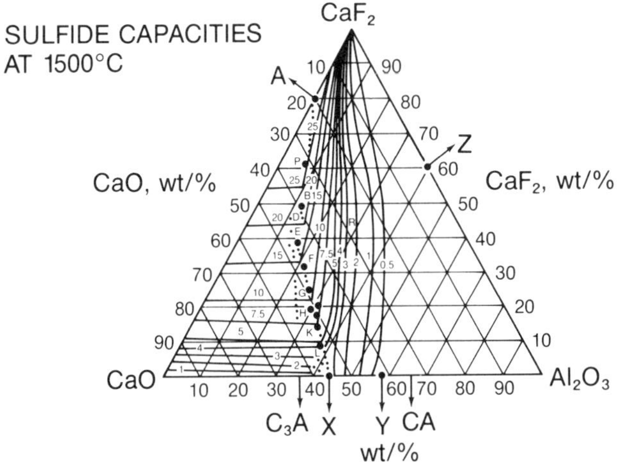

FIGURE 10 Sulfur capacity of slags in the $CaO-Al_2O_3-CaF_2$ system at 1500° C

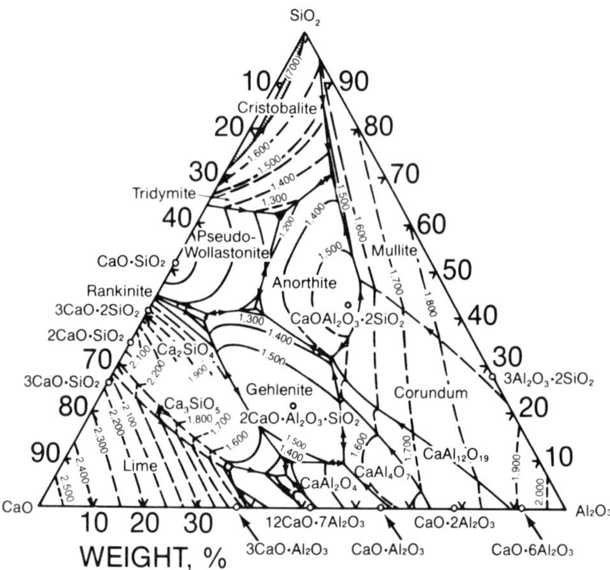

FIGURE 11 $CaO-Al_2O_3-SiO_2$ phase diagram

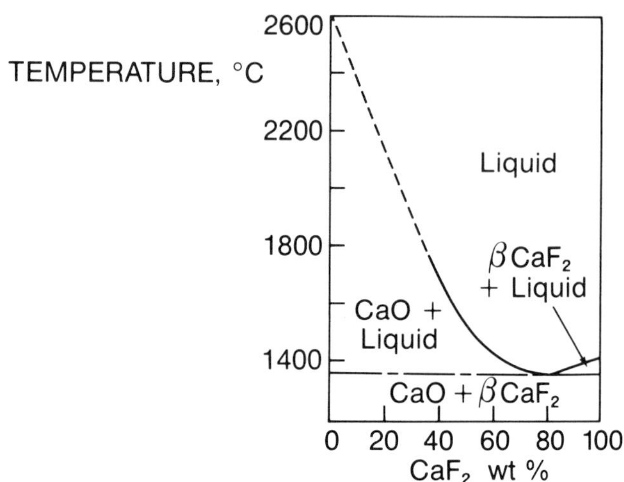

FIGURE 12 $CaO-CaF_2$ phase diagram

with Al_2O_3 contents less than 20% have the highest sulfur capacity even though their melting points are slightly higher. These low-melting, high-sulfur capacity slags have one major disadvantage. They are very corrosive when used in contact with even the best steel plant refractories.

The slags which are formed when lime is fluxed with a combination alumina and silica are depicted in Figure 11.[12] The low melting point areas in the pseudo-wollastonite and gehlenite regions of the diagram are those areas which have been shown to have the highest sulfur capacity, in a similar way to that described previously in the $CaO-Al_2O_3-CaF_2$ system. The use of CaF_2 with slags of this type lowers the melting point even further and would, no doubt, increase their ability to remove sulfur. The ability of CaF_2 alone to reduce the melting point of CaO is shown in Fig. 12.[13] Additives corresponding specifically to the CaF_2-CaO eutectic composition at about 85% CaF_2, and with a melting point of about 1360 C, are used in some desulfurizing systems.

The sulfur capacity of slags in the $CaO-SiO_2$ system is shown in Figure 13 as a ratio between the sulfur in the slag and the sulfur in the metal. It is clear from this diagram, that the higher the ratio of lime to silica, the greater the amount of sulfur which can be retained within the slag phase.[14]

In the multi-component slag systems associated with the manufacture of iron and steel, the magnesia and lime are considered as basic oxides, while the silica and alumina are considered as acidic components. In addition, the oxygen potential of the slag system, as indicated by the iron oxide (FeO) and manganese oxide (MnO) contents of the

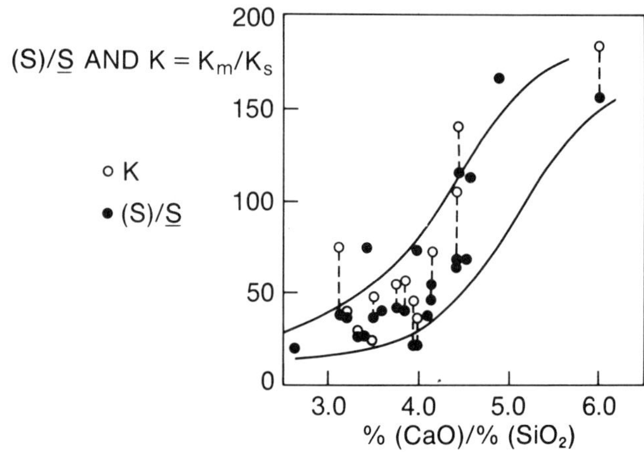

FIGURE 13 Sulfur distribution ratio $(S)/\underline{S}$ as a function of slag basicity

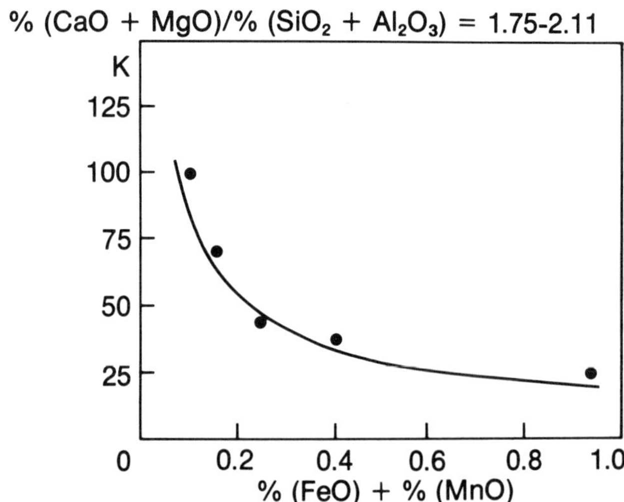

FIGURE 14 Effect of E + MnO on distribution ratio of sulfur

slag, has a large effect on the sulfur capacity of the slags. This strong relationship between oxides and sulfur capacity is shown in Figure 14. It was stated previously that the formation of CaS by reaction with CaO was accompanied by a release of oxygen, which in turn, reacted with metals or metalloids either in the iron or steel or slag. Slags with a high oxygen potential will provide large quantities of oxygen in addition to that released during the formation of CaS. With highly oxidized slags, large amounts of deoxidizing metals and metalloids will be required in order to decrease the oxygen potential in the system to a level which will favor the formation of CaS. Additional details on these factors affecting desulfurization are given in appended papers #7 and #12.

The free energies of formation of many of the oxides[15] encountered in iron and steelmaking are shown in Figure 15. The free energy of formation of CaO is the most negative of all of the oxides mentioned. Although, the free energy of formation of SiO_2 is only slightly more negative than that of MnO which has been shown to interfere with desulfurization, CaO and SiO_2 do form high melting point compounds of great stability. This affinity between the two oxides reduces the activity of the silica in the slag to a fraction of that indicated by the chemical composition. The same is true for lime and alumina. With the exception of alumina and silica, many of the other oxides shown on this Figure with modest values of free energies of formation, interfere with desulfurization.

A further insight into the ability of iron oxide to provide the oxygen which can interfere with the formation of calcium sulfide is shown in Figure 16. When iron oxide is in equilibrium with iron, the equilibrium oxygen content is 0.2% or 2000 ppm. For the purposes of this discussion, the activity coefficient of oxygen is taken as unity. With an activity of FeO, of 0.10, there would be 0.02% or 200 ppm of oxygen in the iron, more than enough to prevent significant desulfurization. Inspection of Figure 16 indicates that only 2 or 3% FeO in a slag consisting of lime and silica will provide sufficient oxygen to inhibit desulfurization.[16] Thus, during iron and steelmaking operations, it is necessary to exercise extreme caution in order to minimize the amounts of those oxides in the slag which have small negative free energies of formation, if significant desulfurization is to be obtained. The combined effects of slag basicity and the strong influence of iron oxide on the sulfur distribution ratio are shown in Figure 17. From this diagram, it is evident that desulfurization is most readily accomplished with fluid, highly basic slags of low oxygen potential.

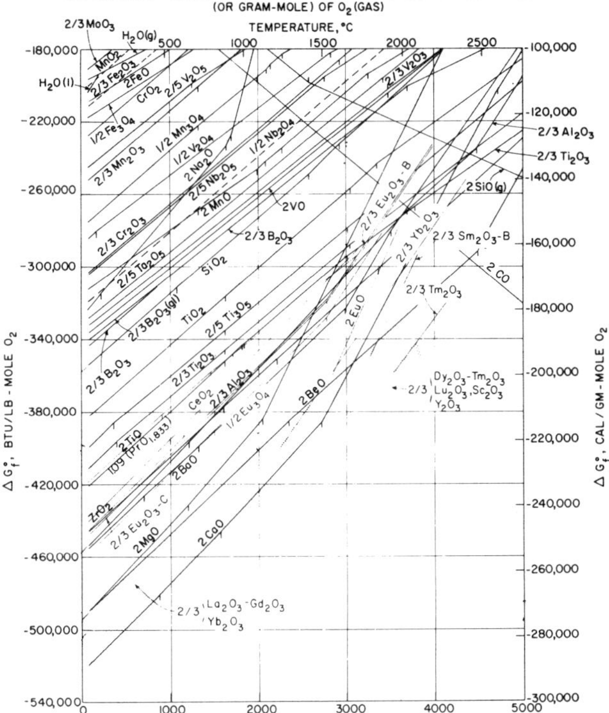

FIGURE 15 Standard free energies of formation of the oxides as a function of temperature

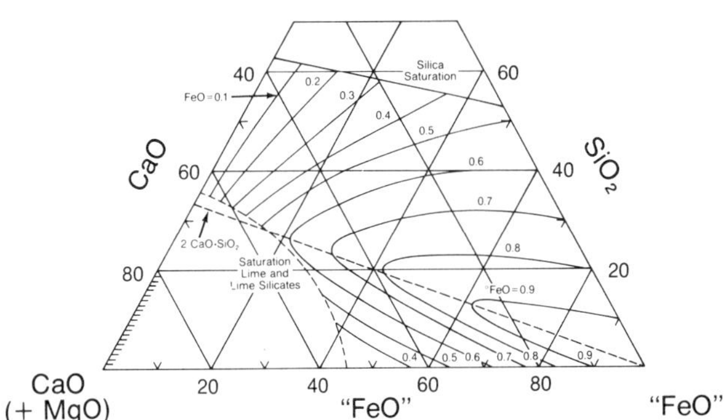

FIGURE 16 Activity of ferrous oxide in CaO-FeO-SiO_2 system at 1600° C relative to pure FeO in equilibrium with liquid iron

In many instances, magnesia and lime have been considered as equally effective components in desulfurizing slags. Recent information has shown, however, (Figure 18) that as little as 10% MgO in an otherwise powerful desulfurizing slag, can reduce the desulfurizing power of that slag by 80%.[17] While the addition of MgO to BOF slags is certainly advantageous from the standpoint of good lining life, it will interfere with the ability of the slag to remove sulfur.

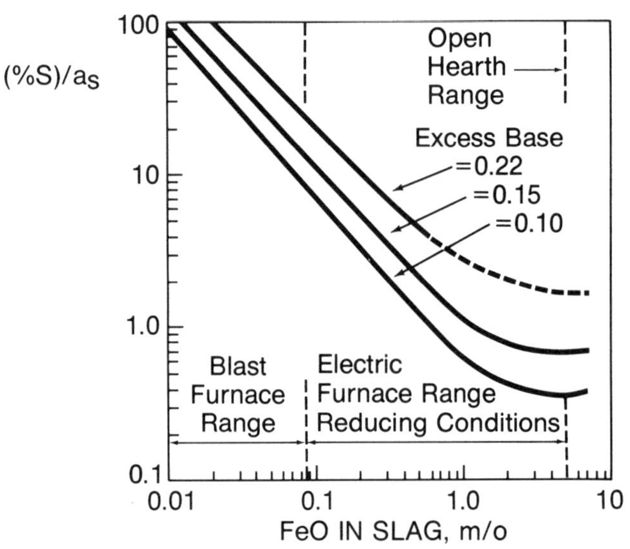

FIGURE 17 Sulfur distribution with complex $CaO-Al_2O_3-SiO_2$ slags at 1500-1700 C

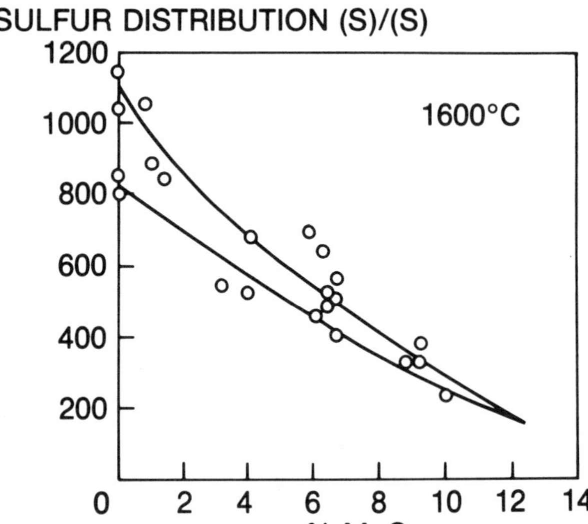

FIGURE 18 Sulfur distribution as a function of the MgO content in slags of the $CaO-Al_2O_3-SiO_2-M_gO$ system at 1600° C

2.6 Reaction Locations for Desulfurization

As mentioned previously, both magnesium and the rare earth elements are soluble in molten iron. It is possible, therefore, that desulfurizing reactions may proceed by precipitation of sulfides from the melt following interaction between the various solute elements present. This may even take place in the absence of high sulfur capacity slags. It has been proposed that calcium also functions in this way. However, there appear to be very few results to support the contention that desulfurization with calcium is mainly direct formation of CaS within the metal bath. In the case of the other alkaline earth metals, with very limited solubility in liquid iron, the desulfurizing reaction must proceed by some other mechanism. In such cases, the presence of high sulfur capacity slags will assist in the desulfurizing process.

The thermochemistry of calcium carbide desulfurization would indicate that the reaction can proceed within the melt itself. However, since it is a solid-liquid reaction with only a small time available for a particle of calcium-carbide to move from a position within the ladle to the surface of the metal, it is unlikely that direct reaction between the metal phase and the rising carbide particle can account for the most of the desulfurization. Thus, with the exception of magnesium and the rare earth elements, it would appear that the most likely location for the desulfurization reactions is at the slag-metal interface. Since this interfacial area is small compared to the total volumes of slag and metal present, rapid desulfurization will only occur if the metal and slag phases can be continuously agitated in order to form new slag-metal interfaces as frequently as possible. The most common method for agitation is by gas injection. In such cases, the gas may be used as a carrier for the reactants or simply as a means of stirring. Other methods which are commonly used for stirring include rapidly rotating paddles immersed in the iron, and shaking ladles. There is much to be said for the simplicity of such stirring systems, and, where ladle configurations permit, the results between the various systems compare favorably. Finally, agitation may also be obtained by electromagnetic stirring. An example of this is to be found in the ASEA-SKF system. The stirring energy provided by such a system, however, appears to be small compared to that obtained with gas bubbling. Some further comments on the various methods available for agitation will be presented in Chapters 3 and 4.

2.7 Summary

Thermochemical data have been presented to illustrate the type of desulfurization reactions which can be expected to take place with magnesium, rare earths, calcium carbide, and high sulfur capacity slags. When slags are used for desulfurization, the base to acid ratio should be high, the melting point low, and the oxygen content of the system as low as possible. In addition, desulfurization will be more efficient when the ladle refractories are made from oxides with a high stability. Finally, since desulfurization in most cases will depend on the transfer of sulfur from the metal to a slag phase at a common interface, the importance of stirring to change that interface as rapidly as possible connot be overestimated. The various aspects of desulfurization which have been described in this section will be used in the following chapters to account for the behavior observed with the different desulfurizing treatments.

CHAPTER 3

DESULFURIZATION OF IRON

3.1 Magnesium Applications

3.1.1 General Comments

In a recent address by the President of the Metals Society, he stated that the trend to lower sulfurs to meet more demanding specifications began in 1970, (Figure 1). Desulfurization with magnesium is one of the methods commonly used for this purpose. Various forms of magnesium are available with different methods of addition. There are certain advantages and disadvantages which are common to all desulfurizing methods based on magnesium.

Advantages:

1) It is a method for producing iron with low sulfur content on a regular, consistent basis.

2) Skimming of blast furnace slag prior to desulfurization is not required.

Disadvantages:

1) The long-term availability of magnesium may be suspect.

2) Slag must be skimmed before charging the iron into the steelmaking furnace. However, this is also true for all other desulfurizing methods used for iron.

3.1.2 Mag-Coke

One system which was readily available in 1970 for desulfurizing molten iron to low sulfur levels was based on the use of coke impregnated with magnesium,[18] (Mag-coke). This system had received wide acceptance in the foundry industry with respect to the production of ductile iron, where it served to decrease sulfur and produce spheroidal graphite. In one method of application, Mag-coke is placed in an inverted graphite basket which contains numerous vent holes. This basket is then attached to a bloom with sufficient weight to counteract the buoyancy of the magnesium and coke. The bloom and basket are plunged into molten iron and allowed to remain in the metal until the reaction associated with the vaporization of magnesium, has ceased. The violence of this reaction can be partially controlled by the size of the Mag-coke, less stirring being obtained with larger pieces. The degree of desulfurization that can be achieved is shown in Figure 19. Further details on this process are given in appended paper #3.

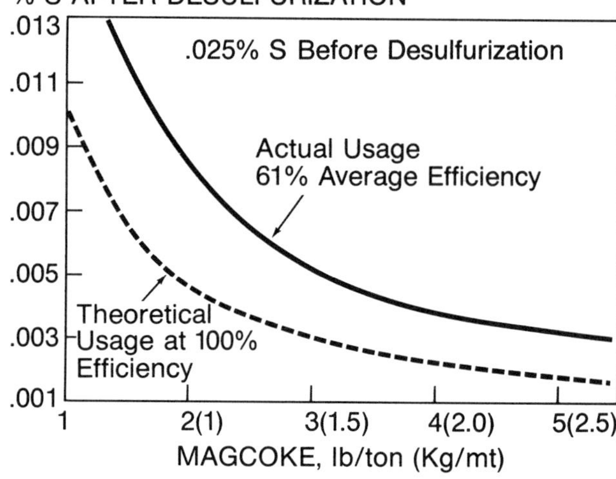

FIGURE 19 Final sulfur content of iron (0.025% sulfur before addition) with increasing additions of Mag-Coke

Advantages:

1) The capital requirements necessary to install this method of desulfurization are relatively small.

2) The method can be used to treat occasional ladles of iron and produce metal with low sulfur contents.

Disadvantages:

1) The treatment is relatively expensive.

2) There is only a single source of Mag-coke.

3) A fume extraction hood system is required.

3.1.3 Lime Mag

The later development in the use of magnesium was the injection of lime and magnesium powder into iron.[19] The injection system used is rather unique in that the magnesium and lime are fluidized in separate feeder systems and mixed just prior to their entry into the lance. One of the other innovations of this system is in the design of the lance. It was found if the lance was straight, the bubbling action due to the carrier gases and the vaporization of the magnesium resulted in rapid erosion of the refractories on the lance. To overcome this problem, a section of the lance about 2 feet from the bottom, was bent through 45 degrees from the vertical. This change in lance design not only resulted in improved lance life, but provided a stirring action that was spread over a wide area of the ladle.

The magnesium in this process is largely responsible for the desulfurization. The

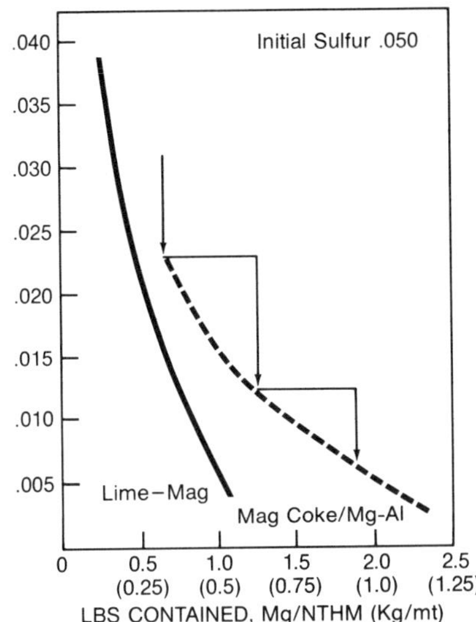

FIGURE 20 Comparison of magnesium consumption for desulfurization of hot metal

lime served mainly as a diluent to control the rate and violence of vaporization of magnesium. As with Mag-coke, it is not necessary to skim the blast furnace slag from the ladle prior to the desulfurizing treatment. However, it is necessary to remove the slag after treatment to prevent reversion of sulfur from the slag into the steel during steelmaking. The results obtained with this method are shown in Figure 20. With the use of this combination of lime and magnesium, the smoke generated is not unduly excessive and there are some shops where this practice is used, without a hood. Further details of this process are given in appended paper #4.

There are several detrimental aspects to this process. The first is associated with the supply of magnesium which is further complicated by the need for fine powders. In addition, these fine powders are pyrophoric when not handled properly, and fatal accidents have occurred. Another problem is in the feeding of lime, which even when extremely dry, compacts while sitting in the container. A satisfactory solution to this problem has not yet been reported. Magnesium-aluminum alloys have been used in place of magnesium powders for this application because they can be crushed more easily and they have some advantages from the standpoint of pyrophoric reactions. However, as shown in Figure 20, the cost effectiveness of the magnesium-aluminum alloy is lower than that with magnesium powder alone.

In summary, the advantages and disadvantages can be stated as follows:

Advantages:

1) The cost of attaining low sulfur levels is relatively small.

2) The process does not require expensive premixed powders.

3) The capital requirements are modest, particularly if no fume extraction hood is required.

4) The process has a high degree of flexibility in that it can be used to treat large tonnages of hot metal, or simply an occasional ladle.

Disadvantages:

1) There is always a danger of fires and explosions if fine magnesium powders are not properly handled.

2) The compaction of lime in the storage container, even in the absence of moisture can lead to operating problems.

3.1.4 Salt Coated Granules

The most recent innovation in the application of magnesium for the desulfurization of iron is the use of magnesium granules coated with a combination of sodium, calcium, magnesium and potassium chloride salts. This new form of magnesium is a more efficient desulfurizer and is safe to handle. The desulfurizing action occurs by the precipitation of magnesium sulfide, and based on the data obtained in appended paper #3 and other trials, the solubility product for the reaction:

$$\underline{Mg} + \underline{S} \longrightarrow MgS$$

is, $\% \text{ Mg} \times \% \text{ S} = 1.5 \times 10^{-4}$

Figure 21 shows the desulfurizing capability of this material compared with two other magnesium processes for removing sulfur from iron. The improved desulfurizing ability of this form of magnesium may be due in part to the solubility of magnesium in chloride salts even at temperatures above the boiling point of magnesium. This effect would minimize the escape of magnesium vapor into the atmosphere, and thus, more magnesium is available for desulfurizing. As a consequence of this behavior, there is a possibility that this process could be conducted without a fume extraction system. Lances of simple design suitable for the process have been developed, and a very simple fluidizing system is used to easily convey the magnesium particles in the gas stream through the piping and lance equipment into the iron. The only disadvantage of this process is the one it shares with all magnesium based desulfurizing processes - potential supply problems. This is further complicated due to the special processing required for manufacture of these salt-coated granules. At the present time, there is only one producer of these granules in North America. Further details of this process are given in appended paper #5.

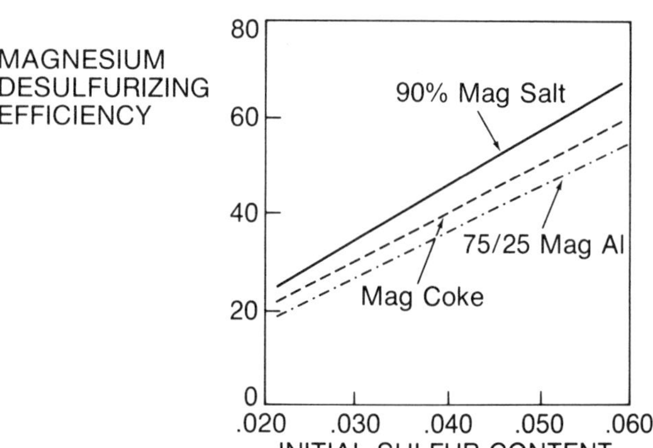

FIGURE 21 Comparison of the desulfurizing efficiency of various magnesium based reagents at 2450° F

In summary then, the advantages and disadvantages can be simply stated:

Advantages:

1) Low sulfur levels can be obtained more efficiently than with any other magnesium based process.

2) Desulfurization can be achieved at a lower cost than with any other magnesium based process.

3) Magnesium exists in a form which is safe to handle.

4) Hood requirements with this operation should be minimal.

5) Injection of the material into molten iron is relatively easy.

6) Because of good flowability of the material, only minimum capital costs are required for the injection system.

7) The system is highly flexible and can be used to treat large tonnages of hot metal or only an occasional ladle.

Disadvantages:

1) This process has the disadvantages common to all magnesium based methods for desulfurization, i.e., availability, and the need to remove the sulfur containing slag before charging the hot metal into the steelmaking vessel.

3.2 Calcium Carbide Applications

3.2.1 General Comments

Production of ductile iron by the foundry industry had started in 1950, and by 1970, when the demand for low sulfur steels had grown to significant tonnages, the other common method of desulfurization used by the foundry industry which was applicable to the desulfurization of pig iron, was the use of calcium carbide. Trials with calcium carbide had been conducted within the steel industry in the 1950's. At that time, however, desulfurized hot metal was added to the open hearths without first skimming the sulfide-containing slag. With open hearths burning high-sulfur fuel oil, this operation showed little promise, and was subsequently abandoned. The foundry industry persevered with the use of calcium carbide and developed reproducible techniques for attaining low sulfur levels prior to the addition of the nodularizing agents such as magnesium. Thus, when the time came for the steel industry to produce low sulfur hot metal, techniques for desulfurization were available.

Referring again to Figure 4, it can be seen that the free energy of formation of CaS is more negative than the free energy of formation of CaC_2. These differences in free energy indicate that calcium carbide will react with sulfur to form calcium sulfide. The earliest applications of calcium carbide for desulfurization were with injection techniques, where it was supposed that the reaction occurred beneath the surface of the bath. However, with later developments, based on the use of mechanical stirring or by agitation obtained with gas bubbling, it was found that a major portion of the desulfurization process occurred at the slag-metal interface. This being the case, skimming of the blast furnace slag prior to desulfurizing treatment is a mandatory requirement. In addition, skimming of the sulfur containing slag and any unreacted calcium carbide after desulfurization, is also necessary. At wet dekishing stations, this has been a major problem due to reaction of the calcium carbide with water. This aspect highlights one other problem common to all calcium carbide desulfurization processes, i.e., the inherent danger in handling calcium carbide. When carbide reacts with water, it forms acetylene which in certain concentrations, is spontaneously explosive. While the safety record with calcium carbide has been exemplory, it is unreasonable to assume that the use of calcium carbide will never result in an explosion.

The supply picture for calcium carbide is good. There are several manufacturers of the material and most of these have now installed grinding equipment producing the fine particles which are most adaptable for injection techniques. Calcium carbide furnaces are similar to submerged arc furnaces for ferro-alloy production and the expansion of this capacity could be less time consuming and less expensive than the installation of electrolytic cells required for the production of magnesium. All of the information

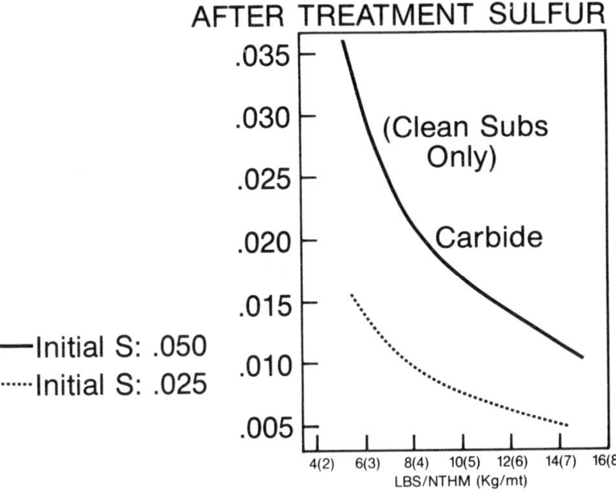

FIGURE 22 A comparison of the calcium carbide requirement necessary to desulfurize from 0.025 and 0.050 sulfur to any required level

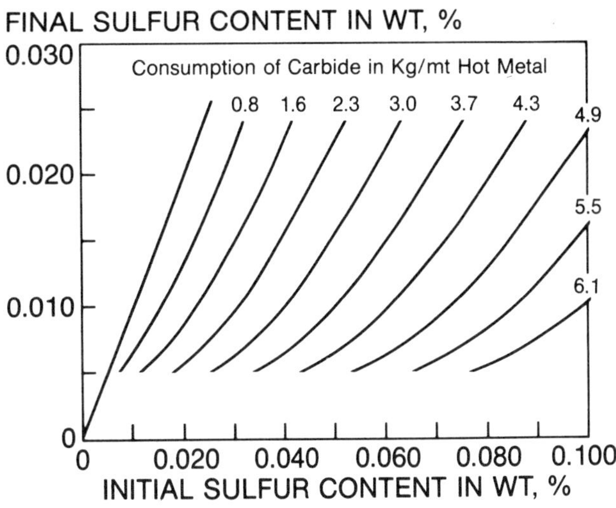

FIGURE 23 Relationship between sulfur contents and the consumption of carbide reagent

presented above is common to all methods based on the use of calcium carbide. The variation and results obtained with different methods of using calcium carbide will be discussed in the following sections.

3.2.2 Injection Aspects

Typical results obtained with the injection of calcium carbide into hot metal submarine ladles,[19] which had been previously skimmed to remove the blast furnace slag, are shown in Figure 22. The shape of the curves in this figure indicates the greater efficiency of calcium carbide at high sulfur levels. For example, seven pounds of calcium carbide per ton (3.5 Kg/mt) will reduce the sulfur content of the iron from 0.025 to 0.010 (0.47 pounds calcium carbide per 0.001% sulfur), but it requires an additional 7 pounds per ton (3.5 Kg/mt) of calcium carbide to go from 0.010 to 0.005% sulfur (1.4 pounds calcium carbide per 0.001% sulfur). For the production of steels with less than 0.01% sulfur, sulfur levels obtainable without first reducing sulfur contents in the iron to 0.005%, will require either steel desulfurization, or positive sulfide shape control. The most effective way to run a blast furnace may result in sulfur contents in the iron greater than 0.050%. At these higher sulfur levels, calcium carbide requirements to obtain hot metal with 0.010% sulfur could be as high as 15 pounds per ton (7.5 Kg/mt) or more. The logistics of injecting that quantity of carbide into the iron could present a problem.

The desulfurizing ability of calcium carbide can be enhanced if the carbide "also contains other components which decompose at hot metal temperatures, thereby evolving a gas."[21] Examples of such components are sodium or calcium carbonate. It is likely that the evolution of gases from the decomposition of the other compounds added with the carbide does not occur completely as the compounds rise through the iron following injection. The release of gas from these compounds within the slag on the surface should result in gentle localized stirring over the bath area, this resulting in increased effectiveness with respect to the desulfurizing process. The improvement obtained with this modified calcium carbide treatment is shown in Figure 23. The data in Figure 22 indicated that it would take 15 pounds (7.5 Kg/mt) of pure carbide to lower the sulfur level from 0.050% to 0.010%. The data in Figure 23 indicate that it will require only about 8 pounds (4.0 Kg/mt) of carbide in the mixture to achieve the same results.

3.2.3 Mechanical Agitation

Since calcium carbide does not melt at the temperatures of iron desulfurization, the action which occurs has to be one between solid carbide particles and liquid iron. Such heterogeneous reactions are notoriously slow, and it is likely, therefore, that desulfurization with carbide occurs in large part at the slag-metal interface. For this reason, procedures have been developed for adding calcium carbide to the surface of the iron after skimming, and stirring with a refractory coated paddle to constantly change the slag-metal interface and achieve desulfurization. The results obtained in Japan with the "KR" stirring method[22] are shown in Figure 24. This graph shows that it takes 4.4 pounds (2.2 Kg/mt) of carbide to go from 0.040% to 0.010% sulfur. For comparison with other methods of carbide

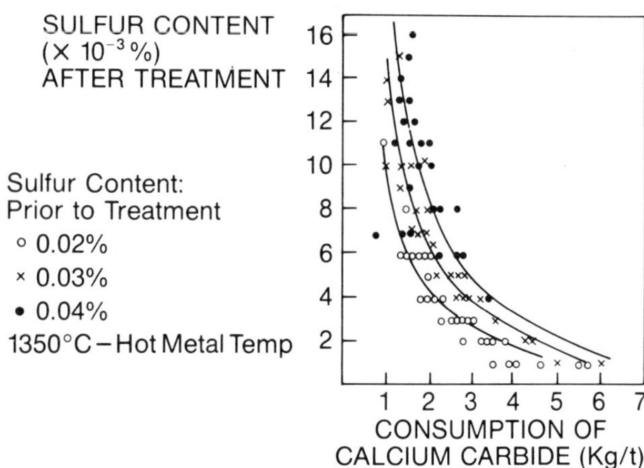

FIGURE 24 Relationship between the consumption of desulfurizing agent and sulfur content of pig iron by the KR process

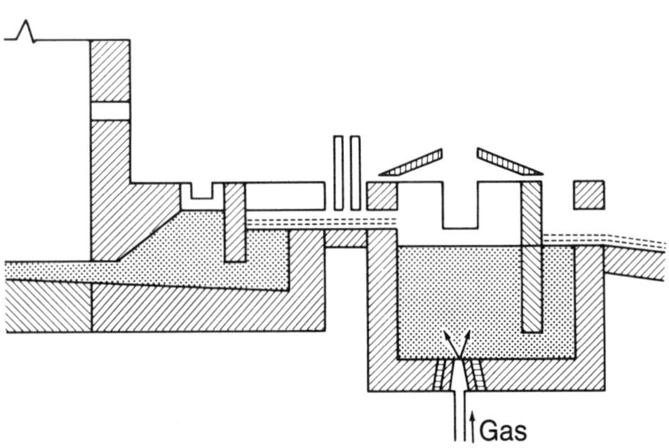

FIGURE 25 A cutaway view of a typical slag skimming and dwell unit for porous plug stirring used with CaC_2

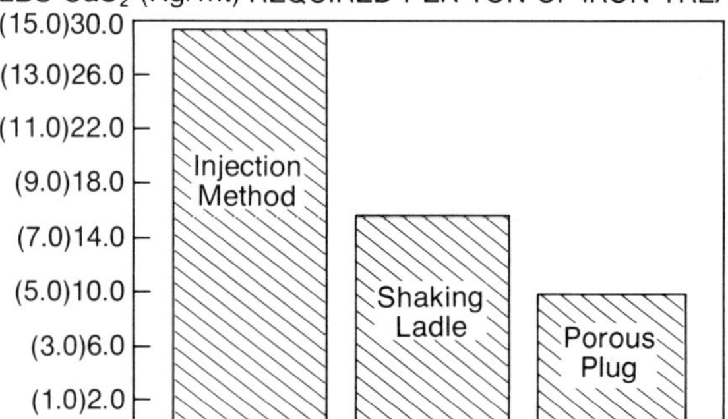

FIGURE 26 A comparison of the effectiveness of various methods of using CaC_2

desulfurization, it can be estimated that it would take 5.5 pounds (2.75 Kg/mt) of carbide to go from 0.050 to 0.010 sulfur with mechanical stirring. This is substantially better than the 15 pounds (7.5 Kg/mt) per ton of carbide required with the injection of carbide alone, or the 8 pounds (4.0 Kg/mt) per ton necessary when injecting carbide with other "decomposing compounds." The KR process utilizes a refractory covered stirring device to achieve mixing. The same effect can be accomplished, however, using shaking ladles or any other device which changes the slag-metal interface rapidly. It is worth noting that these mechanical stirring devices are only applicable in essentially cylindrical vessels. In other words, they would not be appropriate for torpedo cars.

Experience in the foundry industry with mechanical stirring confirms the Japanese data. In one large ductile iron foundry, a ladle lined with 50% alumina brick is used together with a paddle to agitate the iron. Paddle life is about 5-1/2 days and the paddle rotation speed is 120 rpm. Thirty pounds (13.63 Kg) of sodium carbonate is added with the carbide and a stirring time of 12-15 minutes is employed. The sulfur content of the iron is decreased from 0.080 to 0.012% using 12 pounds (6.0 Kg/mt) of carbide per ton of iron. The carbide is sized 20 mesh by down. The results obtained indicate that it would require 7.06 pounds (3.53 Kg/mt) of carbide per ton to lower the sulfur from 0.050 to 0.010%. This is in reasonable agreement with the Japanese observations reported in Figure 24.

3.2.4 Porous Plug Operations

The simplest method for agitating the slag-metal interface to promote desulfurization with carbide is with a porous plug in the bottom of a refractory lined vessel.[23] Such a device is shown in Figure 25. One major advantage of this design is that the spent calcium carbide slag flows continuously from the iron with little or no operator attention. In the foundry where this particular practice is followed, the carbide is added with 0.1 to 0.3% coke breeze, sized 1/2 inch by down so that the products of reaction which are dry and about 1/2 inch in diameter simply roll out of the dwell unit of their own accord. The operating life of the reaction vessel, which is lined with 85% alumina brick, is 300-500 hours. The dwell time of the iron averages 8 minutes, and in this time, the sulfur is reduced from 0.100% to 0.010%. Based on these numbers, the carbide required to go from 0.050% to 0.010% sulfur would be 4.44 pounds (2.22 Kg/mt) of carbide per ton, and the dwell time could probably be reduced proportionally. If the dwell time could be reduced to 5 minutes, then in order to treat 600 tons of hot metal per hour, a reaction vessel capable of holding only 50 tons would be required. This would supply enough iron to run a good sized BOF shop. The use of this porous plug continuous desulfurizing system has been accepted by many large foundries for the desulfurization of iron prior to the addition of magnesium-containing alloys for the production of ductile iron.

In the particular foundry where these porous plug dwell ladles are used, work has also been conducted with injection and shaking ladle techniques for the addition of carbide. Figure 26 shows the carbide requirements for desulfurizing from 0.100 to 0.010% sulfur with the three techniques. Extrapolation of the data to desulfurization from 0.050 to 0.010% sulfur, would indicate that it would require 13.33 pounds (6.67 Kg/mt) of carbide per ton with carbide injection, 7.11 pounds (3.56 Kg/mt) of carbide for mechanical stirring (this is in reasonable agreement with other investigators), and 4.44 pounds (2.22 Kg/mt) of carbide with the porous plug-dwell ladle. It may well be that adaptations of this particular technique could find useful application for desulfurization within the steel industry.

3.3 Sodium Carbonate Applications

Perhaps, the oldest method for desulfurization of iron is the use of soda ash.[24] Experience with this material has been accompanied by several problems. First, desulfurization has been too erratic for the producers of ductile iron to enable them to minimize their nodularizing additions; second, referring back to Figure 4, it can be seen that desulfurization with sodium compounds is bound to be more sensitive to temperature than any other method (notice the steep slope of the free energy curve for Na_2S); third, the slags resulting from the use of sodium compounds are extremely corrosive with respect to all types of refractories; fourth, the use of soda ash results in the formation of fine airborne particles of sodium oxide that are particularly irritating to the skin, especially when the ambient temperature is high and people in the vicinity may be perspiring; and fifth, the reaction of soda ash during desulfurization results in the formation of large volumes of CO which can burn within the dust collecting equipment with disastrous results. The reactions which produce this CO are shown below:

$$Na_2CO_3 \longrightarrow 2NaO + CO_2$$

$$CO_2 + C \text{ (as Kish etc)} \longrightarrow 2\,CO$$

$$Na_2O + 2\,\underline{S} \longrightarrow Na_2S + 2\underline{O}$$

$$\underline{C} \text{ (as Kish etc)} + O \longrightarrow CO$$

Above the ladle, this CO combines with oxygen in the air to form CO_2. With traces of sodium oxide in these gases, this reaction

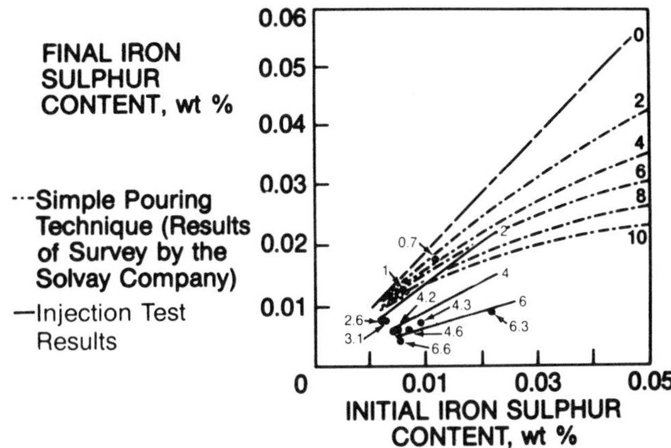

FIGURE 27 Relationship between initial sulfur and final sulfur with soda ash additions

produces a large yellow flame. This flaming action has been the reason that the most recent attempts to use soda ash desulfurizers have been abandoned. The dust collectors installed to handle the irritating sodium oxide particles, contained fiber glass bags, which were destroyed by the high temperature created during the combustion of carbon monoxide.

Despite all the problems with soda ash, it is still used in foundries and certain blast furnace plants. Results obtained both by injection techniques and by merely throwing the soda ash in the ladle are shown in Figure 27. Certainly, the efficiency of the process is improved and the problems associated with refractory erosion, airborne sodium oxide particles, and combustion effects are reduced when soda ash is injected. Close inspection of the data in Figure 25 still indicates, however, that the problem of erratic desulfurization exists, even with injection techniques.

3.4 Desulfurization with Lime and Lime Based Slags

It is a well known fact that lime is the active ingredient in many desulfurizing processes. However, when attempts are made to apply this technology to the desulfurization of hot metal, one of the difficult problems which must be overcome is that most lime based slags have melting points which are higher than the temperature at which it may be desired to desulfurize iron. There are two possible solutions. The first is to use lime powders of such a fine size, for example, under 325 mesh, that there is a large surface area of lime exposed to the iron in order to absorb sulfur. If the lime is too fine, however, part of the addition may simply be carried off into the dust catcher system. In addition, fine lime particles may also agglomerate into coarser aggregates under the slight pressure which exists within the feeder hoppers. If this occurs, it becomes difficult to inject the material due to blockage of the hopper throat and the associated delivery system. This compaction of the lime can occur even in the absence of moisture. Another method for overcoming the problem associated with the high temperature is to lower the melting point of the lime by the addition of other stable oxides and ensure that the temperature of the molten iron is maintained at a high enough level so that at least partial melting of lime containing slag occurs. For effective desulfurization with lime, some element should be present which will react with the oxygen released by the formation of CaS from CaO. In hot metal, this could be carbon while in steel, it would be some other deoxidizer.

When using lime-based slags, it is essential that slag carryover from the blast furnace be removed before desulfurization. In the presence of large quantities of acid slag making materials, such as silica, the desulfurizing power of lime is drastically reduced. It is also further hindered by the presence of small amounts of unstable oxides, such as FeO. As with other desulfurizing processes, it is essential that the sulfur-containing slag be removed from the iron before metal is transferred to the steelmaking vessel.

Figure 28 shows the relative desulfurizing ability of a CaO -5% CaF_2 -5% carbon mixture compared to calcium carbide.[25] In this particular plant, agitation of the melt is accomplished either by impeller stirring or by nitrogen bubbling. Since injection techniques were not involved, lime particles with a grain size of 0.2 -1.0 mm could be used with a minimum of difficulty. The curve in this figure shows that it requires about twice as much of this slag mixture as calcium carbide to accomplish the same amount of desulfurization. However, good grade burnt lime may cost only a fourth as much as calcium carbide. Further details on this process are given in appended paper #6.

The effect of temperature on desulfurization with lime-based slags is shown in Figure 29. These data indicate that if the iron is desulfurized at temperatures of 1450 C (2642 F) or above that the process is effective. If the temperature drops to 1350 C (2462 F), it becomes questionable, and if desulfurization is attempted at still lower temperatures, 1250 C (2282 F), little or no desulfurization takes place. A phase diagram in Figure 11 provides an explanation for the temperature dependence of this system. The eutectic in the CaO - CaF_2 system is at 1390 C (2534 F). With molten iron above this temperature, some liquid CaO-CaF_2, eutectic will be present. This will enhance the opportunity for desulfurization

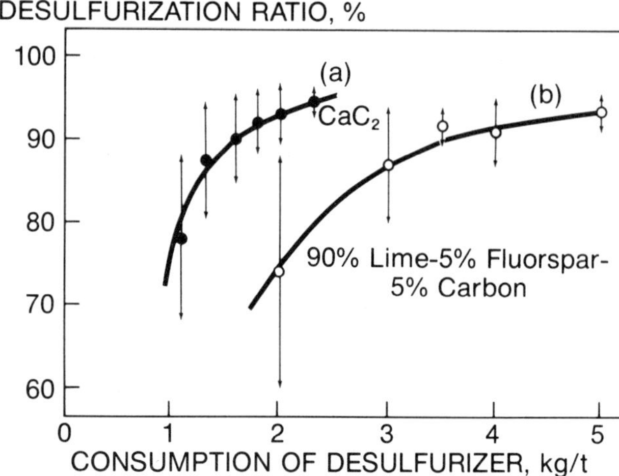

FIGURE 28 Desulfurization ratios with CaC_2 and $CaO-CaF_2$- carbon mixtures

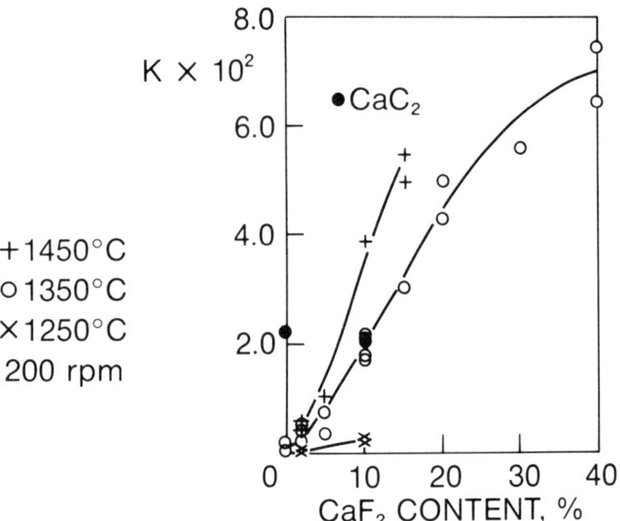

FIGURE 29 Effect of CaF_2 content on apparent desulfurization rate constant

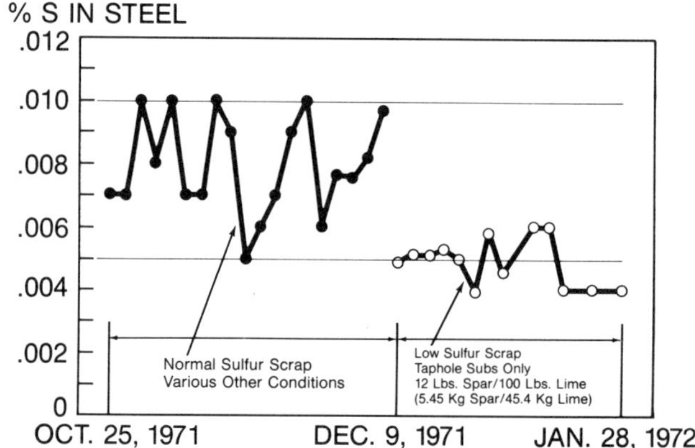

FIGURE 30 Chronological chart of percent sulfur in low sulfur steel heats

by interaction between the two liquid phases, i.e., the molten iron and the molten slag. Below 1390 C (2534 F), the lime will remain solid resulting in a slow heterogeneous reaction. Thus, an important cost consideration when using lime-containing slag to reduce the sulfur content of hot metal, is the cost associated with super heating the iron to the required temperature in order to ensure that the desulfurizing process is effective.

3.5 The Influence of Hot Metal Sulfur on Steel Sulfur Contents

Figure 30 shows the range of sulfur contents obtained in steel with hot metal sulfur levels which average 0.005%, with and without selection of low sulfur scrap in the BOF. In North America, the average scrap charged to a BOF with the hot metal is about 30%. To meet the less demanding applications, such as those indicated in Figure 1, with a regular scrap charge to the BOF, a sulfur maximum in the steel of 0.010% can be maintained provided the hot metal sulfur content has been reduced to 0.005%. This would require relatively large desulfurizer additions and particularly careful attention to detail. Under these circumstances, no further desulfurization of the steel should be necessary. For the most demanding applications requiring final steel sulfur less than 0.005%, further sulfur removal from the steel in the ladle will be required, even with 0.005% sulfur hot metal charged to the BOF. Further details pertaining to steel desulfurization will be discussed in Chapter 4.

CHAPTER 4

DESULFURIZATION OF STEEL

4.1 Preliminary Considerations

In the previous Chapter on fundamental aspects, the activity coefficient of sulfur in conventional pig iron was shown to be about 5. This implies that it is about five times easier to remove sulfur from hot metal than it is from low carbon steel, other factors being equal. For this reason, every effort should be made to remove as much sulfur as possible from the iron so that the amount of sulfur to be removed during steelmaking can be minimized. If sulfide shape control additives are necessary, the amounts required will also be minimized by first using low sulfur hot metal in the steelmaking vessel. In plants with excess blast furnace capacity, steel desulfurization can be eliminated in many cases by the use of well desulfurized hot metal for almost the complete BOF charge. However, under normal conditions, excess blast furnace capacity is a luxury very few steel plants enjoy and the charge to the steelmaking furnace frequently contains significant amounts of scrap. For this reason, some form of steel desulfurization is generally essential.

4.2 Sulfur Removal in Conventional Steel Melting Furnaces

It was shown in the previous Chapter that the sulfur distribution between slag and metal phases is low when the slags contain high quantities of unstable oxides such as FeO and MnO. The slags used in open hearth, convertor and electric furnace melting (particularly with fume extraction through the roof), are all high in these unstable oxides. With the use of heroic measures, most of which adversely affect shop productivity, sulfur levels of less than 0.010% may be obtained. Most of the techniques for sulfur removal under these circumstances, are based on the use of much larger than normal slag volumes and partial or complete removal of the sulfur containing slags. In the BOF, this could mean higher than normal lime charges and more or less complete slag removal during the heat. In the electric furnace, constant feeding of burnt lime or limestone while continuing to flush slag phase, accomplishes the same objective.

In the bottom blown Q-BOP process, this basic convertor is designed in such a way that powdered lime can be carried through the tuyeres into the steel by the oxygen being used for decarburization. Although the oxygen content of this system is high, the lime has an opportunity to fuse in those locations where the oxidation reactions are occurring and some desulfurization should take place. To date, no detailed results have been reported which document the behavior of sulfur in this particular process.

One recent change in convertor practice which will interfere with desulfurization in the vessel is the use of high magnesia slags in order to prolong the life of the basic lining. While there is no doubt that the vessel lining life will be increased by adding magnesia in one form or another with the charge materials, it should be remembered that magnesium oxide will reduce the sulfur distribution ratio between the slag and metal phases. Figure 17 shows that the distribution ratio is decreased by a factor of 5 when the magnesia content of the slag reaches 12%.

When stainless and high alloy steels containing more than 3% nickel are being produced, sulfur removal can be accomplished by the addition of a very convenient nickel-magnesium alloy containing 4.5% magnesium. The density of this particular alloy addition is greater than molten iron, and thus, it sinks slowly within the bath to the hearth of the furnace, permitting a slow release of magnesium. Since magnesium has a greater solubility in steels containing nickel, there is an opportunity for magnesium to dissolve in the metal and subsequently react with sulfur to form magnesium sulfide. The results which can be obtained using this technique are shown in Figure 31, where the sulfur content of the steel has been decreased from 0.010% to 0.005% with an alloy addition of 6 pounds (3.0 Kg/mt) per ton.[26]

Before concluding this section, it should be emphasized that desulfurization in conventional steelmaking furnaces is generally inadequate both from the standpoint of chemistry and also that of economics, to meet modern requirements for low sulfur

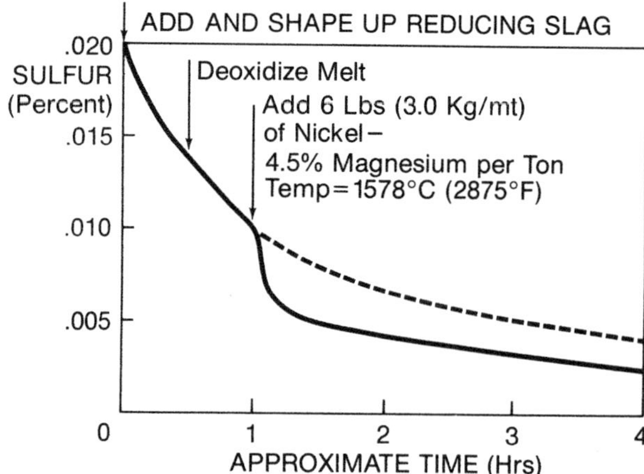

FIGURE 31 Effectiveness of nickle - 4.5% magnesium alloys for sulfur removal

TABLE II Desulfurization in the AOD

VESSEL DESULFURIZATION IN 16-17 TON (14.5-15.45 mt) AOD HEATS

Heat	Grade	% Sulfur Initial	% Sulfur Final	Practice
58073	316	0.021	0.010	No Special Sulfur Treatment
58115	304L	0.022	0.004	400 Lbs (181.8 Kg) Additional Lime in the Reduction Mix – No Other Effort
58031	305	0.037	0.006	Slag Off – 500 (227.3 kg) Lime, 60 (27.3 Kg) CaF_2 – Blow 2 min at 12,000 cfh Ar.
58106	304	0.050	0.007	Slag Off – 600 (272 kg) Lime, 100 (45.45 Kg) CaSi, 50 (22.73 Kg) CaF_2 – Blow 3 min at 12,000 cfh Ar.

steels. The proliferation of various desulfurizing procedures in secondary refining vessels is further confirmation of this fact.

4.3 Sulfur Removal During Secondary Refining

4.3.1 The AOD Process

It has been stated elsewhere "stirring is without a doubt, the most important tool of ladle metallurgy."[27] One of the finest examples of the beneficial effects of violent stirring is to be found in the desulfurization of high chromium stainless steels in the Argon-Oxygen Decarburization Process (AOD).[28] A schematic drawing of an AOD vessel is shown in Figure 32. This violent stirring in association with the high temperatures required for decarburization of high chromium stainless steels can cause severe lining erosion. For this reason, refractory costs are one of the major expenses involved in running an AOD vessel.

The desulfurizing capabilities of the AOD are well illustrated by the data presented in Table II. As mentioned in Chapter 2, the activity coefficient of sulfur in high chromium steels can be considerably less than 1 and may, in fact, approach 0.1. This implies that sulfur removal from high chromium steels is about ten times more difficult than sulfur removal from chromium-free iron. In the light of this fact, the desulfurization results shown in Table II are particularly noteworthy. The various conditions by means of which these results were obtained provide excellent examples of the principles which were described in Chapter 2. The violent agitation obtained within the vessel which was lined with high-magnesia brick, promotes desulfurization in the following ways:

1) High concentrations of CaS are prevented from forming at the slag-metal interface.

2) The high lime slag is well mixed with the molten steel.

3) The oxygen potential of the slag-metal system can be maintained at a low level by the addition of strong deoxidizers - in this case, calcium silicide.

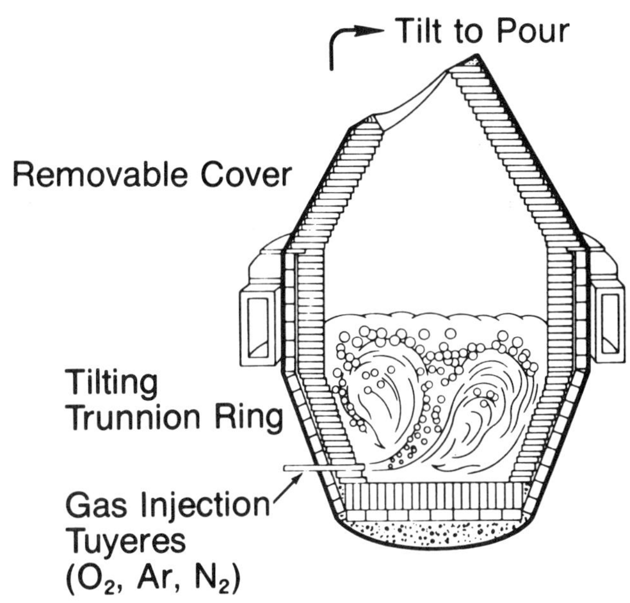

FIGURE 32 Schematic diagram of a typical AOD vessel

The inherent desulfurizing capability of this process is illustrated by the results obtained with heat #58073, in which the sulfur content is decreased from 0.021 to

0.010%. In heat #58115, additional lime is added with the reduction mix, thus increasing the CaO/SiO_2 ratio of the slag. As indicated by the relationship shown in Figure 12, this will result in additional transfer of sulfur from the metal to slag. In this case, the sulfur level was decreased from 0.022 to 0.004%. In heat #58031, excellent desulfurization is achieved by removing the original slag and introducing a $CaO\text{-}CaF_2$ slag which together with additional stirring for two minutes at 12,000 CFH argon, provides metal with a final sulfur content of 0.006% from steel which originally contained 0.037% sulfur. These good results can be attributed both to the beneficial effects of the fluxing agent, CaF_2, as well as the enhanced agitation provided by the argon blowing. The greatest amount of desulfurization was obtained with heat #58106, in which the sulfur content was decreased from 0.050 to 0.007%. This excellent performance was achieved by removing the original slag, forming a new high-lime slag, well-fluxed with CaF_2, adding a strong deoxidizer, CaSi, and blowing for 3 minutes at 12,000 CFH with argon. As shown in Figure 13, the reduction in FeO and MnO (and in this case Cr_2O_3) from within the slag phase, together with strongly deoxidized metal, promotes transfer of the sulfur from the metal to the slag phase.

It will be evident from the above comments that the desulfurization reactions in the AOD vessel are brought to a rapid endpoint by the violent stirring obtainable within the vessel. While the AOD is used mainly for the production of high-chromium stainless steels, the principles illustrated here are applicable to the production of steels for hostile environments or even for the desulfurization of steels for less demanding applications.

4.3.2 The ASEA-SKF Process

A schematic diagram of an ASEA-SKF[29] unit is shown in Figure 33. The electrodes are capable of adding enough heat to compensate for that lost during the time the steel is in the ASEA unit together with any additional superheat required to ensure proper teeming. The induction coils shown on the outside of the ladle supply stirring energy to the steel in order to ensure good mixing of the metal and slag at their common interface. In order to prevent heating of the ladle shell by the induction coils, the shell is made from austenitic stainless steel. Although not shown on this particular diagram, other equipment is available with this particular facility to enable steels to be vacuum degassed.

Conventional ladles are generally lined with bloating fire-brick refractory. However, in this ladle unit superior refractories are used. For example, at the slag line, a magnesite brick lining is used which provides a stable refractory of low-oxygen potential, with the ability to resist attack from high-

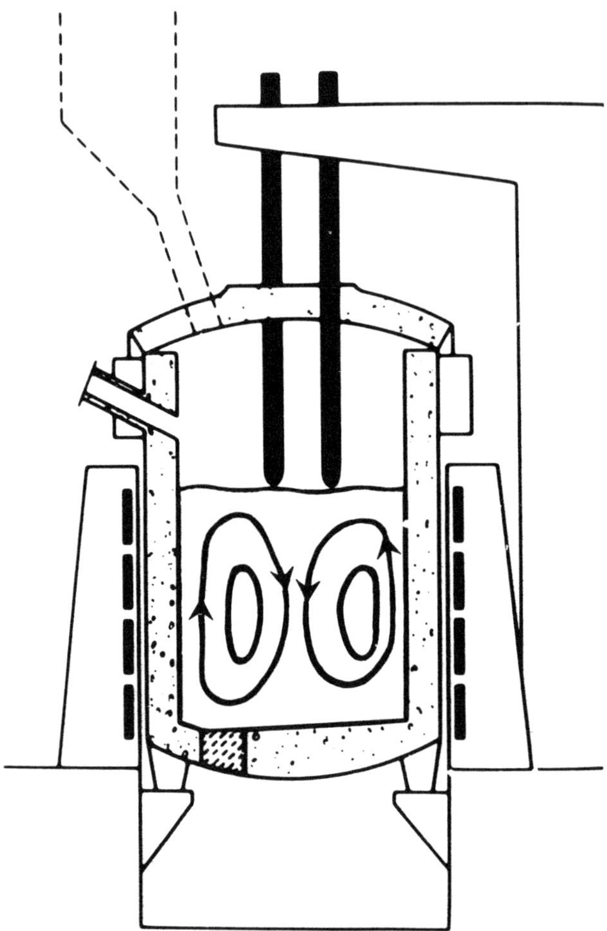

FIGURE 33 Schematic diagram of an ASEA-SKF system

lime slags. The remainder of the ladle is lined with high-alumina refractory which, although not as stable as the magnesite, is sufficiently adequate in this particular location where the walls are less subject to attack by the high-lime slags. With respect to the slags used during this process, efforts are made to ensure a minimum amount of electric furnace slag remains on the ladle. For this reason, the electric furnace is slagged-off and any slag carried over from the furnace is skimmed from the ladle. In this way, only a small amount of furnace slag remains on the ladle during subsequent processing. To obtain good desulfurization of metal, enough CaO, CaF_2 and aluminum is added to the ladle in order to obtain a slag with low oxygen potential and a high basicity index (3.5 to 4.5).

After degassing and slag conditioning, various types of desulfurizing additions may be made. The results obtained with a number of different additions are shown in Figure 34 from which it can be seen that mischmetal additions appear to be the most positive method for reducing sulfur. However, because of their higher price, this may not be the

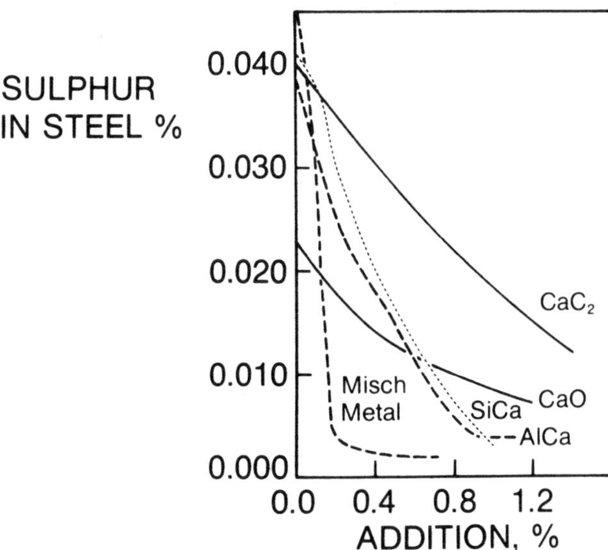

FIGURE 34 Desulfurizing response with various materials in an ASEA-SKF unit

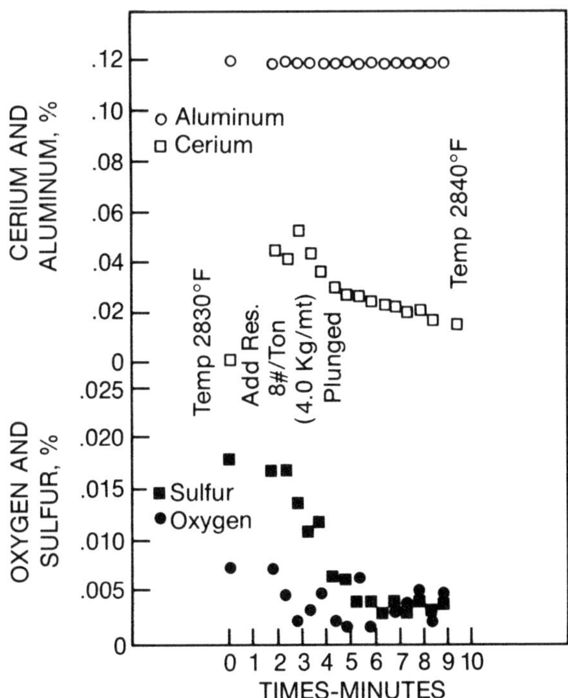

FIGURE 35 Sulfur removal in a magnesia lined three ton induction furnace with rare earths

most economical method. When mischmetal additions of 0.20% are plunged into the metal in the same way as star aluminum would be plunged, sulfur contents in the steel of 0.003% and less are regularly achieved with steel which contained initial sulfur levels as high as 0.015%. The total time required for heating, degassing, slag modification, alloying and desulfurizing is typically about 2-1/2 to 3 hours.

With respect to the mechanism for sulfur removal, the technical literature attributes most of the desulfurization to the precipitation of rare earth sulfides. However, referring to Figure 5, (it will be observed that at 1600 C (2912 F)), it would be necessary to have 0.60% cerium (or rare earths) in equilibrium with 0.003% sulfur to explain this low sulfur content on the basis solely of sulfide precipitation.[30] It would therefore appear that desulfurization is unlikely to be the result of sulfide precipitation alone. From this standpoint, the data provided in Figure 35 may account for the behavior observed during desulfurization in the ASEA vessel. This heat was made in a 3 ton induction furnace with a magnesia lining to which had been added a slag composed of lime with some metallic aluminum to ensure a low oxygen potential. Gentle stirring was achieved in this heat by blowing argon through a submerged lance. When 8 pounds (4.0 Kg/mt) of rare earth silicide [2-1/2 pounds (1.25 Kg/mt) of rare earths per ton] were added together with the aluminum, the oxygen content dropped to 12 ppm, and the sulfur content decreased rapidly from about 0.020% to less than 0.004%.

From Figure 35, it can be seen that after two minutes, the cerium content was 0.050% (0.100% RE), and the sulfur was 0.017%. Using the information given in Figure 5, it can be determined that precipitation of rare earth sulfide should take place. After four minutes, however, the cerium content has decreased to 0.030% (0.060%RE), and the sulfur level had decreased to 0.010%. Under these circumstances, the data in Figure 5 would indicate that the cerium and sulfur contents are both in solution at this point and no further precipitation of rare earth sulfide should occur. It may be, that under the strongly reducing conditions which exist in the presence of cerium, conditions are enhanced for sulfur transfer from the metal to the slag phase, particularly with the stirring achieved by argon bubbling which provides the change in slag-metal interface necessary for rapid desulfurization. Based on these results, it would appear that desulfurization in the ASEA process is a combination of sulfide precipitation and slag-metal reactions.

4.3.3 The Finkl-Mohr Process

The Finkl-Mohr system[31] is very similar in many respects to the ASEA-SKF process, except that all heating, degassing, alloy additions and desulfurization processes are carried out under vacuum. In many instances, the Finkl-Mohr system is operated with conventional bloating fire-brick, however, it is more effective in all respects when the ladles are lined with more stable oxides in the same

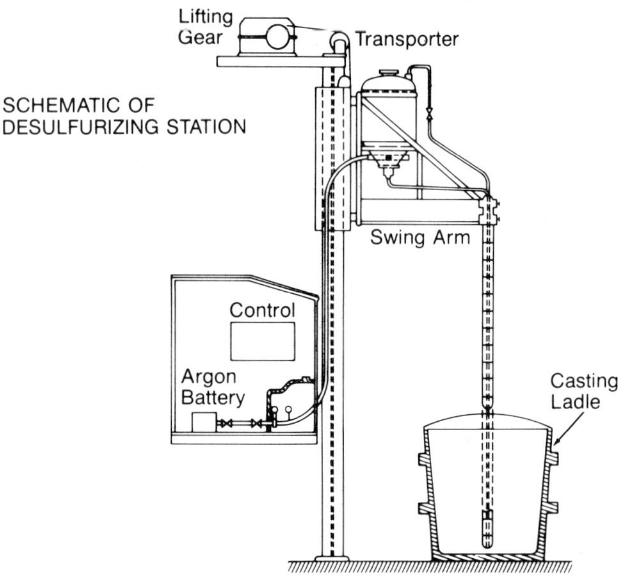

FIGURE 36 Schematic diagram of a TN desulfurizing system

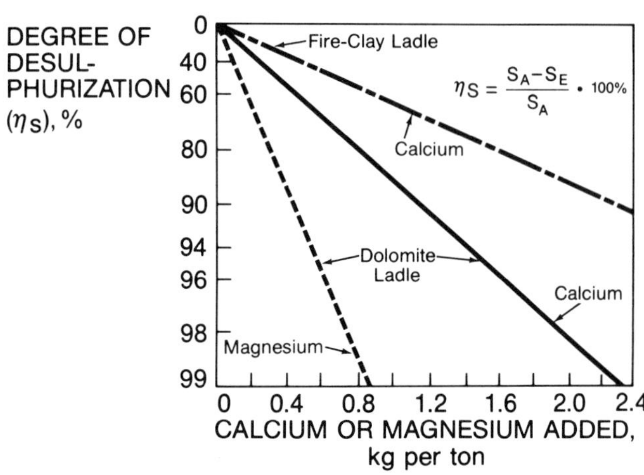

FIGURE 37 Degree of desulfurization as a function of calcium and magnesium additions

manner as those used in either the ASEA-SKF or TN processes. One disadvantage of conducting the process under vacuum at all times is that hopper capacities for additions to be made under vacuum are generally design limited so that the amount of materials which can be added during processing is restricted. In addition, since calcium and magnesium vaporize readily under vacuum conditions, they are unsuitable for use as desulfurizing additions. This implies that desulfurization processes under vacuum will be restricted to those which can be obtained mainly with high sulfur capacity slags. In this respect, the addition of rare earths would ensure that oxygen potentials within the slag-metal system are maintained at a low level, thus favoring desulfurization as discussed previously.

4.3.4 The TN Process

The Thyssen-Niederrhein process (TN)[32] or the Calcium-Argon-Blowing Process (CAB) is shown schematically in Figure 36. In this process, powdered desulfurization reagents such as CaSi, CaC_2 or magnesium alloys are injected to a depth of about 10 feet (3.23m) below the surface of liquid steel in a ladle. A synthetic lime spar mixture is added to the top of the ladle prior to treatment in order to retain the sulfide reaction products which are formed during desulfurization. A ladle cover is installed on the ladle prior to injection in order to minimize splashing of slag and steel and also to facilitate fume collection during the injection process. Optimum desulfurization results are obtained with this process when the steel is fully-killed prior to treatment and steelmaking slag carryover to the treatment ladle is minimized. According to the patent description of the process, the slags used should have a preferred lime/acid ratio of at least 2 to 1 with FeO content in the range of 0.50 to 1.00%. Based on the data contained in Figure 12, this slag should have a distribution ratio between 25 and 30.

Desulfurization results obtained using either CaSi, CaC_2 or magnesium are shown in Figure 37. This diagram also illustrates the superior desulfurization obtained when magnesite or dolomite ladle refractories are used. Sulfur levels of 0.005% can be obtained with this process when the degree of desulfurization approaches the 99% level. The preferred alloy for desulfurization is CaSi. The proponents of this process attribute most of the desulfurization to the precipitation of CaS from the calcium addition. The maximum solubility of calcium in steel at 1600 C (2912 F) is about 0.005% (50 ppm). The data in Figure 6 indicated that rare earth sulfides have about the same free-energy of formation as calcium sulfide. When desulfurization with rare earths was discussed, it was determined that 0.60% rare earths would have to be in solution to be in equilibrium with 0.003% sulfur. It is very likely that a similar amount of calcium would have to be in solution if equilibrium was established between calcium sulfide and molten steel containing 0.003% sulfur. Such high calcium contents have never been determined in steel to which calcium has been added. It would seem more likely that calcium reacts with oxygen dissolved in the metal to form calcium oxide which in turn reacts with sulfur. Stirring to promote slag-metal reactions is accomplished by injection of argon and vaporization of calcium as it rises through the steel. Further details are given in appended papers #8 and #9.

When the TN equipment is used to inject slag making components with CaSi, or slag

TABLE III Comparison of costs of sulfur removal and sulfide shape control with the TN and ASEA-SKF processes

COSTS $/TON LIQUID STEEL

Type of Cost		ASEA/SKF + Vacuum	TN
Refractories	Ladle	2.77	2.13
	Lance	—	0.69
Energy	el Power	1.66	—
	Electrodes	0.89	—
Superheating in LD-Converter		—	1.23
Slag	Lime	0.10	—
	Synthetic	—	0.96
Desulphurizer	REM	9.20	—
	CaSi	—	4.70
Media	Water, Steam	0.43	—
	Argon	—	0.41
Maintenance		0.96	0.64
Labour		1.17	0.30
Diffusion Heating		0.16	0.26
Production Costs		17.34	11.32
Capital Costs at 200.000 Ton/Year		6.65	1.33
Total Costs		23.99	12.65

making components with aluminum, or slag making components alone, sulfur removal to low levels can be attained if the aluminum content is high before the injection process is started.[33] In this way, a low oxygen content in the steel is assured, and this of course, favors desulfurization. The results obtained with different additives are summarized in Figure 38. Based on previous comments, it is possible that if slags which give high sulfur distribution ratios were used with higher tap aluminums, more complete desulfurization might be achieved. While data are available on the use of barium containing silicon alloys for deoxidation purposes, no data are available on the use of these alloys for desulfurization. From information on the deoxidation reactions, there appears to be little or no difference between the use of barium-containing and barium-free alloys.

As mentioned above, stirring is accomplished by means of the carrier gas used to inject the desulfurizing alloys as well as by vaporization of the calcium. For every cubic foot of gas at room temperature, almost 7 cubic feet of gas (0.20 m^3) will be released at the surface of the steel after it is heated to steelmaking temperature. Stirring energy has been computed to be 800 Watts/meter^3 for the expansion of the argon gas and 1200 Watts/meter^3 from the vaporization of the calcium.[27] In contrast, the induction stirring power on the ASEA-SKF unit is only about 100 Watts/meter.^{3} Since the Finkl-Mohr system is stirred only by the bubbling of inert gases, the stirring energy here should be comparable to the 800 Watts/meter^3 component of the TN process.

Although the TN process has been conducted with regular bloating ladle brick, the best and most consistent results are obtained when the process is performed with ladles lined with magnesia refractories. It should be noted, however, that there are several economic penalties associated with the use of magnesia lining:

1) The original cost of the lining is higher than the cost of regular ladle brick.

2) Magnesia refractories are heavier than regular brick and in some cases, this may increase crane loads above safe limits.

3) Magnesia refractories will spall badly if subjected to the thermal cycles commonly used in ladles. For this reason, magnesia ladles are often maintained at temperatures of 1400 F or higher at all times in order to prevent spalling. This necessitates the use of oil or gaseous fuels and could be a costly procedure in today's energy crisis.

The cost of treating steel by the TN process when using CaSi is about half the cost of treating steel to meet the same require-

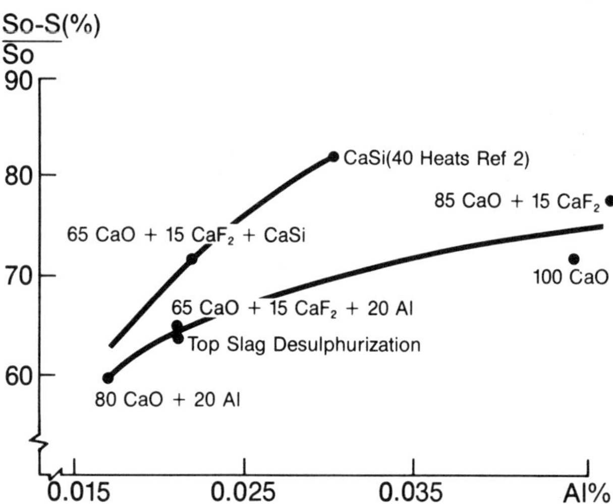

FIGURE 38 Desulfurization as a function of aluminum content before treatment using various desulfirizing slags and slag metal combinations

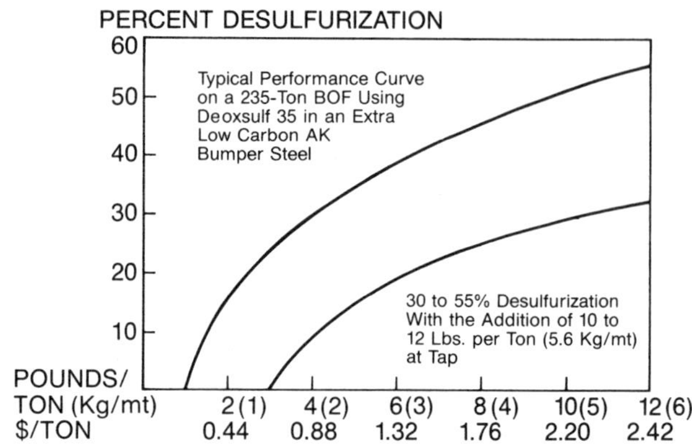

FIGURE 39 Desulfurization with slag additions to the ladle during tapping

ments in an ASEA-SKF unit. The higher cost of the ASEA units are related in most part to the higher capital cost of the ASEA equipment, and the higher cost of the mischmetal compared to calcium-silicon alloys. While calcium carbide is less expensive than calcium silicon, calcium carbide cannot be used for desulfurization except with high carbon steels, since most of the carbon is absorbed by the metal (Table III). When using slags alone for desulfurization, a further reduction in cost of almost $2.00 per ton may be achieved.

Hydrogen and nitrogen are absorbed by the steel when using the TN process. Relative concentrations of these elements are compared in the following Table with the values obtained during treatment of steel in an ASEA-SKF unit:

Process	Nitrogen in wt %
Normal gas purged L.D.	0.006
ASEA-SKF	0.007
TN	0.009

Process	Hydrogen in ppm
ASEA-SKF & vacuum	2.2
ASEA-SKF no vacuum	3.6
TN	3.6

The higher levels of both hydrogen and nitrogen have been confirmed by other operators of the TN process. While these increases in both hydrogen and nitrogen are small, they can be very significant. The higher hydrogen levels increase the possibility of flaking in heavier plates, while the higher nitrogen contents can lead to a reduction in ductility. A particular advantage of the Finkl-Mohr process is that all operations are carried out under vacuum and there is, therefore, a better opportunity to produce steels with lower concentrations of hydrogen and nitrogen.

4.3.5 Desulfurization with Slag Additions

The simplest approach to desulfurization is based on the use of some type of material which can be added to a ladle lined with conventional ladle brick without the use of injection or other special addition techniques.[34] Such methods are being used on a large scale in North America. There are several manufacturers of appropriate additives and one of their patents in this field specifies a preferred composition for one additive as: 70% lime, 15% fluorspar, and 15% metallic aluminum (fine powder form). In order to obtain effective desulfurization, it is stated in the patent that the following conditions should prevail:

1) The slag formed in the ladle should have a high CaO/SiO_2 ratio.

2) The temperature should be sufficiently high to ensure that a fluid slag is obtained with the aid of the CaF_2.

3) The oxygen potential at the slag and steel interface should be maintained at a low value. This is accomplished with the powdered aluminum.

4) Intensive stirring or agitation is beneficial in promoting the desulfurization reactions.

From these comments, it will be apparent that the best results are obtained when the tap hole of the furnace can be appropriately plugged in order to minimize the carryover of furnace slag into the ladle. The results obtained under these conditions in the case of extra low carbon aluminum killed steel are shown in Figure 39. The sulfur reduction in this particular example, is from 0.018 to 0.012%. Although the variability in this desulfurizing process is high, there is, nevertheless, a very significant desulfurization obtained with a minimum cost for materials and no capital expenditures. It is unlikely, however, that this particular approach would be appropriate for the most demanding applications.

4.4 Sulfur Removal During Casting

When rare earth elements are added to molten steel during the teeming operation, it is possible to reduce the sulfur content of the steel from 0.025 to 0.012% with an addition of 0.100% rare earths. The largest user of rare earths in North America, employs this particular technique and obtains desulfurization results of this magnitude. In addition to the reduction in sulfur, the remaining sulfides are predominantly, either rare earth sulfides or rare earth oxysulfides, which are globular and do not deform during rolling. At the present time, only rare earths have been shown to be capable of removing sulfur during the teeming operation.

One of the common methods used for adding rare earth during teeming is to commence the addition as soon as there is a "pad" of metal in the bottom of the mold. The objective is to complete the additions by the time the ingot mold is half full. Under these circumstances, with a rare earth addition of 0.10% for the entire steel ingot, made when the mold is only half full, the rare earth content of the steel at this point could be as high as 0.20%. According to the information in Figure 5, 0.20% rare earths are in equilibrium with 0.006% sulfur at 1550 C (2832 F). It would be expected, therefore, that rare earth sulfides would precipitate from the steel, and aided by the stirring energy generated by the incoming teeming stream, separate out into the surface slag.

4.5 Detrimental Effects of Deviations from Optimum Practice

In general terms, optimum desulfurization is obtained when the oxygen content of both the steel and slag phases are low, the lime silica ratio of the slags is high, the ladle is lined with stable refractories, and the steel is vigorously stirred in order to ensure good contact between the metal and slag phases. While the principles for desulfurization can be stated in relatively simple terms, the execution of these desulfurization principles is not easy. One of the most troublesome features associated with desulfurizing steel in the ladle, is the carryover of high oxygen content slag from the steelmaking vessel. With respect to this particular factor, it has been clearly shown by Stelco[35] that there is a direct relationship between the efficiency of calcium usage (or sulfur removal) and the depth of slag in the ladle, Figure 40. By modifying the tapping practice, Stelco has been able to limit carryover of high FeO slag from the BOF into the ladle to a depth of two inches (5.08 cm), 93% of the time.

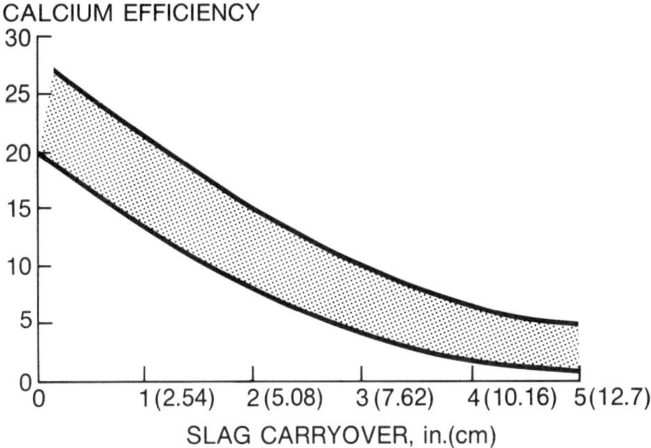

FIGURE 40 Calcium efficiency in desulfurizing in the TN (CAB) process as a function of slag carry over from the furnace

As mentioned previously, optimum desulfurization is obtained in vessels lined with stable refractories, preferably magnesia. The cost of installing these linings and also maintaining them at a temperature high enough to prevent premature failure, is much higher than the cost associated with the use of conventional ladle linings. However, if fire clay ladles are used, the required sulfur removal may not be achieved, and certainly sulfur specifications corresponding to a maximum value of 0.007%, will not be achieved consistently, particularly if there is excessive slag carryover from the BOF. The cost for installation and care of high alumina refractory linings is not as great as that for magnesia linings. However, their silica content ranges from 15 to 25%. Since both silica and alumina are oxides which are less stable than magnesia, desulfurization will not be as effective.

There are innumerable reasons for deviating from optimum practices, but the penalty for deviation will be increased use of desulfurizing materials and a reduction in reproducibility of results. In many cases, it is to be expected that the panalties which are incurred by taking short cuts will outweigh all of the advantages.

CHAPTER 5

INCLUSION SHAPE CONTROL

5.1 Effect of Inclusions on CVN Energy

The ductility of steel, as measured by reduction of area in a tensile test or CVN Energy, increases when the number of second phase particles is reduced and their shape is globular.[36] These second phase particles may be either non-metallic inclusions or carbides. This concept is presented graphically in Figure 41. The most detrimental particles in steels designed for hostile environments are the inclusions, which consist mainly of sulfides. When sulfides, (or oxides) which are plastic at hot rolling temperatures elongate, it is well established that their influence becomes much more pronounced on the physical properties in the transverse direction than on the physical properties measured in the direction of rolling. If the inclusions can be modified so that they remain globular after rolling, then their influence on tests taken in both the longitudinal and transverse direction will be the same. It was stated in the introduction that when CVN impact ductility is high, fracturing along the length of line pipe can be prevented, and if started, it can be stopped. It is necessary, therefore, not only to desulfurize steels which require high ductility, but also to globularize the remaining inclusions. Figure 41 also shows that even if it were possible to produce a steel with no inclusions, the ductility would be deficient if the carbon content was high and in apearlitic or platelike form. For this reason, linepipe steels are ordinarily made with carbon contents of less than 0.10% to meet this requirement.

In the previous Figure, the effect of volume fraction of inclusions on ductility was shown. This same concept is shown in Figure 42, however, in this case, the CVN Energy is shown as a function of the sulfur and oxygen contents of the steel, values with which steelmakers are much more familiar.[37] In the work depicted on this figure, the plates were cross-rolled as much as they were elongated and the CVN values are about the same in both the longitudinal and transverse directions. When the sulfur content is reduced from 0.025 to 0.004%, there is six-fold improvement in CVN values provided the oxygen is less than 10 ppm. If the oxygen content is greater than 25 ppm, there is only a two-fold increase in CVN values for the same sulfur level of 0.004%. In conventional steelmaking the two-fold increase in CVN values with 0.004% sulfur and 25 ppm oxygen is likely. Reduction of oxygen to 10 ppm or less, necessary to achieve a six-fold improvement in CVN shelf energy cannot be obtained with conventional steel plant practices. In most steels, the major inclusion forming component is sulfur and since the amount of sulfur may

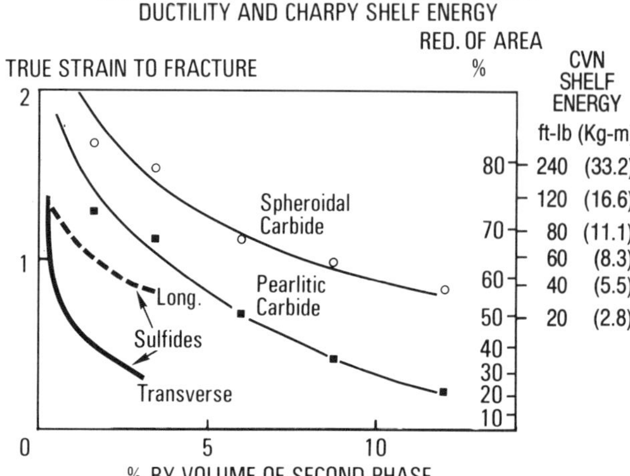

FIGURE 41 The effect of second phase particles on ductility

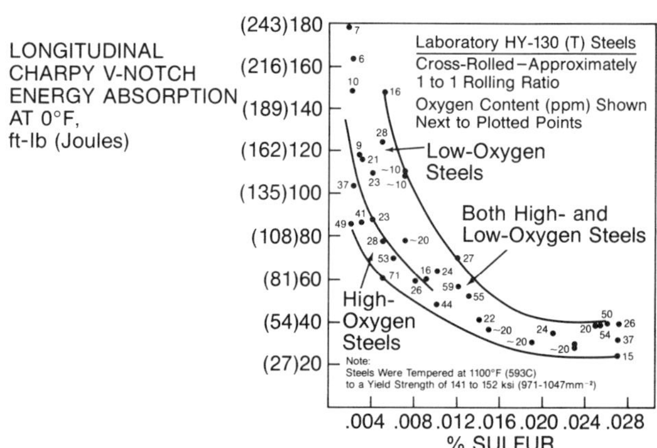

FIGURE 42 Effect of sulfur and oxygen CVN values of quenched and tempered HY-130 plates

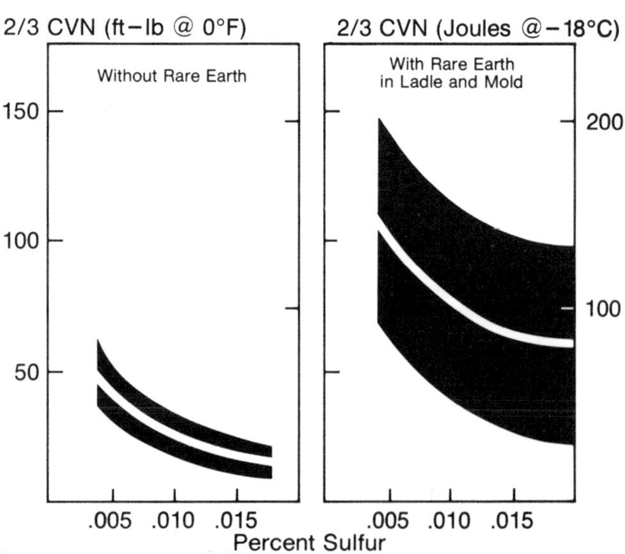

FIGURE 43 Relationship between sulfur and CVN values with and without sulfide shape control with rare earths

be as much as two and one-half times greater than the oxygen content, the effect of sulfur predominates. However, when the sulfur content and oxygen content are equal (0.004% sulfur and 0.004% oxygen), the oxygen content has a decided detrimental effect compared to oxygen values of 0.0006% (6 ppm) where the highest CVN values are obtained.

In the absence of cross-rolling, further improvements in CVN values may be obtained in low sulfur, low oxygen-containing steels by globularizing the remaining sulfides. Rare earths, calcium, zirconium, titanium, and tellurium additions have been used for this purpose. The improvements obtained by globularization of the sulfides, and a comparison of the relative effects with respect to the different additives will be discussed in the following sections.

5.2 Improvements in Toughness with Sulfide Shape Control

5.2.1 Rare Earth Additions

During a 40,000 ton trial at a major North American steel mill,[38] the transverse CVN values of controlled rolled plates were measured for steels with and without rare earth additions. The CVN values obtained with 2/3 width specimens tested at -18 C (0 F), are shown in Figure 43. For normal steelmaking conditions, the oxygen content would not be expected to be less than 0.002% (20 ppm) at any sulfur level. Based on the data given in Figure 42, a two-fold increase in CVN values should be obtained when the sulfur is reduced to 0.005%. Improvements in CVN values from 25 to 50 ft-lbs (33.75 to 67.5 Joules) were in fact achieved. With sulfide globularization by rare earths, CVN values of 60 ft-lbs (81 Joules) at 0.020% sulfur exceed the CVN values at 0.005% sulfur where no attempt was made to globularize the sulfides. At 0.005% sulfur with shape control, the average impact values with rare earths is almost 100 ft-lbs (135 Joules).

The dramatic improvement in CVN values obtained with globularizing additions can be explained when the length of the inclusions found in steels with and without rare earth additions are measured.[39] Data of this type are presented in Figure 44. In this diagram a statistical analysis is given of the length of the inclusions found in the steels. In steel which contained no globularizing additions, 99% of the inclusions were less than 100 microns in length, and 50% of the inclusions were more than 25 microns long. In a companion ingot made from the same heat, to which rare earths were added in the ingot mold, either in the form of mischmetal or rare earth silicide, 99% of the inclusions were less than 20 microns in length and 50% were less than 10 microns long. This reduction in inclusion size readily explains the dramatic improvement in CVN values obtained with the shorter globularized inclusions.

A recent Japanese article[4] has clearly related the average inclusion length with CVN energy, Figure 45. With a mean inclusion length of 25 microns the associated absorbed energy is about 5 Kg-m (47.25 Joules). However, when the average inclusion length is decreased to 10 microns, the average impact energy is almost 10 Kg-m (94.5 Joules).

5.2.2 Calcium Additions

The use of calcium for sulfide shape control has increased rapidly during the last half of the 70's. Improvements in CVN values which can be obtained with calcium are shown in Figure 46. In this example,

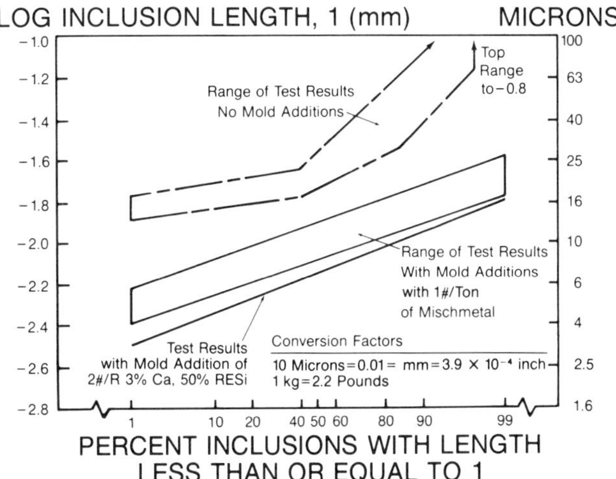

FIGURE 44 Statistical distribution of length of inclusions found in steels with and without rare earths

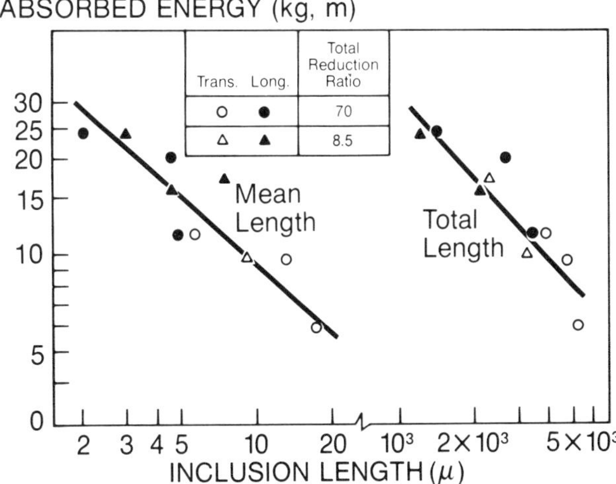

FIGURE 45 Effect of inclusion length on absorbed energy in the longitudinal and transverse direction

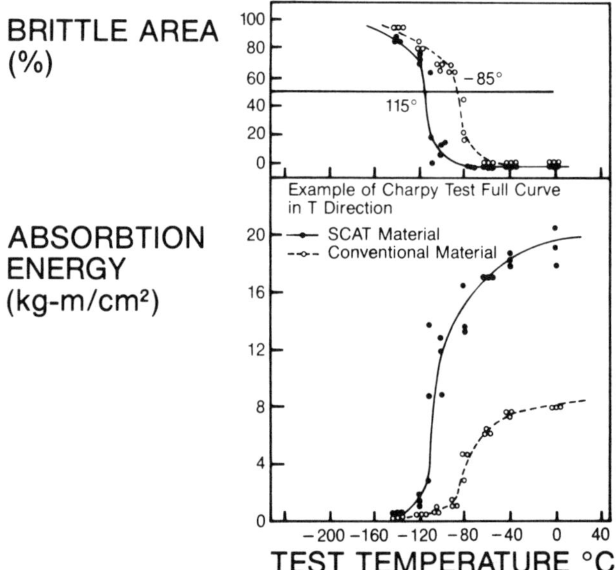

FIGURE 46 Comparison of CVN values in low sulfur steel with and without sulfide shape control with calcium

the calcium was added to the ladle by means of the Sumitomo Bullet Shooting Method,[40] (SCAT-Sumitomo Calcium Aluminum Treatment). CVN values increased from 8 to 20 Kg-m/cm^2 (75.6 to 189 Joules) an increase of 2-1/2 times. The increase in transverse CVN values shown in Figure 43, for rare earth additions, and at similar sulfur levels, was two-fold. When optimum results are obtained with calcium, the improvements are greater than those obtained with rare earths. However, due to the low solubility and high vapor pressure of calcium, it is more difficult to obtain the optimum properties consistently. Data from Oxelosund also confirm this fact.[33]

Process	Toughness-Joules	Standard Dev.
ASEA-SKF (RE's)	96	44
TN (CaSi)	121	59

5.2.3 Zirconium Additions

Sulphide inclusions can also be globularized with zirconium. The effectiveness of zirconium[41] in improving transverse impact properties of pipeline steel together with a comparison between the effects of zirconium and rare earths on tranverse CVN values of the same steel, are shown in Figure 47. Zirconium does not increase CVN values to the same extent as rare earths. In addition, since zirconium has a very strong affinity for nitrogen and will produce nitride precipitates, it will interfere with any heat treating processes which have been designed to precipitate aluminum nitrides. For these reasons, the use of zirconium for sulfide shape control has decreased substantially in recent years.

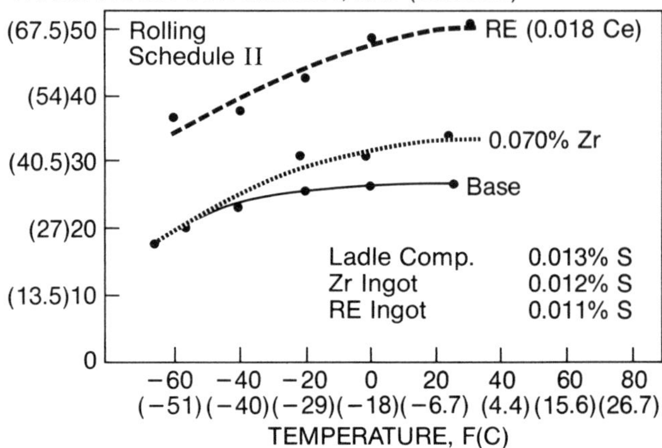

FIGURE 47 Comparison of the effect of zirconium and rare earths on transverse CVN values

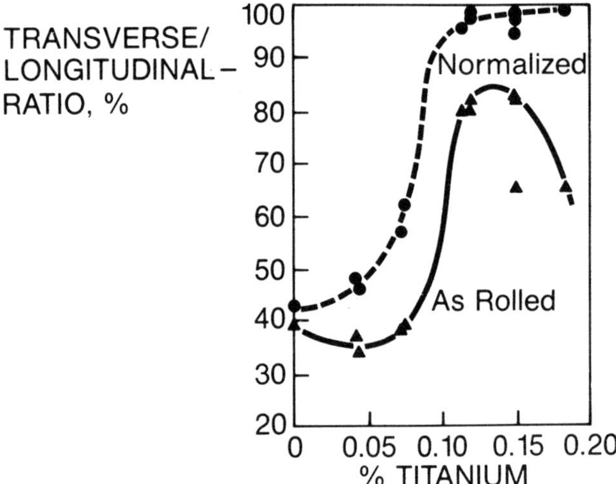

FIGURE 48 Effect of titanium on the transverse to longitudinal ratio of CVN shelf energy on hot strip

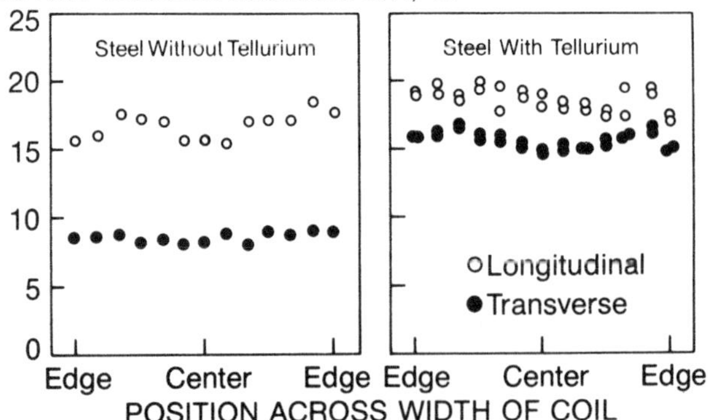

FIGURE 49 Effect of tellurium on the ductility of hot strip measured with a notched tensile test

5.2.4 Titanium Additions

Titanium is another element which will effectively globularize sulfides.[42] In addition to sulfur, titanium has a moderately strong affinity for oxygen and has a greater affinity for nitrogen and carbon than any other element except zirconium. Because of this affinity of titanium for so many other elements, the function of titanium as a sulfide globularizer is difficult to control. This aspect is illustrated in Figure 48 where the ratio of transverse to longitudinal impact properties is shown to reach a maximum of 80% at 0.13% titanium. However, this ratio remains at a maximum value over a very limited range of titanium contents and then decreases rapidly at either slightly higher or lower titanium levels. Titanium steels are noted for their poor toughness in heavier sections and for this reason sulfide shape control with titanium is not used for steels where the plate gauge would be much in excess of 0.25 inches (0.64 cm).

5.2.5 Tellurium Additions

A little known, but effective method for sulfide shape control is by the use of tellurium.[43] Improvements in transverse ductility obtainable with tellurium are shown in Figure 49. Unfortunately, no CVN data are available, but this notched tensile elongation measurement is a good indication of ductility. The effectiveness of tellurium compared to that of zirconium or rare earths has not been reported. Optimum results with tellurium are obtained in the concentration range 0.0025 to 0.006% tellurium in the steel.

5.3 Effect of Sulfide Shape Control Additives on the Composition and Morphology of Inclusions

There would appear to be two predominant mechanisms responsible for sulfide shape control. According to one mechanism, the sulfide shape control additive together with manganese forms a sulfide solid solution which does not deform during hot rolling. In the second case, the non-metallic inclusion consists of a sulfide or oxy-sulfide of the inclusion controlling element. These aspects will be discussed in more detail in the following sections.

5.3.1 Rare Earths

The types of inclusions formed when rare earth elements are added to molten steel have been well documented.[44] Thermodynamic data are available in the literature which predict the sequence of reactions which will take place in molten steel containing oxygen and sulfur. With complete sulfide shape control, inclusions of the type illustrated in Figure 50 are obtained. Due to the strong deoxidizing potential of the rare earth elements, the central core of this inclusion is rare earth oxysulfide. This core is surrounded by a rare earth sulfide. The small amounts of alumina located around the outer edges of the inclusion is not a typical feature. When insufficient rare earth elements are present to form globular sulfides, a combination manganese/rare earth sulfide may form. These combination manganese/rare earth sulfide inclusions are less plastic during hot rolling than sulfide inclusions containing manganese alone. In many cases when the rare earth content of the steel is about 0.05%, the only oxides found are rare earth oxysulfides. With smaller amounts of rare earths present, the rare earths and aluminum may react together to form inclusions of two general types: (RE)$_2$Al O_3

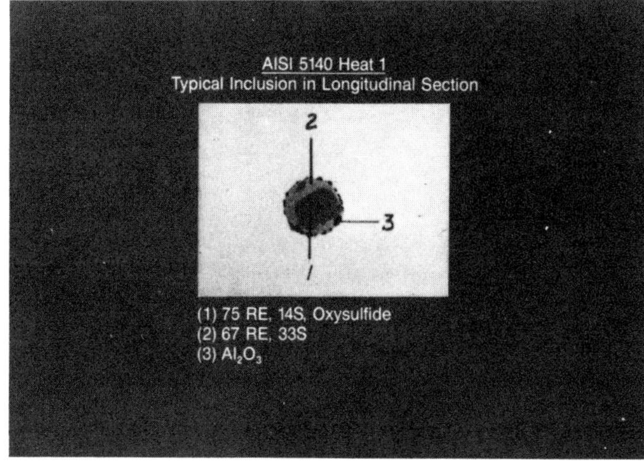

FIGURE 50 Typical rare earth oxysulfide (RE) O_2S inclusion surrounded by a rare earth sulfide (RE)$_2S_3$. Found in a steel with 0.050% rare earths and 0.025% aluminum

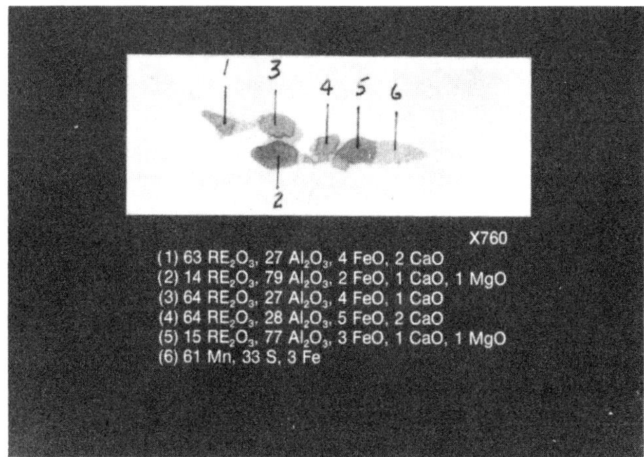

FIGURE 51 Typical (RE) Al_xO_y inclusions found in a manganese sulfide matrix in a steel with 0.020% rare earths and 0.025% aluminum

and (RE) $Al_{11}O_{18}$. Typical examples of these complex oxide inclusions formed in a steel containing insufficient rare earths (less than 0.05%) are shown in Figure 51. Additional information on rare earths is given in appended paper #10.

5.3.2 Calcium

Calcium has the ability to change the properties of manganese sulfides by forming a solid solution with manganese and sulfur. Microprobe analysis of one such typical inclusion taken from a study by Uemura[40] is shown in Figure 52. The core of this inclusion is a calcium aluminate particle which contains little or no sulfur. The sulfur in combination with manganese and calcium is deposited in a ring around the calcium aluminate core which would be the first phase to solidify from the melt. In this particular study, no evidence was presented for the existence of single phase calcium sulfide inclusions. This would suggest that desulfurization with calcium is due to some mechanism other than direct precipitation of CaS.

The chemical composition of inclusions found in steel desulfurized by the TN process using calcium-silicon alloys alone, or other combinations of slags and metals,[32] are shown in Figure 53. The inclusions identified by the number 4 (duplex inclusions with calcium), would appear to be similar to the inclusions shown in Figure 52, where the sulfur is present in a solid solution formed by manganese and calcium. Although other metals and slags can desulfurize steel to low levels, the resulting inclusions are more plastic than those obtained with calcium-silicon alone. This is confirmed by the length to width ratios (L/W) of the inclusions described in Figure 53. This would confirm the necessity for the addition of other sulfide shape controlling elements to those steels where uniformity of properties in both transverse and longitudinal directions is required.

The complexity of the behavior of calcium in desulfurization and sulfide shape control is further illustrated by consideration of the data given in Figure 53. It will be noted that inclusions identified as category 4, appear in all heats examined. Consideration of Figures 8 and 9, particularly along the lime alumina plane, would indicate that inclusions of this type could be partially liquid at steelmaking temperatures and have a sulfur capacity as high as 5. Small droplets of this type could also provide a vehicle for desulfurization. Additional definitive work is required before the mechanisms for sulfur removal and sulfide shape control with calcium can be fully understood.

5.3.3 Titanium and Zirconium

Titanium and zirconium each appear to function in a very similar manner. At lower levels of addition, both elements form complex manganese sulfides containing titanium or zirconium. These sulfides are frequently associated with the carbides and nitrides of the respective elements.[42] With increasing concentration of titanium and zirconium, inclusions with manganese are gradually replaced by nonmetallics of the form $Ti_4C_2S_2$ or $Zr_4C_2S_2$ which appear as grain boundary films. This progressive change in inclusion composition is shown in Figure 54. In Figure 48, it was shown that the ratio of transverse to

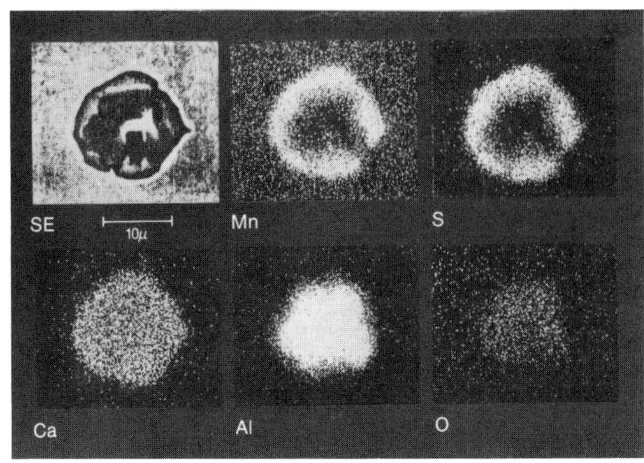

FIGURE 52 Distribution of inclusion morphology obtained with different desulfurizing methods

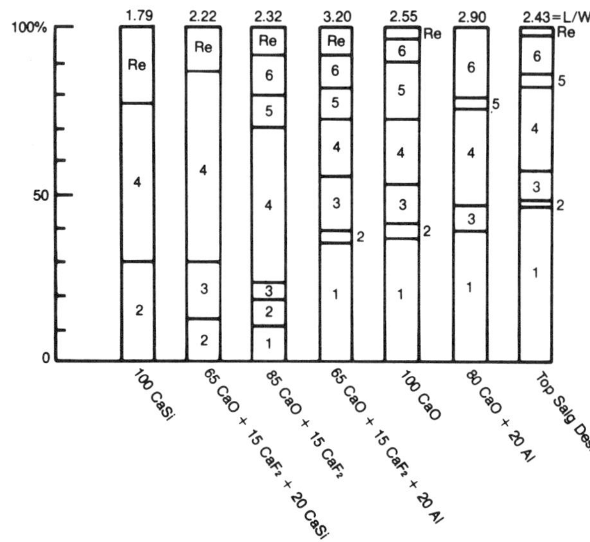

FIGURE 53 Distribution of inclusion morphology obtained with different desulfurizing methods

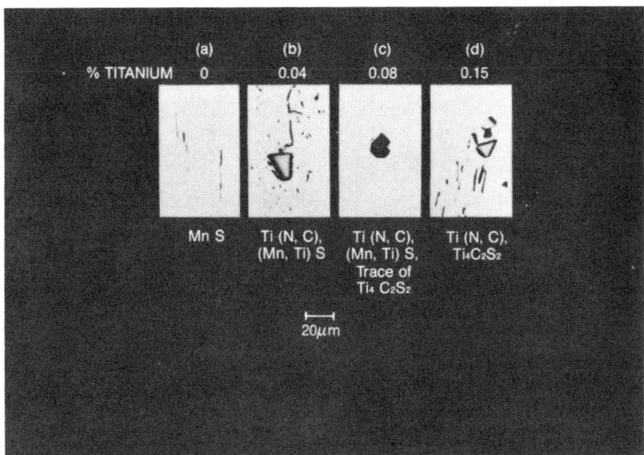

FIGURE 54 Inclusions found in hot-rolled aluminum killed steel containing increasing amounts of titanium

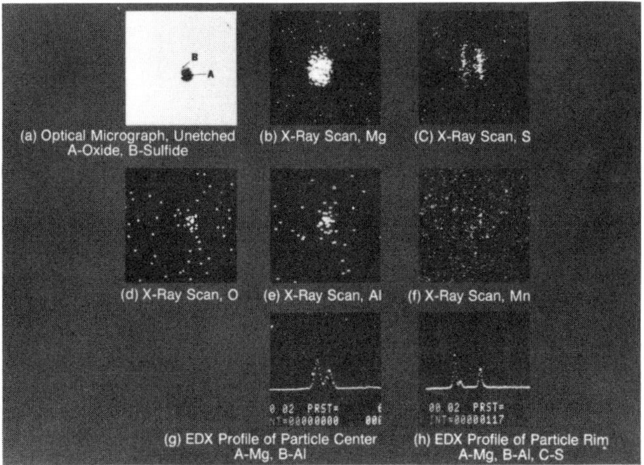

FIGURE 55 Micro-probe data obtained from a steel desulfurized with magnesium

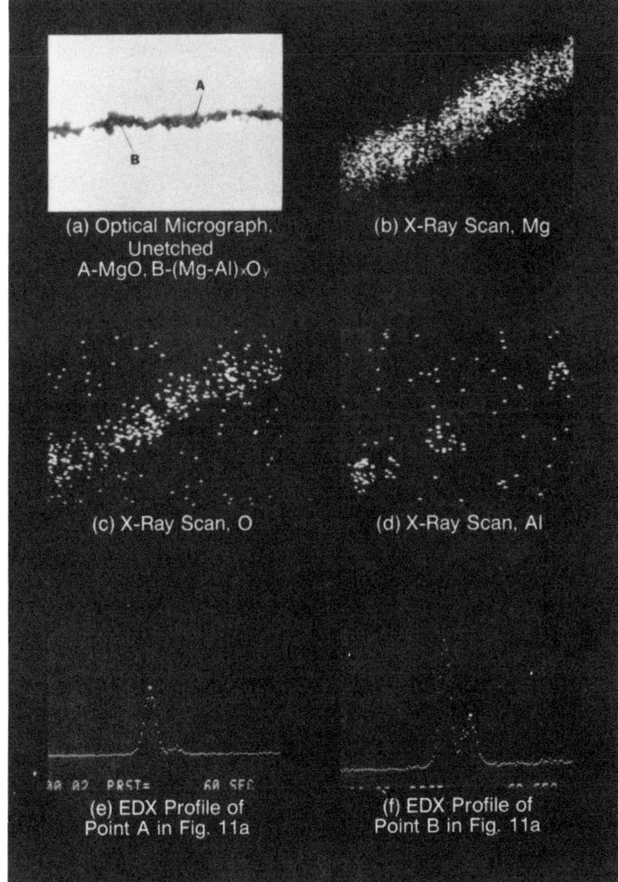

FIGURE 56 Micro-probe data obtained from a typical alumina inclusion that has been modified with magnesia from a magnesium addition

longitudinal CVN shelf energy dropped sharply at 0.15% titanium. This is coincident with the appearance of the $Ti_4C_2S_2$ film type inclusion shown in Figure 54D. Data are available in the literature which show that zirconium carbo-sulfides can cause extensive cracking problems during primary rolling on blooming and slabbing mills.

5.3.4 Magnesium

From thermodynamic data discussed previously it was apparent that magnesium dissolved in molten iron in sufficient quantities would react with sulfur and precipitate magnesium sulfides.[26] Microprobe analysis of the inclusions shown in Figure 55 confirms this behavior. Additional microprobe studies of inclusions found in this steel indicated the presence of MnS - MgS non-metallics. In addition to the effect of magnesium on the sulfides, magnesium can change the composition of inclusions normally formed during aluminum deoxidation. The microprobe data given in Figure 56 indicate oxide inclusions containing a combination of magnesium and aluminum throughout the oxide stringer.

5.3.5 Tellurium

Tellurium would appear to modify manganese sulfides in a manner similar to that of calcium.[43] Microprobe analysis of the inclusion shown in Figure 57 shows the presence of both manganese and sulfur at the ends of the inclusion with some tellurium being distributed throughout.

5.4 Influence of Reoxidation on Shape Control Additives

All of the elements used for sulfide shape control, with the exception of tellurium, are known to be strong deoxidizers. They are therefore susceptible to reoxidation at any stage during the casting operation when steel is exposed to the atmosphere. Various data are available in the literature which describe the detrimental effects of

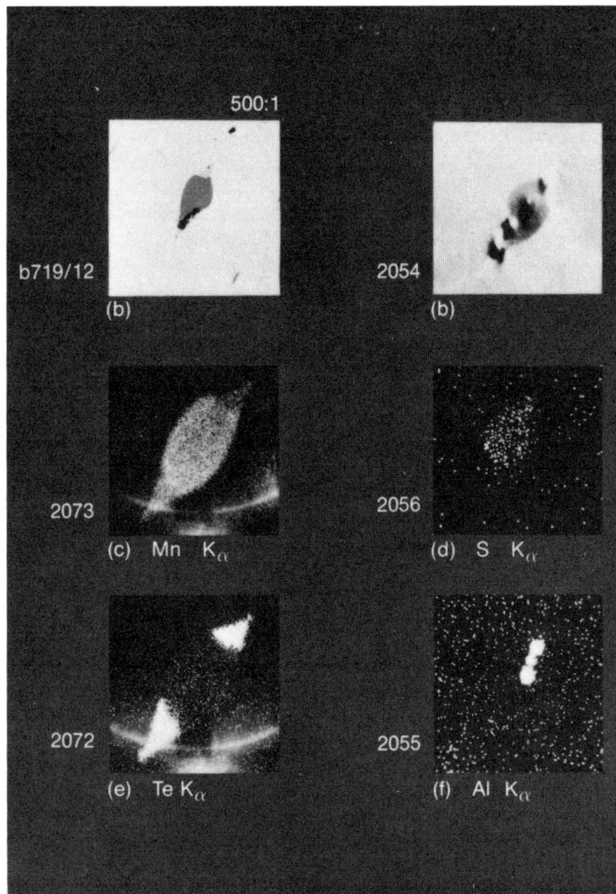

FIGURE 57 Micro-probe data obtained from a typical inclusion found in steel to which tellurium had been added

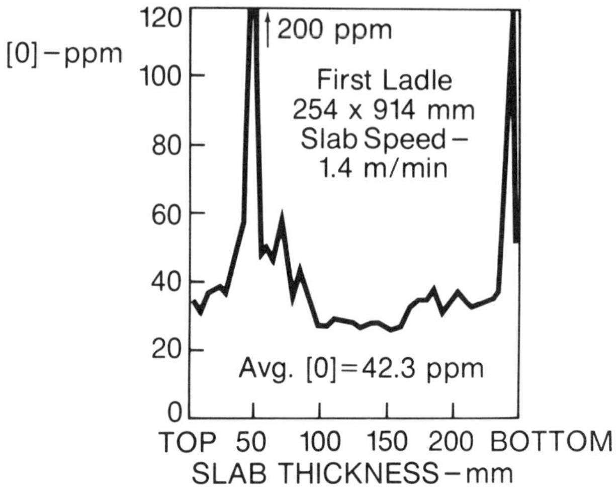

FIGURE 58 Oxygen profile across an aluminum killed slab cast with incomplete protection against reoxidation

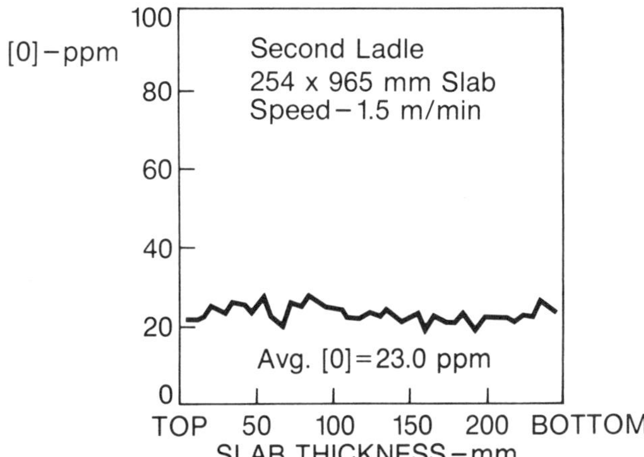

FIGURE 59 Oxygen profile across an aluminum killed slab cast with complete shrouding between ladle and mold

reoxidation on aluminum, rare earth and calcium additives. While aluminum is not used to control sulfide shape, its behavior with respect to oxygen is similar to that of the sulfide shape control elements. In order to illustrate the effects of reoxidation, some observations which have been reported with respect to aluminum, are included below. In addition, since sulfide shape control additives are generally always added to steel which has been aluminum deoxidized, the behavior of aluminum during pouring operations, is a good indicator of the extent to which reoxidation has occurred. This subject is discussed in detail in appended paper #11.

5.4.1 Aluminum

The concentration of oxygen below the surface of a continuously cast slab with imperfect protection against reoxidation is shown in Figure 58 to be in excess of 120 ppm.[45] The average oxygen content of the steel was only 42.3 ppm. When the pouring system was modified to prevent reoxidation of the pouring stream between the ladle and the tundish and tundish and mold, the oxygen

FIGURE 60 Examples of reoxidation of rare earths with and without protection against reoxidation

content was decreased by about half to 23 ppm, and was uniformly distributed across the slab. The low oxygen levels obtained with good shrouding are illustrated in Figure 59.

5.4.2 Rare Earths

The sulfur prints taken from two small ingots cast from the same heat of steel containing rare earths, are shown in Figure 60 to illustrate the effect of reoxidation.[46] The ingot on the left was poured through the atmosphere without protection. As a result, the rare earth oxysulfides formed by reoxidation can be seen concentrated below the shoulder of the hot-top in quantities which would cause severe problems in practice. The companion ingot shown on the right of Figure 60, was protected against reoxidation during teeming by the addition of two pounds (.91Kg) per ton of Teflon (CF_4) chips during casting. The Teflon created an inert atmosphere within the mold and around the stream so that reoxidation products were prevented from forming.

5.4.3 Calcium

The benefits expected from steels which have been desulfurized with calcium-containing alloys, can be lost due to reoxidation effects. During the discussion of the TN process in Chapter 4, a Table was given which indicated that there was an increase in nitrogen content of the steel, which is good evidence that reoxidation has occurred. The rate of dissolution of oxygen in steel is considerably greater than that of nitrogen so that any increase in nitrogen content would be accompanied by a substantial increase in oxygen content. This oxygen will react with the dissolved calcium and if excessive reoxidation occurs, there will be insufficient calcium left to maintain complete sulfide shape control. This aspect is illustrated in Figure 61. This figure shows that when calcium-containing heats are stirred for periods longer than necessary, an increased opportunity is given for contact with the atmosphere, and impact properties are reduced. In the example illustrated here, an extra stirring time of only two minutes was sufficient to bring about a decrease in the impact properties of the steel by 23% (from 150 Joules to 115 Joules).

It is to be expected that steels treated with titanium or zirconium will also suffer detrimental effects if precautions are not taken to prevent reoxidation. A detailed discussion of reoxidation and the influence of pouring stream character on steel cleanliness and consequently inclusion shape control, is given in appended paper #11.

5.5 Influence of Inclusion Morphology on Steel Hydrogen Tolerance

With the production of steels containing lower residual concentration of the surface active elements, oxygen and sulfur, as well as fewer non-metallic inclusions, it is becoming increasingly apparent that the absorption of hydrogen by molten steel and slag during ladle treatments and casting operations, can have serious implications with respect to steel properties during subsequent processing, as well as behaviour in service. Although all of the factors which influence the behavior of steel with respect to hydrogen, are not yet fully understood, it would appear that the total number of inclusions present as well as the shape of the inclusions are important parameters which strongly influence the ability of a particular steel to tolerate hydrogen.

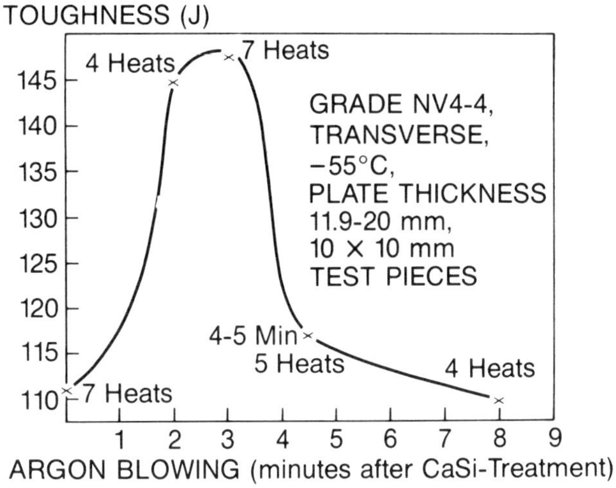

FIGURE 61 CVN values as a function of time of argon is blow after Ca-Si treatment

In Section 4.3.4 on the TN process, values were given for the hydrogen content of steel produced by the ASEA-SKF process compared with that produced using the TN process. The change in levels of hydrogen content during both of these processes is illustrated in Figure 62. At Oxelosund where both of these processes are used to desulfurize steel to equivalent sulfur levels, the importance of the lower hydrogen content obtained with the ASEA-SKF process, (2.2 ppm) compared to that found with the TN process (3.6 ppm), may be judged from the fact that steel from the latter process is not used for heavier plate grades because of the increased risk of flaking.

The detrimental effects associated with cleaner steel, i.e. steel containing fewer inclusions, is shown on Figure 59. In the case of the aluminum killed steels containing no sulfide shape control elements, the detrimental effects of hydrogen seem to be less pronounced. On the other hand when the sulfur and oxygen contents of steel are decreased by the addition of rare earth, or in the case of electro-slag refined steel with

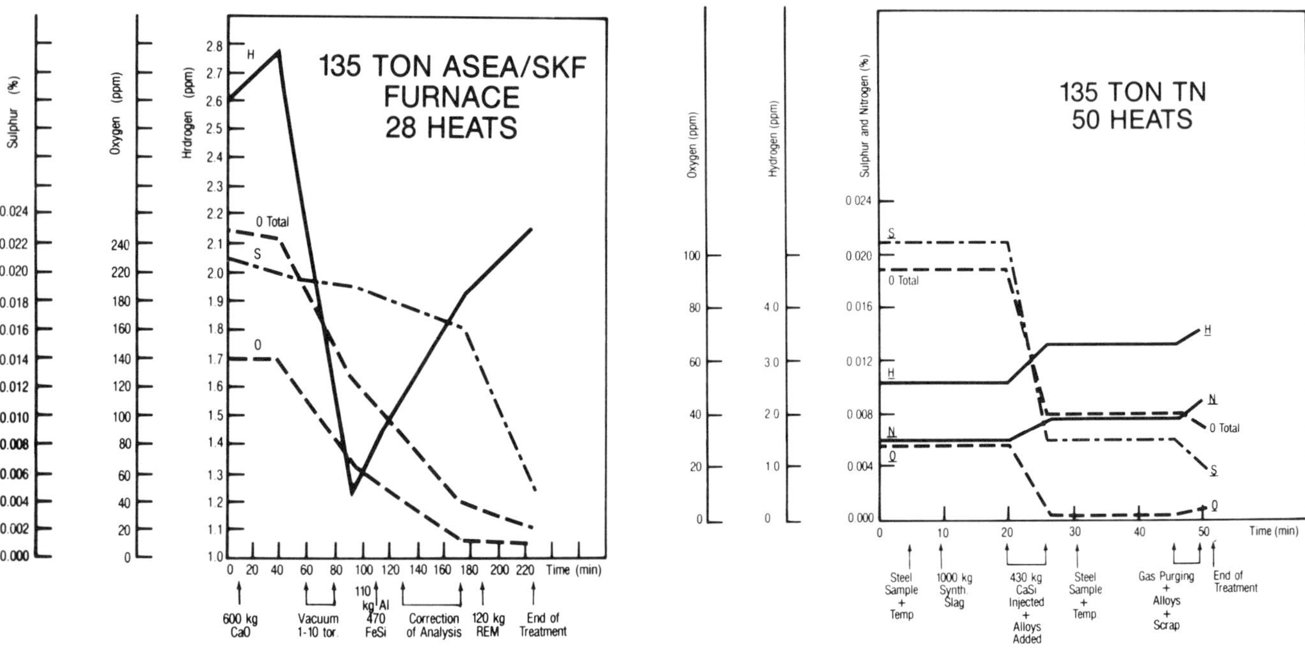

FIGURE 62 Hydrogen, oxygen and nitrogen contents of steels made in the ASEA-SKF and TN processes

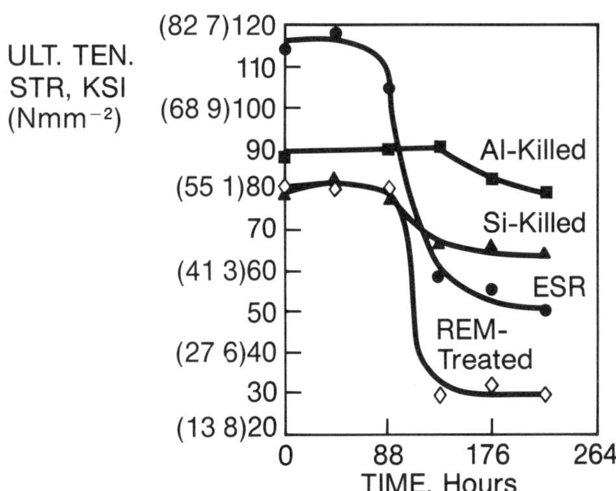

FIGURE 63 Effect of hydrogen charging of steels with various deoxidation practices on their tensile strength in the through thickness direction

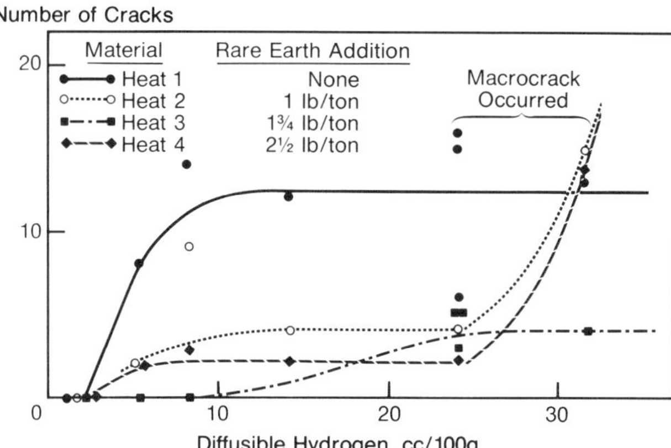

FIGURE 64 Effect of rare earth treatment on the hydrogen cracking tendency of four steels

superior cleanliness, the volume fraction of inclusions to which the hydrogen can diffuse and thus accumulate in innocuous pockets, is substantially decreased. When all of these reservoirs for hydrogen are filled, cracking and loss of strength can occur as is shown in Figure 63.

In addition to the number of inclusions, or reservoirs, available to act as hydrogen sinks, there is good evidence that the shape of these inclusions has a significant influence on the ability of the steel to tolerate hydrogen.[48] This is shown in Figure 64.

When simulated heat affected zones (HAZ) of weldments were charged with hydrogen, a reduction in cracking was observed with those steels where the inclusions had been globularized by additions of rare earth compared to steel of the same composition containing unmodified manganese sulfides which elongate during hot rolling and create sharp notches in the steel matrix. At all levels of rare earth additions, cracking was much less severe at hydrogen concentrations much higher than that at which the steel with no rare earth additions began to exhibit severe cracking. This beneficial effect of sulfide

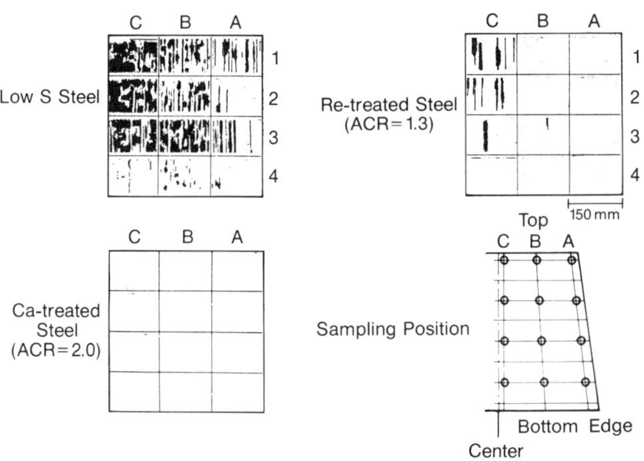

FIGURE 65 Comparison of the susceptability to hydrogen induced cracking between low sulfur and rare earth or calcium containing steels

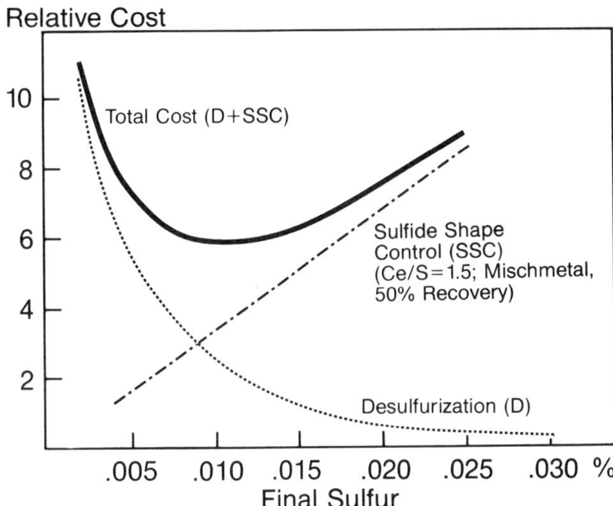

FIGURE 66 Relative cost of obtaining desired properties with desulfurization and desulfurization in combination with sulfide shape control

shape control is illustrated again[49] in Figure 65. In low sulfur steel with no sulfide shape additive, severe cracking was observed with the British Petroleum (BP) cracking test which charges the specimen with hydrogen in a corrosive solution. This behavior is attributed to the presence in the steel of elongated, thin plate-like sulfides. When the sulfur present in such a steel is globularized with calcium, or partially globularized with rare earths (in this case, the residual rare earths were insufficient to ensure complete globularization), the sharp notches are eliminated or at least reduced in number, and the cracking tendency of this steel is decreased.

5.6 Relative Costs for Desulfurization and Sulfide Shape Control

With steels for service in hostile environments, there are occasions when specifications call for maximum values in CVN energy. There are many more cases however, where sufficient improvement in ductility can be achieved with less strenuous measures in order to produce steels with CVN values high enough to meet demanding applications. The relative costs[50] of these various levels of CVN improvement using a combination of desulfurization and sulfide shape control (SSC) are shown in Figure 66. From Figure 43, it can be seen that at 0.010% sulfur, together with sulfide shape control, the average CVN value is 75 ft-lbs (101.3 Joules). With no sulfide shape control and 0.005% sulfur, the average CVN value is only 50 ft-lbs (67.5 Joules). The data shown in Figure 66 indicate that the cost of obtaining steels with 0.010% sulfur together with sulfide globularization is approximately equal to the cost of producing steels with 0.005% sulfur with no shape control. However, with the former practice there is a 50% improvement in impact properties. This represents an economical solution to what could otherwise be a difficult problem.

When average CVN values of 100 ft-lbs (135 Joules), are required, this can be accomplished by decreasing the sulfur level to less than 0.005%, together with sulfide shape control, as illustrated in Figure 43. However, from Figure 66 it can be observed that the combined cost of desulfurization and sulfide shape control will be approximately double that of the cost required to produce steels with 0.010% sulfur and sulfide shape control. Thus, to obtain a 50% improvement in CVN energy, the costs involved are increased by 100%.

Similar curves showing the improvements to be expected over a range of sulfur contents with the other elements capable of controlling sulfide shape have not been prepared because of the lack of available data.

CHAPTER 6

CONCLUDING COMMENTS

1. Sulfur can be removed from either blast furnace hot metal or from steel by the following methods:

   a) The addition of metallic elements which have a high affinity for sulfur, for example, magnesium or rare earths

   b) The addition of reactive compounds such as calcium carbide or soda ash

   c) The use of slags containing lime

2. The mechanisms for sulfur removal may involve precipitation of sulfides within the metal phase by direct reaction between dissolved sulfur and those metallic elements which are soluble in liquid iron. This mechanism applies specifically to those metals which have a reasonable solubility in liquid iron, for example, magnesium and the rare earth elements. Sulfur may also be removed by reaction at the interface of a solid additive, or as is more frequently the case, at interfaces between the molten slag and metal phases.

3. The efficiency of sulfur removal by a slag phase is enhanced by the following factors:

   a) Slags with a high base to acid ratio

   b) Slags with a low melting point and high fluidity

   c) Slags with a low oxygen potential

4. All methods of desulfurization are enhanced if the interfacial area between the reacting phases is continually renewed by vigorous agitation of the melt using gas purging, mechanical mixing, or electromagnetic stirring.

5. The above basic principles apply both to the desulfurization of iron and the desulfurization of steel.

6. Since the activity of sulfur in iron is about five times greater than that of sulfur in steel, sulfur removal from blast furnace iron is more easily accomplished than sulfur removal from steel.

7. By operating the blast furnace to produce high-sulfur hot metal, and desulfurizing the iron by an appropriate external treatment, energy requirements are decreased, and productivity of the blast furnace is increased.

8. Desulfurization of hot metal outside the blast furnace can be accomplished by any of the following treatments:

   a) Mag-coke

   b) Lime-mag

   c) Salt-coated magnesium granules

   d) Calcium carbide

   e) Soda ash

   f) Lime-based slags

9. For optimum sulfur removal from hot metal, carryover of blast furnace slag into the desulfurizing vessel should be minimized. After desulfurization, the sulfide-rich slag must be prevented from entering the steelmaking furnace, otherwise, sulfur will revert to the metal phase during the refining process.

10. While fluid, highly basic slags can be generated in the primary steelmaking vessel, the high oxygen potential of the system does not favor extensive sulfur removal, unless expensive time consuming actions are taken to reduce sulfur by the use of double slag techniques. As in the case of the blast furnace, favourable conditions for sulfur removal can be more readily established outside the primary furnace.

11. Examples of secondary refining processes which can be used to desulfurize steel are listed below:

    a) AOD

    b) ASEA-SKF

    c) Finkl-Mohr

    d) TN

    e) Special Ladle Slag Additives

12. For efficient utilization of desulfurizer additives, the oxygen content of the metal phase should be as low as possible. In blast furnace iron, since the metal is essentially carbon saturated, both the concentration and activity of oxygen are maintained at a low level. For desulfurization of molten steel in secondary refining vessels, oxygen concentration and activity are kept low by ensuring the steel is fully killed before the addition of desulfurizing agents.

13. Effective desulfurization of steel in secondary reaction vessels, will only be accomplished if carry-over of high FeO slag from the steelmaking furnace is prevented. In addition, the reaction vessel should be lined with high stability refractories of low oxygen potential, i.e. they should contain the minimum amount of silica.

14. Desulfurization of steel can be accom-

plished during teeming by addition of rare earth elements to the mold. This has the advantage that interaction between the highly deoxidized steel and refractory materials is avoided.

15. Even when sulfur has been reduced to low levels by various desulfurizing treatments, further improvements in ductility can be achieved by globularizing the remaining sulfides.

16. The elements most commonly used for controlling the shape of sulfide inclusions are as follows:

    a) Rare earths

    b) Calcium

    c) Zirconium

    d) Titanium

    To a lesser extent, magnesium and tellurium have also been used for sulfide shape control.

17. In general, the elements used to control the shape of sulfide inclusions have a greater affinity for oxygen than for sulfur. This implies that the steel must be protected against reoxidation during teeming, otherwise the shape control properties of the sulfides may be lost with sulfur finally being associated with manganese rather than the shape control elements.

18. The effect of lower sulfur and globularization of sulfides on the behaviour of hydrogen in steel is complex. With cleaner steel, i.e., fewer inclusions present, the steel would appear to be more susceptible to problems associated with hydrogen, such as cracking and loss of ductility. On the other hand, when elongated manganese sulfides are replaced with globularized sulfides, the ability of the steel to withstand hydrogen induced problems, is enhanced.

19. Decisions concerning the extent to which desulfurization should be done in hot metal or steel, and whether or not sulfide shape control is appropriate, must be made on a plant to plant basis. Additional factors which must be considered are:

    a) Mechanical properties required in the final product

    b) Availability of capital for installations

    c) Availability of desulfurizing materials

    d) Availability of space necessary to conduct the operation

References

1. G. W. van Stein Callenfels, "The Changing Pattern of Technology in Large Scale Steelmaking", Ironmaking and Steelmaking, Volume 4, 1979, pp 145-154.

2. Proceedings Micro-Alloying 75, Union Carbide Corporation, 1977.

3. F. B. Pickering, "High-Strength Low-Alloy Steels - A Decade of Progress", Micro-Alloying 75, pp 9-30.

4. T. Ikeshima, "Recent Metallurgical Aspect of Linepipe and Oil Country Tubular Goods", Transaction ISIJ, Volume 19, 1979, pp 583-594.

5. R. F. Potocic and K. G. Leewis, "Desulfurizing: Sulfur Control Plus More Production", Symposium on External Desulfurization of Hot Metal, McMaster University, Hamilton, Canada, 1975, Editor: W. K. Lu.

6. J. P. Orton, "The Importance of Low Sulfur on Processing and Properties of Steel", Symposium on External Desulfurization of Hot Metal, McMaster University.

7. John Chipman, "Chemical Behavior of Sulphur in Iron and Steelmaking", Metals Progress, Volume 62, No. 6, pp 97-108.

8. K. A. Gschneidner, Jr. and Nancy Kippenhan, "Thermo-Chemistry of the Rare Earth Carbides, Nitrides and Sulfides for Steelmaking", IS-RIC-5, Rare Earth Information Center, Iowa State University, Ames, Iowa 50010.

9. F. C. Langenberg and J. Chipman, "Equilibrium Between Cerium and Sulfur in Liquid Iron", Trans. AIME Vol. 212, p 290.

10. E. Spire, et al., "Ladle-Metallurgy in Steelmaking", Proceedings Electric Furnace Conference, Volume 32, 1974, pp 222-230.

11. F. D. Richardson, "Physical Chemistry of Melts in Metallurgy", Volume 1, pp 291-327, Academic Press, New York & London, 1974.

12. BOF Steelmaking, Volume 2, p 104, ISS, AIME, 345 E. 47th Street, New York, New York.

13. N. Tsuchiya et al., "Limit of Desulphurization in Blast Furnace, and External Desulphurization by Sintered $CaO-CaF_2$", Symposium on External Desulphurization of Hot Metal, McMaster University.

14. H. Ushiyama, "Ladle Furnace Process in Japan", Ironmaking and Steelmaking, 1978, Volume 3, pp 121-134.

15. K. A. Gschneidner, Jr. et al., "Thermo-Chemistry of the Rare Earths", IS-RIC-6,

Rare Earth Information Center, Iowa State University, Ames, Iowa.

16. BOF Steelmaking, Volume 2, p 116.

17. K. Koch and G. Kaestle, "Phase Diagrams And Thermo-Dynamic Activities in Slags", Proceedings of the International Conference on Physical Chemistry and Steelmaking, Versailles, France, October, 1978, pp 78-83.

18. W. H. Duquette, et al., "Low Sulfur Steel (0.005%S) From Mag-Coke Desulfurized Hot Metal", Proceedings Open Hearth Conference, 1973, Volume 56, pp 79-91.

19. P. J. Koros et al., "The Lime-Mag Process for Desulphurization of Hot Metal", Iron and Steelmaker, 1977, Volume 4, No. 6, pp 34-40.

20. A. M. Smillie and R. A. Huber, "Operating Experience at Youngstown Steel with Injected Salt Coated Magnesium Granules for External Desulfurization of Hot Metal", Iron and Steelmaker, 1979, Volume 6, No. 6, pp 20-28.

21. W. Meichsner et al., "Desulfurization of Hot Metal by Injection of Calcium Carbide Based Mixtures", Proceedings Open Hearth Conference, Volume 56, 1973, pp 113-120.

22. H. Kajioka, "External Desulphurization at MSC", Symposium on External Desulphurization of Hot Metal, McMaster University.

23. G. McGlothlin, "Continuous Desulfurization of Iron in High Production Foundries by the Porous Plug Process", AFS Transaction, 1977, pp 5-8.

24. B. Bahout et al., "External Desulfurization of Iron by Pneumatic Injection of Soda Ash in 200t Ladles", Ironmaking and Steelmaking, Volume 4, 1978, pp 162-167.

25. T. Ohya et al., "Desulfurization of Hot Metal with Burnt Lime", Proceeding National Open Hearth and Basic Oxygen Steel Conference, Volume 60, 1977, pp 345-356.

26. D. J. Wilson et al., "Desulfurization, Deoxidation and Sulfide Shape Control with Nickle-Magnesium", Proceedings Electric Furnace Conference, Volume 33, 1975, pp 196-209.

27. B. Tivelius and T. Sohlgren, "Secondary Steelmaking by the ASEA/SKF and the TN Process: A Comparison", Iron and Steel Maker, November 1978, pp 30-39.

28. P. A. Tichauer, "AOD, A New Process for Steel Foundries", Proceedings Electric Furnace Conference, Volume 35, 1977, pp 174-176.

29. N. Grevillius et al., "Operational Experience of the ASEA/SKF Ladle Furance Process at Bofors Steelworks", Proceedings Electric Furnace Conference, Volume 28, 1970, pp 57-62.

30. W. G. Wilson, "Advantages of the Use of Rare Earths", Proceedings Electric Furnace Conference, Volume 35, 1977, pp 42-49.

31. T. J. Wayne, "The VAD Process", Proceedings Electric Furnace Conference, Volume 36, 1978, pp 82-88.

32. E. Spetzler and J. Wendorff, "Injecting Alkaline Earth Into Steel Melts and Its Effect on the Steel Properties", Proceedings Open Hearth Conference, Volume 58, 1975, pp 358-377.

33. B. Tivelius and T. Sohlgren, "The Influence of Different Desulphurizers on the Process Parameters and Properties of Solid Material During TN Ladle Treatment", Iron and Steel Maker, November 1979, pp 38-46.

34. L. Luyckx, U. S. Patent 4142887, "Steel Ladle Desulfurization Compositions and Methods of Steel Desulfurization".

35. J. D. Young et al., "CAB Treatment of Carbon Steels at STELCO", Symposium on the Ladle Treatment of Carbon Steels, McMaster University, Hamilton, Ontario, Canada, 1979.

36. T. Gladman, et al., "Effects of Second Phase Particles on Strength, Toughness and Ductility", BSC/ISI Conference, The Effects of Second Phase Particles on the Mechanical Properties of Steel, Scarborough (UK), 1971.

37. L. F. Porter, US Steel, Applied Research Laboratories, Monroeville, PA Private Communication.

38. R. Ackert and A. Crozier, "Improving the Notch Toughness of Commercial and High Strength Steels", CIM Conference of Metallurgists, 1973, Quebec City.

39. A. J. DeArdo et al., "A Comparison of the Effectiveness of Rare Earth Silicide Alloys and Misch Metal Additions in Steel", Proceedings NOH-BOSD, Conference, 1980. To be published.

40. T. Uemura et al., "Development and Effect of Calcium Adding Technology into Molten Steel", Proceedings Open Hearth Conference, Volume 59, 1976, pp 457-478.

41. W. G. Wilson, "Results from Various Methods of Adding Rare Earths", Proceedings Electric Furnace Conference, Volume 31, 1973, pp 154-161.

42. L. Meyer et al., "Titanium as a Strengthening and Sulfide-Controlling Element in Low Carbon Steels", Processing and Properties of Low Carbon Steel, TMS, AIME, 1973.

43. A. Massip and L. Meyer, "Verbesserung der Mechanischen Eigenschaften Warmgewalzter Bausthal-Flacherzeugnisse durch Sulfidformbeeinflussung mit sehr geringen Tellurgehatten", Thyssen Technische Berichte, Heft 2/78.

44. W. G. Wilson, et al., "The Use of Thermodynamics and Phase Diagrams to Predict the Behavior of Rare Earths in Steel", Journal of Metals, May 1974, pp 14-23.

45. B. C. Whitmore, "Oxide Segregation in Aluminum-Killed Strand Cast Slabs", Iron and Steel Maker, October 1978, pp 34-38.

46. R. J. Leary et al., "Effects of Adding Rare Earth Silicides, Aluminum and Cryolite to Molten Steel", U. S. Department of the Interior, Bureau of Mines, RI-7091, March 1968, p 28.

47. R. Pishko et al., "The Effect of Steelmaking on the Hydrogen Attack of Carbon Steel", Metallurgical Transactions, Volume 10A- No. 7, July 1979, pp 887-894.

48. W. F. Savage, "The Effect of Rare Earth Additions on Hydrogen Induced Cracking of HY-80 Weldments", Sulfide Inclusions in Steel, Published by American Society for Metals, Metals Park, Ohio 44073, pp 233-251.

49. Y. Nakai et al., "Development of Steels Resistant to Hydrogen Induced Cracking in Wet Hydrogen Sulfide Environment", Transactions of the Iron and Steel Institute of Japan, Volume 19, No. 7, 1979, pp 401-410.

50. L. Luyckx and J. R. Jackman, "Current Trends in the Use of Rare Earths in Steelmaking", Proceedings Electric Furnace Conference, Volume 31, 1973, pp 175-181.

APPENDED PAPERS

1. "Recent Metallurgical Aspect of Line Pipe and Oil Country Tubular Goods", T. Ikeshima, Vice President, Sumitomo Metal Industries, Ltd., Transactions ISIJ, Vol. 19, 1979.

2. "Desulphurizing: Sulphur Control Plus More Production", R. F. Potocic, Supervisor, Operating Practice, Iron Production, K. G. Leewis, Research Investigator, Dominion Foundries and Steel Limited, Symposium on External Desulphurization of Hot Metal, McMaster University, Hamilton, (1975) Ed. W. K. Lu.

3. "Low Sulphur Steel (.005% S) From Mag-Coke Desulfurized Hot Metal", W. H. Duquette, Asst. to Supt., Steelmaking Dept., Bethlehem Steel Corp., N. R. Griffing and T. W. Miller, Engineers, Steelmaking, Bethlehem Steel Corp., Proceedings Open Hearth Conference, 1973.

4. "The Lime-Mag Process for Desulfurization of Hot Metal", P. J. Koros, R. G. Petrushka and R. G. Kerlin, Jones and Laughlin Steel Corp., Pittsburgh, Pa., Iron and Steel Maker, Vol. 4, No. 6, 1977, pp 34-40.

5. "Operating Experience at Youngstown Steel with Injected Salt Coated Magnesium Granules for External Desulphurization of Hot Metal", A. M. Smillie and Richard A. Huber, 62nd National Open Hearth and Basic Oxygen Steelmaking and the 38th Ironmaking Conference in March 1979, Detroit.

6. "Desulfurization of Hot Metal with Burnt Lime", T. Ohya, et al., Nippon Steel Corp., Himeji, Japan.

7. "Ladle-furnace Process in Japan", H. Ushiyama, General Manager, Technological Administration Dept., G. Yuasa, Manager, Shibukawa Works, and T. Yajima, Manager, Technological Administration Dept., Daido Steel Co., Nagoya, Japan, Proceedings Secondary Steelmaking, Metals Society, London, May 1977.

8. "Secondary Steelmaking by the ASEA/SKF - and the TN-Process: A Comparison", B. Tivelius and T. Sohlgren, 37th Ironmaking Conference, Chicago, April 1978.

9. "The Influence of Different Desulphurizers on the Process Parameters and the Properties of Solid Material During TN Ladle Treatment", B. Tivelius and T. Sohlgren, Ladle Treatment of Carbon Steel Symposium, May 1979 at McMaster University.

10. "Thermodynamics and Phase Diagrams to Predict the Behaviour of Rare Earths in Steel", W. G. Wilson, D. A. R. Kay and A. Vahed, Journal of Metals, May 1974, pp 14-23.

11. "Control of Inclusions in High-Strength, Low-Alloy Steels", A. McLean, Professor, University of Toronto, D. A. R. Kay, Professor, McMaster University, Micro Alloying 75 - Symposium Proceedings, Union Carbide Corporation, 1977.

12. "Physical Chemistry of External Desulfurization of Molten Iron and Steel", R. D. Pehlke, ISS-AIME Steelmaking Proceedings, 1978, Vol. 61, pp 511-514.

APPENDIX

TECHNICAL PAPERS ON DESULFURIZATION AND SULFIDE SHAPE CONTROL

Table of Contents

RECENT METALLURGICAL ASPECT OF LINE PIPES AND OIL COUNTRY TUBULAR GOODS

Toshio Ikeshima

Vice President
Sumitomo Metal Industries, Ltd.

This paper originally appeared in Transactions Iron & Steel Institute of Japan, Volume 19, 1979

I. INTRODUCTION

It is a privilege for me to have the Nishiyama Prize this time, which I feel is an esteemed prize, and which I understand is usually given to the researchers and engineers who have completed and achieved excellent and superb academic works. What I have done as my research activities in the laboratory is very small and I am rather overwhelmed with deep appreciation.

I joined Sumitomo Metal Industries, Ltd. immediately after graduation from the university, about 40 years ago. The first ten years I spent as a research engineer and then took a position of a manager responsible for research and development for almost all of the rest of my service years with Sumitomo. What concerned me most during those years as a research manager of the corporation is to find out good research subjects or themes for young competent research engineers. Let me explain a little more elaborately on this matter. The first step of the job to select good research subjects or themes is, as shown in Fig. 1, to know and point out exactly what is really needed by the customers and required by the production departments in the manufacturers. Therefore, we must assess properly how important the needs are and when the needs are exactly required.

We not only hear in person or directly what the customers want but also have to predict the customer needs from various sources of information on historical or long-term future trends of social factors. Our judgement to find out whether or not the research themes selected for possible solutions to the customer needs will really meet and satisfy the company's policy, or whether or not such research themes are within the capability of problem solution by research engineers as involved. Even though the research themes so selected are very important, their research work might spend a longer time period. If the themes are in cases beyond the capability of research engineers available at that time, we have to start the work on such research themes in some cases from the beginning such as education and training to obtain the research work force. As you notice, the most important assignment to the research manager is to find out good research themes or subjects and extract the capability of the research engineers to the full or maximum extent.

Next, I would like to ask the research people to apply a lot of fresh ideas to their research work. Better research themes are not a simple continuation, extension or combination of what is done in the past. I would rather like to define them as the new kind of research field going to be developed. I believe and I hope that the ability of a researcher can be increased by challenging all the time what is unknown.

Accordingly, I can say that a good research manager is a kind of coordinator who can link the needs from the customers or the society with a new research activity.

In order to give a concrete illustration of what I mentioned now, I would like to speak on my experience; how the needs for petroleum resource development have given the impacts on metallography and metallurgy in terms of the steel pipe material. The energy issues are very much debated these days and I hope that what I am going to speak might be of any help to you in deepening your understanding of the relation between iron & steel and petroleum resources.

II. OIL RESOURCE DEVELOPMENT AND REQUIREMENTS FOR OCTG (OIL COUNTRY TUBULAR GOODS)

1. Development of Oil and Natural Gas Wells and the Requirements for OCTG

General future development of oil and natural gas resources will be directed towards the followings:

1) Development of deep and high pressure oil or gas wells
2) Development of oil and natural gas wells including hydrogen sulfide and carbon dioxide
3) Development of offshore oil and natural gas wells
4) Development of oil wells in arctic areas

These days the number of deep oil and natural gas wells of 6000 meters or more are increasing and some exceeding 10,000 meters. Among these oil and natural gas wells, some have the atmospheric pressure of 20,000 psi or more and the temperature exceeding 200°C at their well bottom. Another recent tendency is that we have to drill more wells producing oil or gas with much corrosive gas contents of hydrogen sulfide or carbon dioxide. Further well development efforts are required increasingly in the offshore (Photo 1) or areas with cold and hard weather conditions such as the North Sea and Alaska.

The following is a brief listing of the performance parameters essential to steel pipes for these oil or natural gas well:

1) High strength
2) High collapse resistance
3) Corrosion resistance
4) Sealing properties of joints
5) Reliable quality

OCTG are connected with each other by mean of threaded joints or couplings and they are

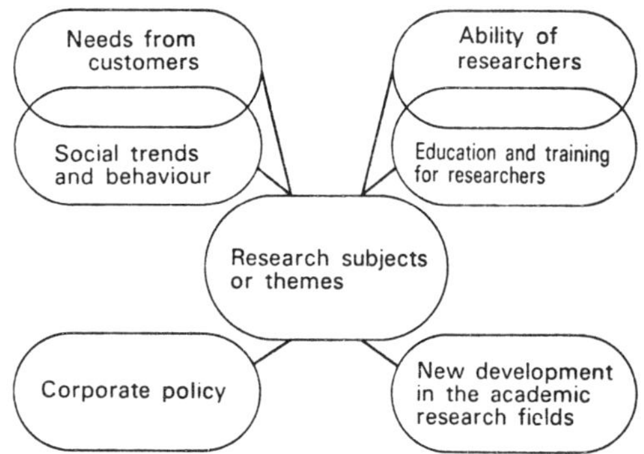

Fig. 1 - Assignment of research managers-Discovery of proper research themes

Photo 1 - Offshore drilling work

Photo 2 - Oil pipeline under construction in Alaska

lowered from the onland or offshore rigs down towards the oil or gas reservoirs. Therefore, the upper pipe should have sufficient strength to withstand the weight of all the steel pipe strings connected to and hung down by it. Because of the significant external pressures which will be acting on the oil pipes laid down in places nearly close to the bottom of the well, the oil pipes are to be designed to withstand pressure stresses. As and when the pipes are used for wells which produce the oil with so much corrosive gaseous components, such pipes are of anti-corrosive steel. The threaded joints or couplings to connect steel pipes are required not only to withstand the weight of all the steel pipes connected or hung down but also to have improved sealing quality so that the oil or natural gas which passes through the inside of the oil pipes may not leak out from such joints or that any liquid outside the pipes may not come in through such joints.

If the string of pipes is broken midway for such reasons as the inadequate strength of steel or defects in the pipe body or threads, all the well facilities and millions of dollars put into the development of such wells will have gone out in vain. Therefore, it is necessary to conduct sophisticated inspections and tests including non-destructive tests for the quality assurance and reliability of OCTG.

2. General Trend of Oil and Gas Transportation, and Pipeline Requirements

As mentioned just above, the types of oil or natural gas wells have changed over the time and so does the transportation of oil or natural gas produced out of those wells. The transportation of oil or natural gas resources will have to meet the following requirements:

1) Increased transportation load
2) Transportation in arctic areas (Photo 2)
3) Transportation along the ocean bottom
4) Transportation of oil or gas heavily contaminated with hydrogen sulfide and/or carbon dioxide

Oil or natural gas exploration projects have been conducted or planned on a significant scale and it eventually required or will require the much larger quantities of oil or gas to be shipped by pipeline facilties. In some cases such pipeline facilities go through the areas with cold and hard weather conditions or along the deep sea beds. The pipelines in other cases have to carry the oil and natural gas with highly corrosive gaseous components. Therefore the pipelines are to meet the design specifications to cope with the followings:

1) Increased diameter and wall thickness
2) Increased strength
3) Increased toughness
4) Corrosion resistance
5) Weldability

The necessity of transmitting oil or natural gas over a longer distance by larger pipelines has made it realized to develop and apply the larger outside diameter pipes. The maximum out-

Table 1 - API specification and examples of steel, heat treatment for each grade[2-4]

Spec.	Grade	Y. S. kpsi (kg/mm²)	Steel	Heat treatment
5A	H-40	40 (28.1) ~	C-steel	As
	J-55	55 (38.7) ~ 80 (56.2)	"	"
	K-55	"	"	"
	N-80	80 (56.2) ~ 110 (77.3)	C-Mn	Q-T or N-T
5AX	(T/B) P-105	105 (73.8) ~ 135 (94.9)	C-Mn-Cr-(Mo)	Q-T
	(C/S) P-110	110 (77.3) ~ 140 (98.4)	"	"
5AC	C-75	75 (52.7) ~ 90 (63.3)	C-Mn	"
	L-80	80 (56.2) ~ 95 (66.8)	C-Mn-Mo	"
	C-95	95 (66.8) ~ 110 (77.3)	C-Mn-Cr-Mo	"

Note As: As rolled, Q-T: Quenched and Tempered, N-T: Normalized and Tempered

side diameter available so far was only 48 inches (1,219.2 millimeters), but it came up to 56 inches (1,422.4 millimeters) for the pipelines recently developed by Soviet Russia. Transportation of oil or natural gas at higher pressures has required the thickness of pipes to increase from 1.000 inches (25.4 millimeters) up to as much as 1.250 inches or (31.8 millimeters). Strength required for the pipelines is also on the move to higher values; from the API grades available in the past in the range of X-52 and X-60 as a maximum up to X-65 or X-70 in recent years.

The recent trend toward the increased installation of pipelines in arctic areas as well as for wells producing corrosive oils and gases has respectively necessitated the development of steel with increased low temperature toughness and increased resistance to corrosion by H_2S, etc.

If we try to satisfy all these requirements as I mentioned, the steel material very often tends to be of less weldability. In most cases the line pipes are connected by welding. The weldability of the steel material is the important requirement.

In the past years, various research and development efforts have been exercised to meet the requirements which came out from time to time for the OCTG and line pipes. I would like to introduce a couple of typical R & D efforts and experiences.

III. RESEARCH AND DEVELOPMENT OF SPECIAL PURPOSE OCTG

1. High Strength Casing and Tubing

Figure 2 shows the historical trend of drilling depth of wells in U.S.A. The maximum depth as recorded in the 1930's was some 3,000 meters. However, in the 1970's there have been an increasing number of productive wells beyond 7,000 meters deep. In wildcat project, it can go as deep as 10,000 meters. As the number of deeper wells increases, the requirement of quality to OCTG has been significantly more severe and the kinds of specifications ruled by API are increased and widely accepted in many countries. Given in Table I are API grades,[2-4] for OCTG example of type of steel and heat treatment.

In old days, the grades in H-40 or J-55 grade were most commonly used and their yield points were 40,000 psi (28.1 kilogram per square millimeters) or 55,000 psi (or 38.7 kilogram per square millimeters). However, the higher grade such as P105 or P110 with the yield point of 105,000 psi (or 73.8 kilogram per square millimeters) or more are being used with the increased drilling depth.

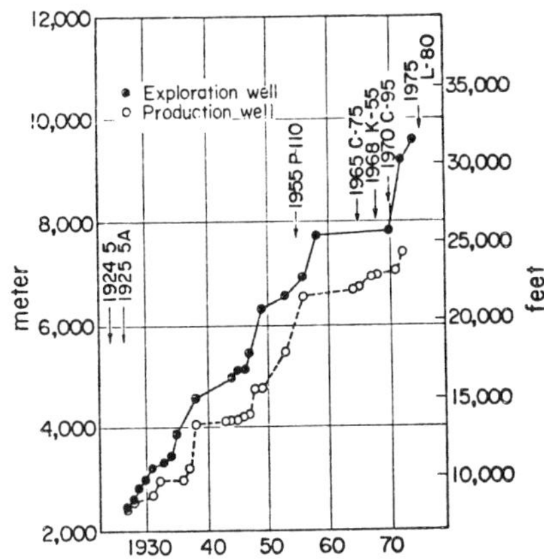

Fig. 2 - Maximum depth changes of exploration and production wells in U.S.A. and API new specification establishment[1]

At some wells deeper than 5,000 meters, high strength OCTG pipes with the yield point of as high as 150,000 psi (105.4 kilogram per square millimeters) have become necessary and used practically, although API has not yet specified for this requirement.

Such pipes are made of Cr-Mo-V steel and the heat treatment is conducted by quenching and tempering. Quenching property is the vital element to get the high strength steel and many efforts have been exercised for the careful examination of chemical compositions of high strength steel and improvements of heat treatment technology. Specially elaborate attention was paid to the prevention of (quenching) crack of high strength steel during quenching, the improvement of low temperature toughness, and the uniformity of quality of the steel after heat treatment.

Photo 3 - Result of collapse test of OCTG

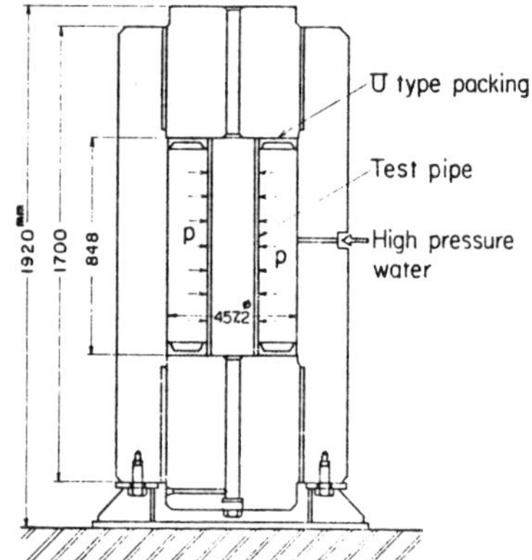

Fig. 3 - Collapse testing devices of OCTG

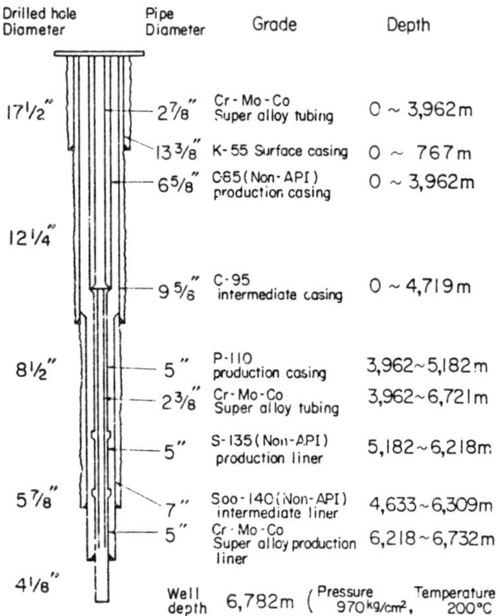

Fig. 4 - String design of very deep sour gas well in Mississippi, U.S.A.[9-12]

2. High Collapse Casing

One of the various types of OCTG is "casing" which is usually applied to the outer side of the drilled hole. The casing tube is exposed to the external pressures caused by the earth or ground surrounding such pipe and to the external pressure by the "completion fluid" to control formation pressure and minimize formation damage. Therefore, the casing has to be designed to withstand such external pressures.

If the casing is short of the pressure-resistant properties, the collapsing shown in Photo 3 will eventually take place. The deeper the wells go down, the stronger the external pressures will be on the casing to be used in places deep to the bottom of wells. Therefore, the endeavors have been made successfully to develop the high resistant to collapse casing. (Fig. 3)

In order to obtain a higher resistance to collapse with the casing, we had to improve the strength of the steel material for the casing and at the same time we had to look into the appropriate shape of the casing with due consideration of thickness deviation and ovality. In addition, any measures were to be devised to reduce to a minimum the residual stress of the casing which develops while the pipes are manufactured.

Therefore, various basic research programs were planned and implemented for the purpose of more improved quality control at pipe manufacturing factories and for assessing the influencing factors by the collapse test facilities especially installed.[5]

3. Resistant to S.S.C.C. (Sulfide Stress Corrosion Cracking)

The problem with the development of high pressure gas wells with high content of hydrogen sulfide is sulfide stress corrosion cracking in the steel material.

Estuary areas of the Mississippi River in the United States have been well known since many decades ago for the natural gas reservoirs which contain about 30% of hydrogen sulfide.[6-12] We were awarded a research contract for the development of casing usable for the well in this area and spent years for the purpose.

As is shown in Fig. 4, the depth of the wells in this area is some 22,000 feet (6,700 meters) and high strength OCTG were planned to be used. However, very high strength steel is by far unfavorable for the purpose in terms of the possible crack of breakage to be caused by the sulfide stress corrosion cracking. Therefore, it becomes necessary that the material design of casing should be implemented by increasing some more thickness to the pipes by so doing to give the appropriate strength at a certain level.

From necessity, we have successfully developed the steel material with good resistance to SSCC which is fit to produce the casing with the outside diameter of 6 5/8 inches (or 168.3 millimeters), the thickness of one inch and a quarter (31.8 millimeters) and the yield point of some 90,000 psi (63.3 kg per square millimeters). These pipes are made of Cr-Mo steel and are heat treated in the same way as with other types of OCTG. However, it has to be noted that is the hardness of steel is so high it tends to cause easily cracking by SSCC due to the hydrogen sulfide, and that if the hardness of steel is lowered in the other way round the steel will become short of strength. Therefore, the heat treatment should be strictly controlled so as to obtain the hardness of steel within a certain narrow range of values. As things so stood, various experiments were conducted to prove the most appropriate heat treatment method and to make the metallographic features as

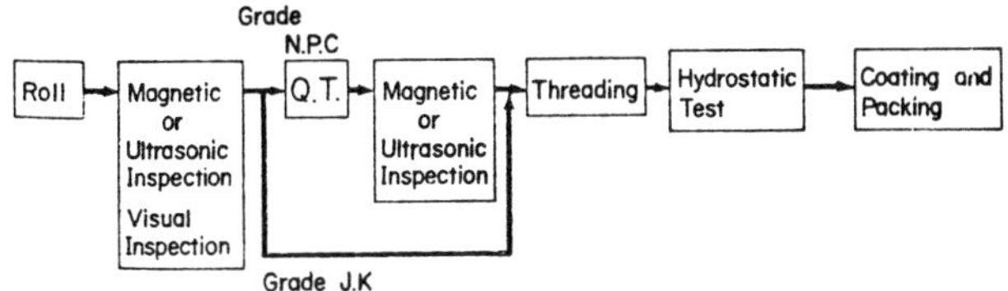

Fig. 5 - Example of inspection system of OCTG

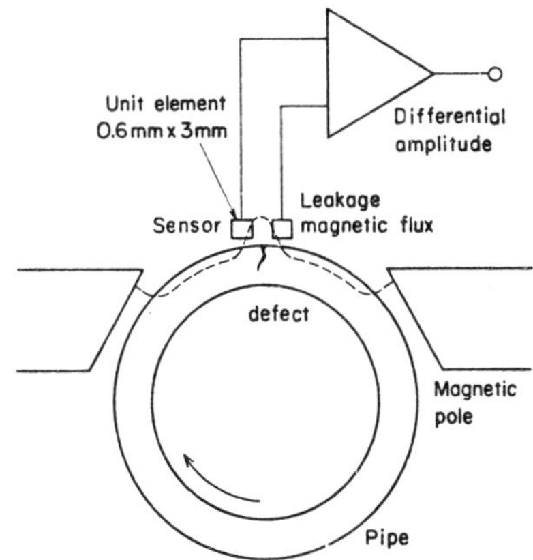

Fig. 6 - Mechanism of non-destructive inspection devices, SAM

uniformly as possible despite the fact that the pipes are thicker.

4. Reliable Quality of Products

As the conditions for application of OCTG are getting more severe and stringent, the reliable quality of the pipes will become the most growingly important and major requirements. Highly sophisticated non-destructive inspection devices have been developed and employed for detecting the material defects or heat treatment cracks of the high strength OCTG.

Figure 5 shows an example of the OCTG inspection system. Under this system, the high grade pipes will be inspected by automatic magnetic and/or ultrasonic inspection on as-rolled stage, and then the pipes are quenched and tempered and finally they are placed again under automatic and/or ultrasonic inspection; piece by piece and all over the length of pipes. One of such testing devices we have developed is the automatic electromagnetic inspection devices which we refer to as "SAM", and Fig. 6 shows the fundamental principle of these devices. The principle applied to these devices is to make use of a phenomenon that the magnetic flux of any magnetised test pipes should show the leakage flux at the defects. In the traditional way, the magnetic particles are sprinkled over the pipes for checking the defects visually. However, the use of Hall elements has made it possible to replace the checking the defects by visual inspection and to do it automatically. The smaller the sensor, the smaller defects the automatic devices can detect. Therefore, the functions of SAM depend largely upon the small-sized sensor available. When we were working on these automatic electromagnetic inspection devices, we happened to learn by chance of the development of a new product by SONY Corporation called as "SONY magnetic diode (SMD)". We readily submitted our request to SONY Corporation for the permission of use of magnetic diode for sensors and met their acceptance. This could serve as an important moment which could lead to the successful development and completion of SAM. And I was honored by Ohkochi Award for this work.[13,14]

IV. RESEARCH AND DEVELOPMENT OF LINE PIPES FOR AREAS WITH COLD WEATHER CONDITIONS

1. Research on Line Pipe Fracture at Low Temperatures

Now, I would like to talk on the line pipes.

Line pipes are used for transporting oil and natural gas in large quantities and therefore their outside diameters are as large as a meter or so but the pipes themselves are not so thick. However, the pipelines required for any oil or natural gas projects would amount to a huge aggregate total quantity or weight, as illustrated by TAPS (Trans-Alaska Pipeline System) in the United States which made use of total 500,000 tons of steel pipes.

It becomes necessary to employ a mass production system and there might possibly develop some limiting factors or constraints with the production process which requires the heat treatment. At the same time in case that the line pipes are to be installed and erected in the areas with cold and hard weather conditions, various considerations will have to be made with the possible rupture of line pipes at lower temperature levels.

There are lots of experiences and studies available so far with respect to the steel plate rupture at lower temperatures, but there still remains much to be studied and solved with the steel pipes.

Let me have this opportunity to explain the fracture characteristics of line pipes; namely the brittle fracture in general terms and the unstable ductile fracture which is special with the steel pipes.

When it matters with the oil pipelines, we must think of the brittle fracture. In case of the natural gas pipelines, we have to look into the brittle fracture and also the possibility that the unstable ductile fracture might spread over a long distance.[15] If the unstable ductile fracture spreads over a long distance of pipelines, a large quantity of gas would gush out and might possibly cause a big fire accident. Even if it takes place, we have to make endeavor to limit its spread to a minimum.

Figure 7 compares the cracking velocity and the pipeline pressure reduction in case of the fracture of pipelines. As and when the fluid in the pipelines is the oil, the pipeline inside pressure will reduce rapidly at a time of pipeline rupture and its reduction velocity will range from 1,500 to 1,200 meters per second. However, the velocity of pipeline cracking to be caused by the brittle fracture is rather slow

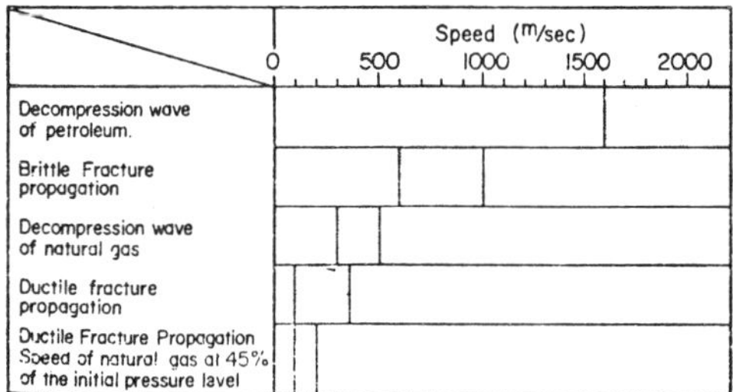

Fig. 7 - Comparison between fracture speed in line pipe and speed of decompression wave

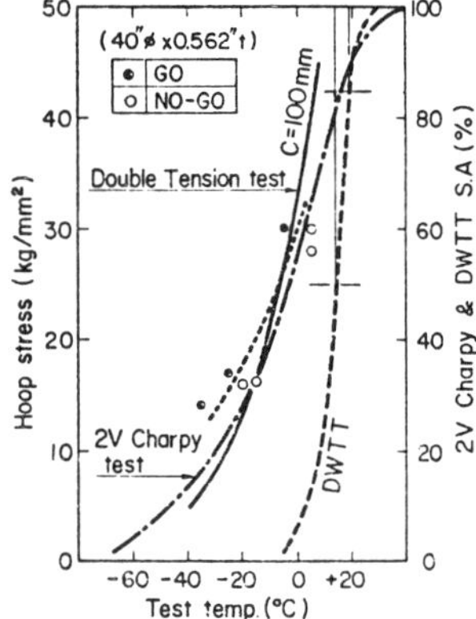

Fig. 8 - Comparison between results of low temp. pipe burst test and those of 2V Charpy test, DWTT and double tension test for specimen

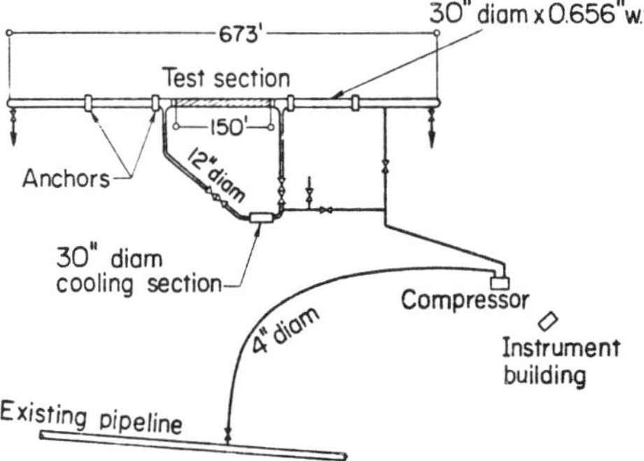

Fig. 9 - Athens test site

and at a level of some 600 meters per second and the pipeline cracking will not continue so long. Therefore, it will become necessary and important to thoroughly probe and study the factors which might cause the cracking or fracture of pipelines for the purpose of manufacturing the fracture initiation-resistant oil pipelines.

In case the pipeline fluid is in the gaseous form, the pressure reduction velocity is lower than brittle fracture speed and more cracking will develop in the oil pipelines.

Given that 55% of gas is leaked out of the pipelines, then the ductile fracture may spread over a longer distance because the pressure reduction velocity is at times slower than ductile fracture propagation of the pipelines.

Photo 4 - Burst test of a single actual size pipe

Photo 5 - Result of unstable ductile fracture propagation test

One of the problems of manufacturing gas pipelines is the production of steel material which can prevent brittle fracture and the unstable ductile fracture and it is also necessary to set up the criteria by which the pipeline fracture characteristics can be proved.

We have conducted tests to burst fracture many types of line pipes under various conditions. Photograph 4 shows one of such tests we made at the open space near our Hazaki Research Center.(16)

1. Arrest of Brittle Fracture Propagation

Figure 8 is the summary of the results of the burst fracture tests at lower temperature levels by using actual size specimen of steel pipes and the laboratory small size toughness tests. Open small circles denote the pipes which can prevent the brittle fracture propagation because of their brittle fracture characteristics. Dotted lines are the curves showing the marginal values of the arrest of the brittle fracture propagation. These curves give the conditions which are very similar to those as obtained by double tension tests conducted at the laboratory facilities. Therefore, it follows that the brittle fracture of the steel pipes at lower temperature levels could be reasonably assumed or predicted by double tension testing. I can conclude that we have the good established criteria for attesting the spread of brittle fracture.

2. Arrest of the Unstable Ductile Fracture Propagation

On the other hand, the unstable ductile fracture has never been experienced with the steel plates. Accordingly, various rupture tests have been planned and implemented in recent years in many countries of the world. In Japan, we were unable to secure the appropriate locations to conduct such tests and we decided to request that we might use the facilities at Athens, Ohio owned by Battelle Memorial Institute of the United States. Figure 9 shows the general plain view of Battelle's test facilities. A pipeline as long as some 200 meters was placed with the test steel pipe arranged in the middle part of said pipeline. Natural gas was supplied from the commercial pipeline system running nearby the test area and compressed by the compressor unit to a high pressure level to fill in the test pipeline and maintain a given pressure. Experiments were conducted during the cold winter months to obtain the required temperature levels. Photograph 5 shows the test result.

We have repeated study works to define the criteria for the prevention of the unstable ductile fracture propagation. Many of the researchers in various countries report that there might be a likely correlation between the arrest of the unstable ductile fracture propagation and the energy values of Charpy at the testing temperature of line pipe burst fracture tests.

However, the logical or theoretical analysis of the line pipe fracture mechanism has not yet been established as to the minimum level applicable to absorption energy values of Charpy or other details, and much has to be studied by researchers and scientists in many countries.

2. Development of High Toughness Line Pipes

In any case the line pipes to be used in the areas with cold and hard weather conditions are required to have high toughness at lower temperature levels. Table 2 shows the typical example of the levels of absorption energy values for the line pipes. Higher energy values are requested for actual application year after year.

Table 2 - Example of the specification of 2V Charpy energy in line pipe

Year	Item	Assurance of 2V Charpy energy
1969~	TASS (Alaska)	−10°C, 30 ft-lb (4.2 kg·m)
1972~	BP (North Sea)	−10°C, 45 ft-lb (6.2 kg·m)
1973	AGL (Australia)	0°C, 50 ft-lb (6.9 kg·m)
Plan	CAGSL (North America)	Minimum temperature of 100% shear area, 80 ft·lb (11.2 kg·m)

Several means are considered to improve the low temperature toughness of line pipes, which include the followings:

1) Chemical composition: low carbon, low sulfur, addition of chemical element(s)
2) Plate rolling: low temperature heating, controlled rolling
3) Welding technique: welding material, welding condition
4) Heat treatment: quenching and tempering

In terms of the chemical compositions of line pipes, lower carbon content is preferred but it also causes decreased strength to some extent, which should be corrected by the addition of appropriate chemical elements such as Nb or V. Lower sulfur content improves the low temperature toughness of the line pipes. In case of the controlled rolling, strict temperature control is required for heating the slab at low temperature and for rolling the plates at the narrow range of temperature.

It is necessary to select carefully the welding material with high low-temperature toughness features and weld the material under the conditions with least input heat. Heat treatment method is in some cases applied to seamless steel line pipes for improving their low temperature toughness. However, the heat treatment is still in the testing stage as far as the large diameter welded steel pipes are concerned.

I would like to introduce to you a couple of research efforts or results to improve the low temperature toughness of line pipes.

1. Studies on Chemical Composition

Shown in Fig. 10 is the relation between the sulfur content of line pipe material and the impact values of material test specimen. As the sulfur content is lowered, the impact value increases sharply and so does by far when the material is treated by calcium.(18) This is because of the inclusion control in the steel material, which could give significant influences on the impact value. As is seen from Fig. 11, shorter length of A-type inclusion or reduced total number of A-type inclusion will lead to higher impact values. Lower sulfur content in the steel serves

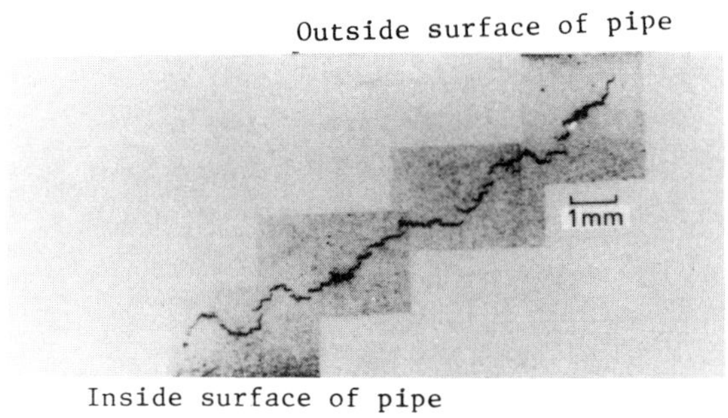

Outside surface of pipe

1mm

Inside surface of pipe

Photo 6 - Example of hydrogen induced crack

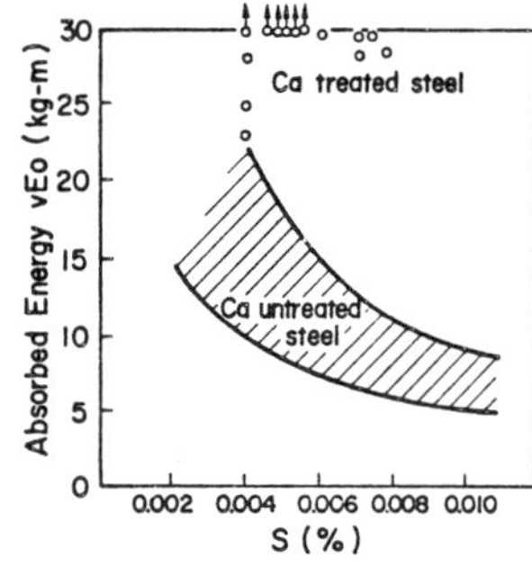

Fig. 10 - Effect of calcium addition on absorbed energy of 32°F in transverse direction of API X-65 grade line pipe

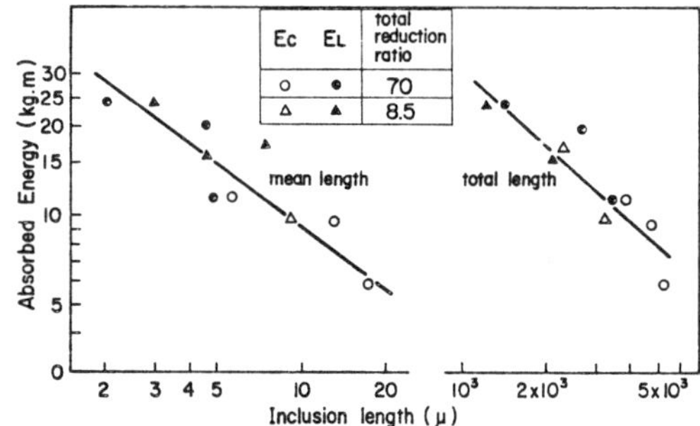

Fig. 11 - Effect of inclusion-length on absorbed energy in transverse and longitudinal direction

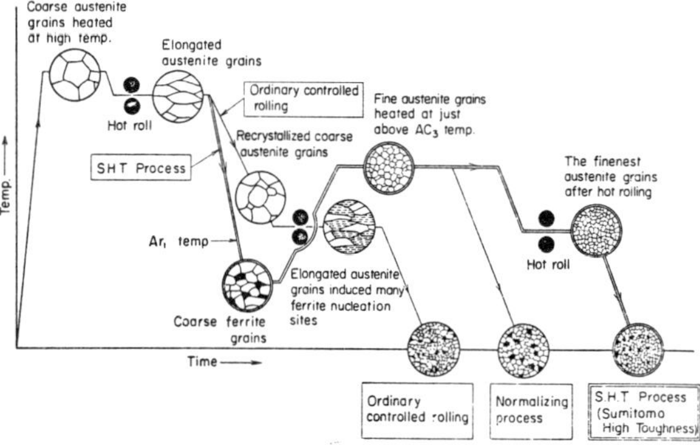

Fig. 12 - Schematic diagram of microstructure changes by SHT, ordinary controlled rolling and normalizing method

to reduce the number of inclusion. The calcium treatment of the steel can prevent the length of the inclusion from becoming longer even if the steel is rolled. This is because the inclusion is shaped into spherical forms by calcium treatment.(19)

2. Studies on Controlled Rolling

Please refer to Fig. 12 which illustrates a method of improving low temperature toughness of line pipes by controlled rolling methods.

Under the standard or conventional controlled rolling, the steel material is first heated in the heating furnace to have the austenite structure. Then, the heated steel is rolled and the austenite is deformed under this process. Such deformed austenite is then recrystallized before being rolled again at or around the transformation point, and turns into the austenite particles with many ferrite nuclei. Such austenite particles will have more fine particle structure as it gets cooled down.(20)

We devised further process steps to have a more complete structure of the steel. We have cooled the steel material as once rolled, down to the temperature levels below the transformation point to obtain the ferrite nuclei and then heated it up again. At this stage, the steel structure turns into the fine grain austenite. The steel material is then again put into the rolling mill at and around the transformation point in order to obtain more complete and finer austenite grain.

We call this new controlled rolling system as the SHT method. The method is a new hot steel processing engineering as established by applying the changes of steel structure at or around the transformation points and its relations with mechanical properties.(21)

Figure 13 gives the mechanical properties obtained by the SHT method. Figure 14 shows the layout arrangement of thick plate steel factory employing SHT method. The SHT can make it possible to reduce the average grain diameter smaller than the normalizing method and to lower the Charpy fracture transition temperature and maintain more strength even with the same steel material used.(22)

V. RESEARCH AND DEVELOPMENT OF ANTI-CORROSION LINE PIPES

1. Research on Anti-corrosive Features of Line Pipes

Another thing to note in recent years is that there is a growing concern over the anti-corrosion of the line pipes.

We have a record of some leakage accident years ago with certain offshore pipeline which took place shortly after the start of the pipeline operation. Examinations of the damaged portion of line pipes disclosed that the leakage was caused by the step-wise cracks developed across the thickness of the pipes. This accident prompted many scientists to try to probe the causes of these cracks and to initiate many researches and studies on the pipe corrosion characteristics in a very short time. Research efforts are now centering upon the following:

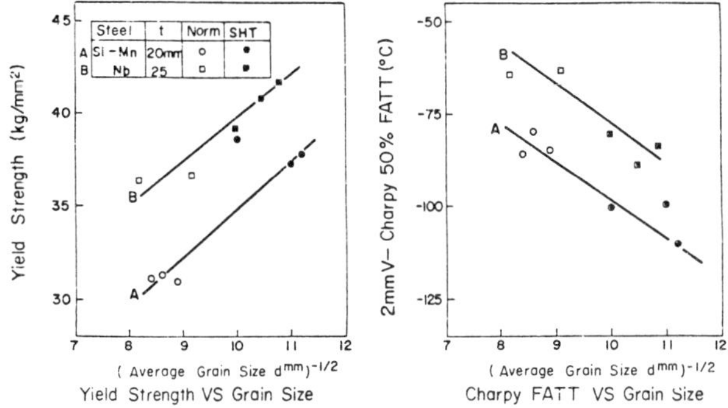

Fig. 13 - Comparisons of mechanical properties of SHT process with normalizing process

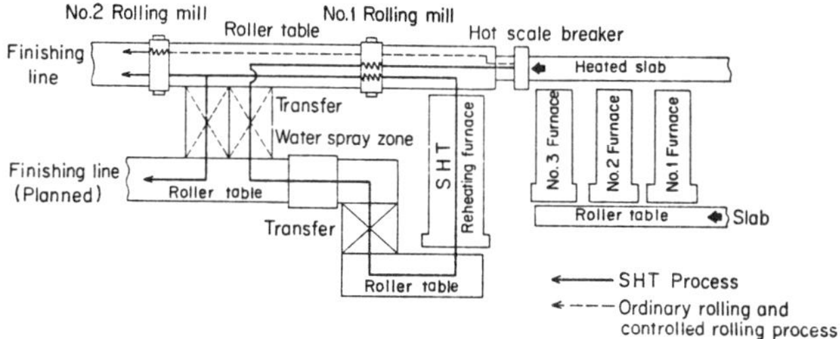

Fig. 14 - Factory layout SHT method rolling process

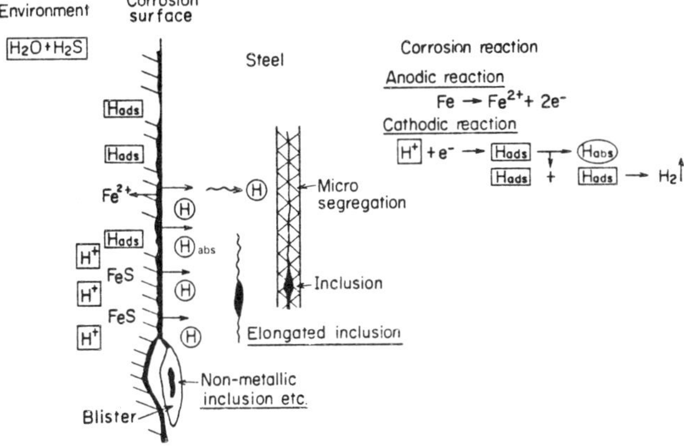

Fig. 15 - Mechanism of hydrogen induced cracking

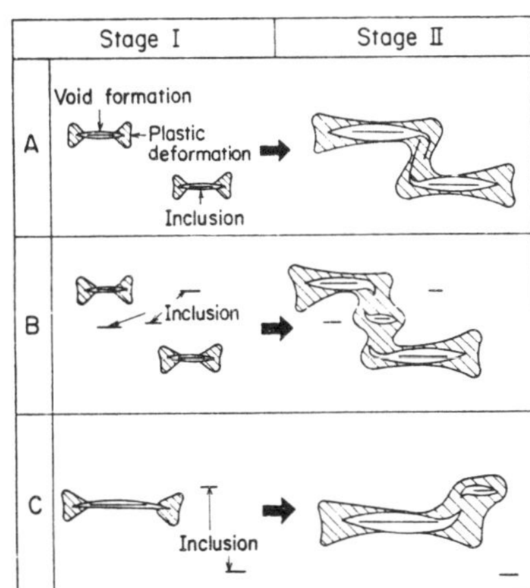

Fig. 16 - Schematic representation of step-wise cracking process

1) Investigation of causes of pipe corrosion accident: analysis of hydrogen-induced cracking
2) Development of corrosion test methods
3) Development of anti-corrosion steel

As I mentioned before, the oil or natural gas recently produced tends to contain more hydrogen sulfide. The hydrogen sulfide may make some chemical reaction with the steel on the inside surface of the pipelines and produce hydrogen atoms. Then the hydrogen atoms get into the steel structure and concentrate around the inclusion and become the hydrogen molecules. Increased pressure by the hydrogen molecules around the inclusion may cause to induce a crack at the ends of the inclusion. This is the mechanism of the so-called "hydrogen-induced cracking" which has been revealed to date. Figure 15 illustrates this crack inducing mechanism and Fig. 16 shows the growth mechanism of step-wise cracks. The plastic deformation at the ends of the inclusion caused by the increasing pressure of the hydrogen around the inclusion may expand, link to other cracks and form into a string of step-wise cracks.[23,24]

In order to develop the steel which could prevent the hydrogen-induced cracking and to further advance the researches on this subject, it was necessitated to reproduce in the laboratory a relation which is found between the crack growth on the actual pipelines and the environmental conditions under which such pipelines were placed, or to find out any testing method which could apply to the correlation of such cracking phenomena.

We then launched a joint research project together with an oil company, one of the important customers of our corporation. We submerged the test line pipe specimen in the artificial sea water saturated with hydrogen sulfide, in order to conduct the experiments to develop hydrogen-induced cracks. Through these experiments we could establish a method to detect the corrosion characteristics or feature of the steel material in terms of the frequency of the growth of cracks.[23,24]

2. Development of Anti-corrosion Line Pipes

During the course of the research and development of the steel material which hardly suffers from hydrogen-induced cracking, it was revealed that the countermeasures good for high toughness steel at lower temperature levels will also work for the purpose of developing the steel in question.

1) Lower sulfur content
2) Addition of copper
3) Conglobation of the inclusion
4) Prevention of abnormal steel structure formation

As is seen from Fig. 17, the frequency of hydrogen-induced cracking will decrease as the sulfur content in the steel becomes small and it will come down sharply at sulfur content of 0.005% or less, just in the same manner as with the case of improvement in the low temperature

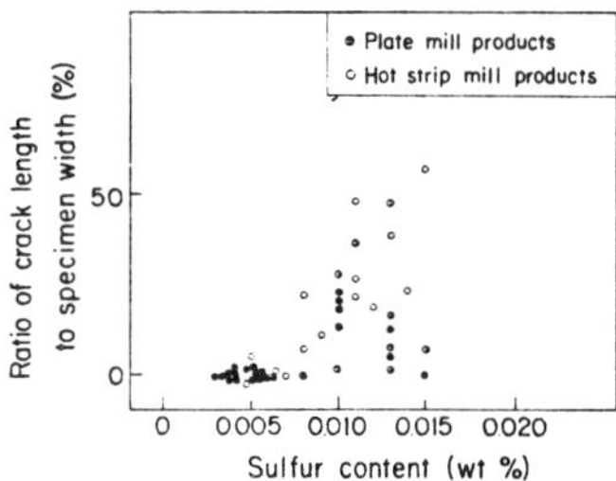

Fig. 17 - Effect of sulfur contents in steel on hydrogen induced cracking

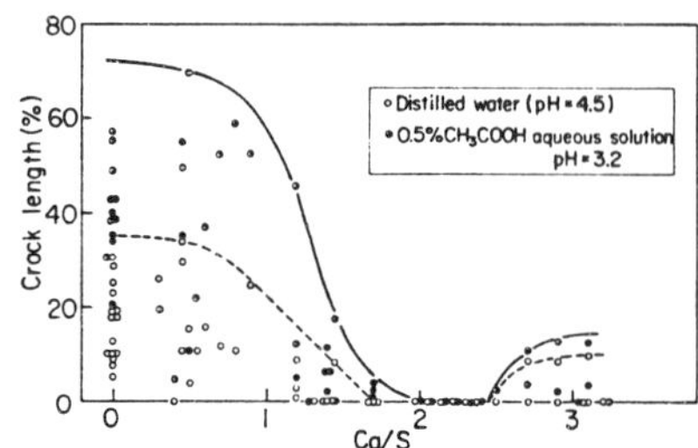

Fig. 18 - Effect of Ca/S on crack length of steels

toughness of the steel. This is because lower sulfur content reduces the number of the inclusion. The lower the sulfur content, the better steel material can be obtained which is insensitive to crack development.

In the near future the sulfur content requirement will be as low as 0.001% or less and the production of steel material with such low sulfur content will become feasible.

The next countermeasure is possibly to add copper to the steel. Copper-bearing steel will react with the hydrogen sulfide and form a coating of copper sulfide all over its surface which serves to prevent the hydrogen from coming into the steel structure. With copper added some 0.30% to the steel, it is noted clearly that the hydrogen penetration will decrease significantly.

However, the effects by copper addition appear only when the acidity, an important element to the corrosion environment, is relatively weak or at pH values of 4.5 or more. As and when pH value comes below this level, the copper sulfide coating on the surface of steel begins to dissolve and does not stand the hydrogen penetration.

Furthermore, the spheroidization of the inclusion will become necessary at the lower pH conditions. As I explained with regard to the development of high toughness steel, the steel has to be treated by calcium. Figure 18 shows a relation between the frequency of crack growth and a ratio of calcium to sulfur. As a ratio of calcium to sulfur comes near to a certain given value, the number of cracks is apparently beginning to decrease.[25,26]

VI. Conclusions

With the progress of petroleum resource development, various requests for the improvement of steel pipes were raised and our industry exercised utmost efforts to cope with them.

Significant impacts of the petroleum development upon the metallurgical engineering field can be summed up as below:

1) Manufacturing technology of extremely lower sulfur content steel
2) Manufacturing technology of the steel with less A-type inclusion
3) Quantitative analysis of the inclusion
4) Thermo-mechanical treatment
5) Micro-alloy steel
6) Hydrogen-induced cracking and stress corrosion cracking
7) Corrosion of steel by high pressure humid carbon dioxide

We were successful in developing a technology to produce low sulfur content steel which had never been imagined. We established a method to manufacture the steel with less A-type inclusion and a technology to shape the inclusion into a spherical form and to prevent such conglobated inclusion from extending long even after the steel being fed into the rolling mill.

In the meanwhile I came to recognize that we have to develop the concept and means to, if I may be allowed to use big words, assess quantitatively the relations between the inclusion and the mechanical properties of the steel. I often explain to the young research engineers that metallurgy has not been felt to be completed as the theoretical science. Because the description of metallurgical science is always qualitative. If you can devise any system of computer-based calculations to prove the metal characteristics will become justified as a science.

In this respect, I think that the fact we can now produce the steel with less inclusion could give any moment or incentive to prompt the quantitative analysis of the mechanical and other properties of the steel. I hope the researchers attitudes go towards such direction in the future.

In the field of the process heat treatment, we could open a new academic and disciplinary dimension to process the steel at or around the transformation point. And here we face new issue that we can change significantly the characteristics of steel as and when this newly developed controlled rolling system is applied to micro-alloy steel which is defined as the steel bearing small amount of niobium, vanadium, or titanium of 0.1% or less.

Today, as far as corrosion problems are concerned, I mention just the hydrogen-induced cracking, although there are many other corrosion problems which must be solved. Especially, the high pressure and high temperature corrosion at the oil wells as deep as 10,000 meters or more where the pressure goes beyond 1,000 atmospheric pressure will possibly make a new issue and a new dimension of research work.

I was rather in a hurry to cover the various research and development topics related to the oil and gas projects and steel pipe requirements. However, what I want to convey is that it is necessary for all the scientists or researchers to think all the time of the possibility how to pose the social needs before the research engineers who want to find solutions to such needs and how to link such arrangement with further advancement of science or academic learning. And what is required for any research manager is to maintain the perspectives as broad as possible and coordinate the social needs and academic development in a better direction.

I do not say that I could do it well. I rather reflect upon what I could not do to my full satisfaction. It was a very poor talk, but I do appreciate this opportunity to speak before such a distinguished gathering. Let me thank you again for honoring me with the esteemed Nishiyama Prize.

REFERENCES

1) World Oil, (1978), Feb. 15.
2) API Spec. 5A.
3) API Spec. 5AX.
4) API Spec. 5AC.
5) The Sumitomo Search, (1976), No. 15, May.
6) J. Petroleum Technology, (1972), June.
7) Oil & Gas J., (1975), Jan. 6.
8) Oil & Gas J., (1975), May 12.
9) World Oil, (1975), June.
10) J. Petroleum Technology, (1976), June.
11) Petroleum Engineer, (1977), March.
12) Private Information
13) The Sumitomo Search No. 17, May, 1977
14) T. Hiroshima: Energy Technology Conference, ASME, Sept., (1977).
15) R. J. Eiber: 4th Symp. on Line Pipe Research A.G.A., (1969).
16) T. Tanaka, the late M. Fukuda, I. Takeuchi and T. Koga: Tetsu-to-Hagane, 64 (1979), 958.
17) W. A. Maxey and R. J. Eiber: Research Report on Ductile Fracture Experiments, Battelle Memorial Inst., (1973), May 4.
18) S. Hasebe and T. Tanaka: AWRA Symposium, May, (1973).
19) M. Fukuda, T. Hashimoto and Y. Kitagawa: The Sumitomo Search, (1975), No. 14 Nov.
20) M. Fukuda, T. Shashimoto and K. Kunishige: The Sumitomo Search, (1973), No. 9, May.
21) T. Tanaka, N. Nozaki, K. Bessyo, M. Fukuda and T. Hashimoto: The Sumitomo Search, (1978), No. 19, May.
22) E. Miyoshi, T. Tanaka, N. Nozaki and M. Fukuda: ASME Publications, 77-Pet-61
23) E. Miyoshi, T. Tanaka, F. Terasaki and A. Ikeda: ASME Publications 75-Pet-2.
24) A. Ikeda, Y. Morita, R. Terasaki and M. Takeyama: 2nd International Congress on Hydrogen in Metal, H2-4A7, June, (1977)
25) M. Kowaka, F. Terasaki, S. Nagata and A. Ikeda: The Sumitomo Search (1975), No. 14.
26) A. Ikeda, F. Terasaki, M. Takeyama, I. Takeuchi and Y. Nara: Corrosion/78, (1978), June.

DESULPHURIZING: SULPHUR CONTROL PLUS MORE PRODUCTION*

R.F. Potocic, Supervisor, Operating Practice,
Iron Production

K.G. Leewis, Research Investigator,
Research Department

Dominion Foundries and Steel, Limited
Hamilton, Ontario

This paper originally appeared in Symposium on External Desulphurization of Hot Metal, McMaster University, Hamilton, Canada, 1975, pp. 2-1 - 2-23.

INTRODUCTION

In the past ten years, burden preparation has been the major factor increasing furnace production. (1,2,3,4,5,6) But the resulting advantages will decrease as raw material quality continues to diminish. Pellet suppliers have difficulty maintaining quality when increased tonnage is needed. To make higher hot metal tonnages, the difference between pellets and total Fe units required must be met with raw ores. Today's lump ores are inferior to pellets. They are soft, have a lower Fe content and higher slag content, and in many cases, they decrepitate in the furnace stack. Further gains in blast furnace production must be obtained through changes in furnace operating practices. For years, blast furnace operators have known that burdening for a less basic slag will increase productivity. (7) A leaner flux rate results in an increased solubility of bosh alkalies. The lean flux rate translates into economic advantages.

1. A more permeable, smoother operating furnace.

2. A lower coke rate.

This furnace operation results in a higher sulphur iron. Quality demands on steelmakers have discouraged blast furnace operators from operating more economically at a lower flux rate. (8) The resulting high (0.045%) hot metal sulphurs cannot be used in basic oxygen furnaces without external desulphurizing to reduce sulphurs to approximately 0.020% S.

The desulphurizing technique adds a reagent to the iron in the torpedo car. The reagent reacts with the sulphur, and the insoluble sulphide floats into the slag. The slag and iron must be separated before charging to the basic oxygen furnace (B.O.F.). At Dofasco, the separation was carried out by using tapholes in the torpedo cars to retain the slag while the clean iron was tapped into the transfer ladle and then into the basic oxygen furnace. On the return trip to the blast furnace, the sulphur-rich slag is dumped into a wet dekishing station (Fig. 1).

There are many desulphurizing techniques and reagents available: Carbide or MgAl injection, mag-coke plunging, runner additions, etc. (9,10, 11,12,13)

Dofasco chose to use mag-coke because of its:

1. Simplicity and ease of operation.

2. Small capital investment.

3. Immediate installation.

With this desulphurizing facility, it was possible to determine if the combination of a less basic blast furnace operation and external desulphurization would produce economic savings. This paper describes the production trial.

Lower Flux, High Sulphur Blast Furnace Operation

Prior to 1974, Dofasco normally produced a monthly average of 0.021% S iron for the basic oxygen furnaces. This excellent low sulphur level was obtained at a high coke rate. Several small trials were conducted to gain experience and confidence with high sulphur operation. The main experiment consisted of a two-week, high sulphur period bracketed before and after by base and transition periods. (TABLE I)

To reduce the number of process variables, a constant burden was charged to the trial blast furnace, the 2000 T/day #2 blast furnace. The flux rate was lowered from 163 lbs/NTHM to 103 lbs/NTHM to achieve a 1.00 basicity. The thermal balance was maintained through pellet adjustment. The major furnace parameters, silicon and

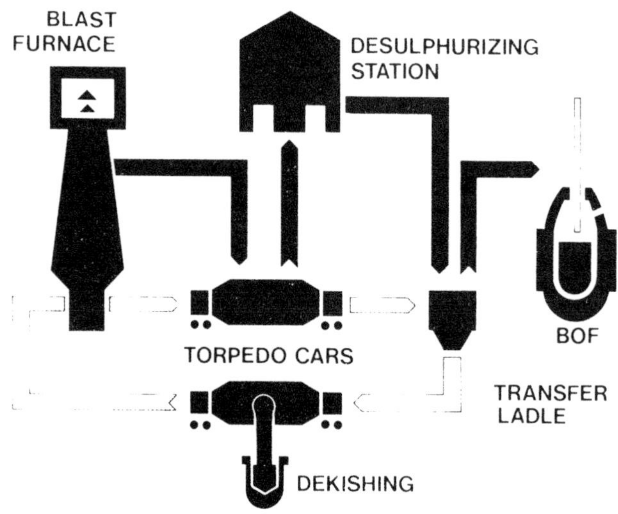

Fig. 1 - Hot metal movement

*Symposium on External Desulphurization of Hot Metal, McMaster University, Hamilton, Canada (1975) Ed. W-K Lu

TABLE I

#2 BLAST FURNACE HIGH SULPHUR TRIAL

		High Sulphur (Sept 16-30 1973)	Average Base Period (Oct 1973)	Corrected Base Period (Constant Wind)
Daily Production	NTHM	2,165.2	1,903.7	1,988.4
Carbon Rate	#/NTHM	939	978	1,005
Total Metallic Burden	#/NTHM	3,101	3,151	3,145
Limestone & Dolomite	#/NTHM	103	163	160
Hot Metal Si	%	1.04	1.01	1.04
S	%	.043	.021	.021
Temp.	^{o}F	2,715	2,752	2,752
Slag Volume	#/NTHM	400	441	441
Bases to Acid Ratio		1.02	1.17	1.17
Wind Rate At Turboblower (Avg.)	SCFM	76,700	71,500	76,700
Wind	SCF/NTHM	51,011	54,081	55,546
Wind	SCF/# Carbon	54.32	55.27	55.27

manganese, were held constant at approximately 1.02% Si, 4.75% C and 1.42% Mn. Once the burden changes were reflected in the hot metal analysis, normal cast-to-cast adjustments were used to sustain a constant hot metal analysis.

The flux rate was adjusted to produce 0.045% ± 0.005% S. This sulphur level was chosen because it was one "mag-coke plunge" above the desired hot metal average of 0.021% S. During the two-week, high sulphur test period, the average sulphur before desulphurizing was 0.043%. The low sulphur base periods averaged 0.021% S. Furnace control was sufficient to maintain consistent Si and Mn, before, during and after the high sulphur period. Silicon was controlled by varying the hot blast moisture. Mn was regulated by adding or subtracting Wabush pellets. The average runner temperature for the high sulphur period was about 40° F below the low sulphur period. At constant hot metal temperature, lowering the slag basicity will raise the hot metal % Si. The actual changes were a 37° F drop in temperature and a 0.03% increase in Si. The leaner burden gave a lower slag volume and a lower basicity. The reduced basicity allowed for twice the solubility of bosh alkalies. (14) Alkali loading during this trial was held constant at 2.5 to 3.5 lbs K_2O per net ton of hot metal (NTHM). If 60% or more of the charged alkali is removed in the slag, no alkali accumulates in the furnace. The alkali removal was 58% for low sulphur versus 80% for the high sulphur period during this trial. The ability to run with a lower slag basicity may allow Dofasco to consider less expensive raw materials with previously unacceptable sulphur and alkali concentrations.

The capacity of the turbo-blower for #2 blast furnace limits the amount of wind blown. The product of wind volume and blast pressure is directly proportional to the horsepower available. The wind rose from 71,500 scfm to 76,700 scfm on high sulphur practice. This 7% increase was due to the lower blast pressure. The lower blast pressure meant the furnace permeability had increased.

The carbon rate dropped 76 lbs/NTHM, from 1,005 lbs/NTHM on low sulphur burden to 939 at high sulphur practice. This increase in furnace efficiency was the result of:

1. Decreased stone rate (from 160 to 130 lbs/NTHM)

2. Lower iron temperature (from 2752°F to 2715°F)

Fig. 2 - Plunging apparatus

Fig. 3 - Typical graphite bell failure

MAG-COKE ADDITIONS

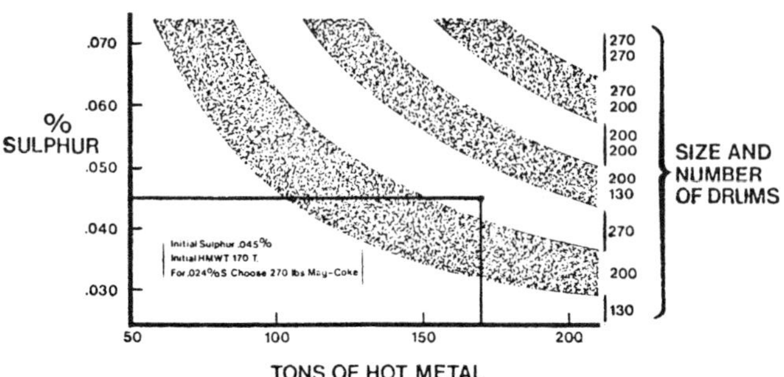

Fig. 4 - Mag-coke additions

3. More permeable furnace (from 71,500 to 76,700 SCFM)

4. Better alkali removal (from 60% to 80% removal)

There was a 13% increase in hot metal production from 1,904 tons to 2,165 tons during the two-week high sulphur trial. The extra tonnage was the result of the greater wind rate (4.3%) and the reduced coke rate (8.8%). The runner system was easier to maintain. Less runner scrap remained after casting. Unfortunately, the maintenance on the trough and skimmer increased because of the increased tonnage and iron fluidity.

Flint's method was used to compare the high sulphur operation to the base period (TABLE II). (7) The reduced coke rate produced a fuel savings of 76 lbs. carbon/NTHM. The reduction in fluxes saved 60 lbs. flux/NTHM. The wind blown was reduced 3,000 CF/NTHM. The efficiency improved to 54.32 from 55.27 CF/lb. of carbon.

Desulphurizing

A variety of techniques and materials were tried at Dofasco. Magnesium wire and powders of lime, calcium carbide and lime magnesium were injected with argon. Calcium cyanamide was added in the runner system. Galag disks, Desulf-x, calcium-magnesium alloys and DOMAG X45 briquettes were all plunged with varying degrees of success. Mag-Coke was the simplest to operate and required little capital expenditure. (9) The "mothballed" #1 blast furnace provided the crane rail spur and the floor space for a pilot desulphurizing plant. Located between the operating furnaces and the basic oxygen furnace shop, next to the torpedo car scales, the old blast furnace provided a near ideal location. The sulphur concentration and weight of iron were known before the torpedo cars were spotted. This pilot plant eventually desulphurized 60% of the plant iron production, a total of 128,000 tons of hot metal desulphurized per month.

In simple form, the mag-coke plunging apparatus consists of a 9-ton weight suspended by chains from a crane, a graphite adapter to bolt the graphite to the weight, a graphite connector to obtain the proper depth of plunging, and a reaction bell of graphite which is pinned to the bottom of the graphite stem (10) (Fig. 2). Connectors average 535 plunges varying between the maximum 750 and a minimum of 270 plunges. The graphite bells ranged from 1 to 72 plunges but averaged 23. The average life varied with daily tonnage treated; the higher the tonnage, the higher the hardware life. There was no preheating of the graphite assemblies at the pilot plant. Quick turnarounds kept the assembly heated and minimized thermal shock. The typical bell failure as seen in Figure 3 is caused by metal solidifying between the bell and stem joint before the graphite fully contracts. Eventually, the top of the bell is peeled away from the stem. Typical connector failure is a

TABLE II

SUMMARY OF HIGH SULPHUR TRIAL

I. BLAST FURNACE

1. Fuel Savings	76# Carbon/NTHM
2. Flux Savings	60# Flux/NTHM
3. Wind Savings	3000 SCF/NTHM
Or Increased Production	270 NTHM/Day

II. DESULPHURIZING

1. Mag-Coke Consumption	1.8#/NTHM
2. Bell & Connector Cost Plus Overhead	30% of Mag-Coke Reagent Cost

3. Rebate .003% S from .018% S vs. .021% S.

III. NET SAVINGS

$$\frac{\text{Cost of Desulphurizing}}{\text{Blast Furnace Savings}} \times 100\% = 69\%$$

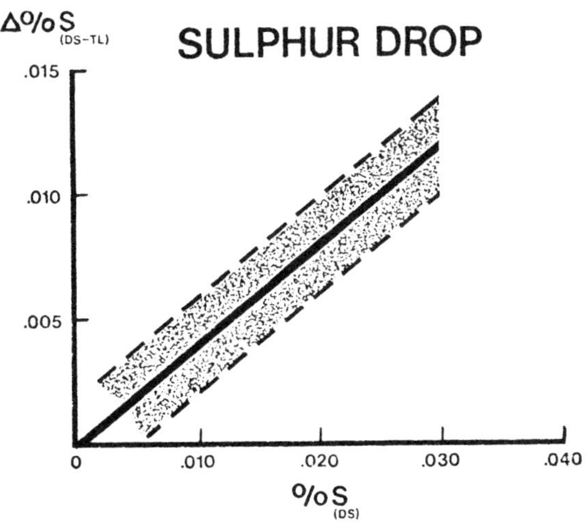

Fig. 5 - Sulphur drop during transport (desulphurizing station - transfer ladle) versus (transfer ladle)

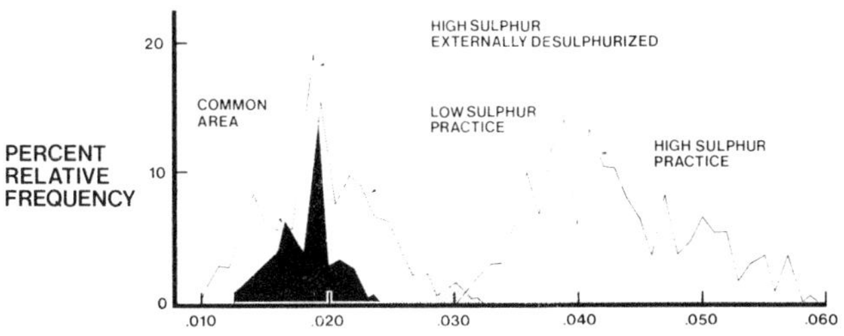

Fig. 6 - Effect of external desulphurization on sulphur range

result of oxidation of the graphite. Fracture easily occurs across the reduced diameter.

Refractory bells have averaged 36 plunges. Refractory connectors last 75 to 100 plunges. Refractory assemblies were uneconomical and their high maintenance made them uncompetitive. A solution, at this time, seems to make refractory bells to fit the conventional graphite connectors.

At the pilot plant, the locomotive can spot only one torpedo car at a time. Before loading, each can of mag-coke is opened to check to see if there is moisture absorbed by noting the absence of metallic lustre. Damp material is always explosive. The plunging apparatus is lowered over the appropriate 130, 200 or 270 drums of mag-coke (Fig. 4). The mag-coke is nailed inside the bell with 10" spikes. The bell with the mag-coke is aligned over the mouth of the torpedo car. The mag-coke is plunged as rapidly as possible to within 6" of the bottom of a newly-lined torpedo car. This immerses the top reaction holes about 3' below the iron slag interface. During the 8-minute reaction time, the weight is suspended from the crane rather than set on the torpedo car. This allows the crane cables to absorb the violent vibrations of the reaction and increase the life of the graphite. The reaction completed, the apparatus is raised. Any residual coke is poked from the bell into the torpedo car. The complete assembly is moved to the other end of the floor and allowed to cool while the locomotive spots the next car, and the second can of mag-coke is readied.

TABLE III

HOT METAL SULPHUR QUALITY

		Base Period	High Sulphur Period	
			As Cast	After Mag-Coke
Average	(%S)	.021	.043	.018
Mode	(%S)	.020	.038	.019
Range	(%S)	.013-.032	.030-.059	.010-.024

TABLE IV

TYPICAL CHANGE IN TORPEDO CAR SLAG

	Before Desulphurizing	After Desulphurizing
S	1.1	1.9
FeO	4.6	3.3
MgO	8.9	12.3*

* The magnesium compounds (MgO and MgS) in the slag are automatically converted to MgO.

The two-week high sulphur period averaged 1.76 lbs. mag-coke/NTHM to lower the average cost sulphurs from 0.043% to 0.023% at the desulphurizing station. A drop in sulphur concentration was noticed during the time of transport from the desulphurizing station to the basic oxygen furnace shop (Fig. 5). The sample obtained immediately after desulphurizing contains MgS still floating from the iron into the slag. When the pins are crushed and run on the Leco, a higher sulphur results. Figure 6 shows the frequency polygons for the sulphur analysis of both the desulphurized and non-desulphurized iron in the B.O.F. transfer ladle. The two distributions are compared in TABLE III. Desulphurizing produced a 0.002% lower average in the 25% smaller range. The common area was 35% of the total. The improvement in hot metal quality visible to the left, was 25% of the total area. The desulphurized metal was 0.024% S or less; a figure many steel men still wish for.

Theoretical calculations predict a temperature loss of 12° F on plunging. Actual thermocouple measurements indicate a cold bell will cause the hot metal to lose 20° F and a hot bell 10° F. The average temperature loss was 13° F.

Because of the good skimmer practice in our cast house approximately 3000 lbs of slag gets into the torpedo cars. Slag samples are difficult to get and may not be completely representative. The typical change in torpedo car analysis can be seen in TABLE IV. Slag sulphurs doubled. All the magnesium in the slag whether MgS or MgO was converted by the X-ray fluorescent spectrometer to MgO. The reaction products MgS and MgO account for the difference.

A taphole is used to separate the torpedo car slag from the hot metal during reladling in the Melt Shop (Fig. 7). The importance of slag-metal separation must be strongly emphasized. Example: After desulphurizing, the slag in the torpedo car contains at least 90 lbs. of sulphur. If half this slag is added with the iron to the 120 T.

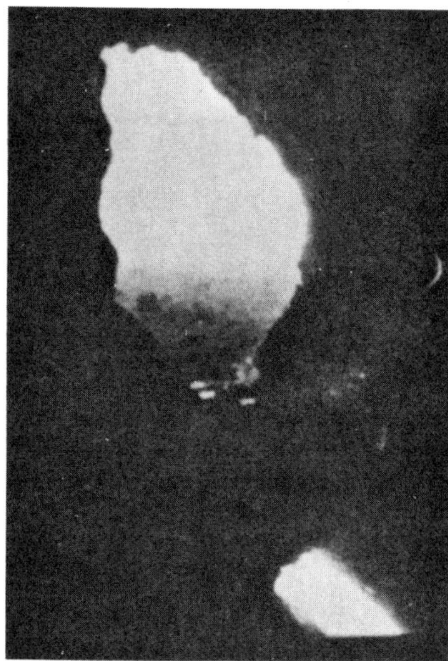

Fig. 7 - Slag metal separation

Fig. 8 - Dekishing operation

B.O.F. transfer ladle, the 48 lbs. of hot metal sulphur to be charged into the B.O.F. will be increased, almost 200%, to 93 lbs. of sulphur. Although Dofasco's taphole separation arrangement is not perfect, it is better than existing methods.

The residue left in the torpedo car, a mixture of slag, kish and iron, is poured carefully into water at the dekishing station (Fig. 8). The torpedo car is clean and free of residue when respotted at the cast house. The product slag at the dekishing station appears granular rather than liquid, but the quantity of graphite mechanically mixed in the kish makes observation difficult.

Desulphurizing does not contribute to accelerated torpedo car lining wear. Splash buildup in and on the mouth of the torpedo car is a problem and this is periodically cleaned. To prevent excessive splash buildup from the desulphurizing reaction, the torpedo cars are filled only to 90% capacity leaving one foot of free board. Sometimes the large lumps of mag-coke residue become frozen in the torpedo car tapholes after separating the hot metal in the basic oxygen furnace shop. These lumps are poked out at the dekishing station or, if stubborn, lanced at the pig casting machine.

Cost Comparison

The cost of desulphurizing was determined over the month of September 1973 (TABLE II). The consumption of mag-coke for the 0.043% to 0.018% sulphur drop was 1.8 lbs./NTHM. The graphite assemblies, labour and overhead came to 30% of the Mag-Coke Reagent Cost. Subtracting a rebate for producing sulphurs 0.003% below the base period average of 0.021% S, the total desulphurizing cost came to 69% of the savings to the blast furnace. Over the past year, the price of magnesium has more than doubled. This has drastically increased the cost of mag-coke desulphurizing, and it is now approaching the measurable blast furnace savings. Methods to decrease the cost of desulphurizing are being investigated. This escalating cost is the main reason for eventually desulphurizing by carbide injection technique.

Conclusions

1. The extra iron production from high sulphur operation amounted to 270 NTHM per day or 13%; 4% from increased wind rate and 8% from lower coke rate.

2. The cost of desulphurizing was 69% of the savings to the blast furnace (September 1973)

3. The average hot metal sulphur distribution to the basic oxygen furnace shop improved.

Recent Developments

The success of this short trial period indicated the need for 100% high sulphur iron production. The pilot plant at #1 blast furnace desulphurized 136,000 NTHM/month. Recommissioning of #1 blast furnace for the needed production forced this temporary station to move. A new desulphurizing facility was designed and constructed.
This building is located in the blast furnace area between the basic oxygen furnace shop and the hot metal scales. This position guarantees that proper weights and sulphur concentrations are known before desulphurizing. The twenty-minute trip to the basic oxygen furnace shop will ensure that the MgS inclusions will have time to float into the slag. The building is essentially two crane aisles over two parallel torpedo car tracks. The east aisle was designed for mag-coke plunging and the west for calcium carbide injection. Eventually four cars, two in tandem per side, can be simultaneously desulphurized. Calcium carbide appears to offer the cheapest and most flexible operation, based on

reagent costs. The high capital investment and long lead time for installation of the carbide process dictated an interim method be used. Mag-coke was chosen because of our past experience. Each aisle was designed to handle a minimum of 8,000 NTHM/day. The mag-coke half is presently desulphurizing Dofasco's total iron production of 7,700 NTHM per day. The injection system will permit evaluation of 0.060% S to 0.080% S levels on blast furnace operation. Benefits to the basic oxygen furnace shop of lowering sulphurs to 0.015% S or less can also be evaluated. The building's flexibility will also allow desulphurizing to specialty steel quality. Mag-coke will be the backup if difficulties are experienced with the carbide system.

Acknowledgments

The authors would like to thank the management of Dofasco for permission to present this paper and their colleagues for the enthusiastic support received. Throughout the programme, this group effort of the production and research departments was smoothly integrated into the regular production schedule with the cooperation of the blast furnace personnel especially Mr. J. Holditch and Mr. S. Durkacz.

Bibliography

1. Strassburger, J. (ed).: Blast Furnace Theory and Practice, Gordon and Beach Science Publishers, New York, NY, (1969), Page 77.

2. Hasegawa, T.: Third International Symposium on the Iron and Steel Industry, Brazilia, Brazil (14-21), October, 1973.

3. Ashton, J. and Holditch, J.: Homogenized Oil Injection at Dofasco, Iron Making Proceedings Vol. 34, Toronto, 1975.

4. Mantegazza, A. and Corignani, M.: Importance of Burden Beneficiation on Blast Furnace Operations, Iron Making Proceedings, Vol. 30, Pittsburgh, 1971.

5. Bates, M.D.: An Aerodynamic Model of the Blast Furnace. A description of the Model and its Application with Special Reference to the Projected Behaviour of Formed Coke, presented at the 3rd C.C. Furnas Memorial Conference, November 1972, Buffalo, U.S.A.

6. Anon.: "Metals and Materials" (October 1974) Page 386.

7. Flint, R.V.: "Effect of Burden Materials and Practices on Blast Furnace Coke Rate", American Iron and Steel Institute, presented (September 1961) at Chicago Regional Meeting.

8. Forster, E., Klapdar, W., Richter, H., Spitzler, E., and Wendorff, J.: "Deoxidation and Desulphurization by Blowing of Calcium Compounds into Molten Steel and its Effects on the Mechanical Properties of Heavy Plates" Stahl und Eisen (1974), May, (474-485).

9. Stanlake, R.C., and Strathdee, B.A.: "Desulphurizing Iron with Mag-Coke", presented at International Magnesium Assoc. Annual Meeting, Cherry Hill, N.J. (May 1973).

10. Duquette, W.H., Griffing, N.R. and Miller, T.W.: "Low Sulphur Steel (0.005% S) From Mag-Coke Desulphurized Hot Metal" presented at N.O.H. and B.O.F. Conference, Cleveland, April 8, 1975.

11. Dastor, P.N. and Cameron, R.W.: "Desulphurization of Hot Metal by use of Calcium Carbide Injection", Regional Technical Meeting, American Iron and Steel Institute, Chicago, October 1973.

12. Carmichael, I., and Brunger, R.: "Desulphurization of Iron: Practical and Economic Considerations", 40th Iron Making Conference Proceedings, Scarborough (1973), British Steel Corporation Research and Development.

13. Gilby, S.W.: "External Desulphurization of Hot Metal Using Magnesium", Ironmaking Conference Proceedings TMS-AIME, Vol. 32 (1973), pp. 133-141.

14. Ashton, J.D., Gladysz, C.V., Holditch, J. and Walker, G.H.: "Alkali Control at Dofasco", Ironmaking Conference Proceedings TMS-AIME, Vol. 32 (1973), pp. 60-72.

LOW SULFUR STEEL (.005% S) FROM MAG-COKE DESULFURIZED HOT METAL

W. H. Duquette
Sparrows Point Plant
Bethlehem Steel Corporation

N. R. Griffing
and
T. W. Miller
Homer Research
Bethlehem Steel Corporation

This paper originally appeared in the Open Hearth Proceedings, Vol. 56, 1973, p. 79.

Demand for steel products with improved performance, such as plates with better through thickness properties, higher notch toughness, and cleaner surface, has grown steadily in the last few years. Concurrently, there has been increasing data in the literature (1) indicating that lower sulfur content is one of the prime requisites for attaining improvements in product performance. In late 1971 a program was initiated at the Sparrows Point Plant aimed at developing a desulfurization process capable of producing steel with sulfur contents as low as .005% S. A review of the prior art together with sulfur material balance calculations had shown that an effective procedure would be one consisting of desulfurization of hot metal followed by additional sulfur controls in the steelmaking furnaces. It was also apparent that any effective method would entail a considerable increase in operating costs and penalties over normal steel production. Capital cost considerations and a desire for operating simplicity led to the installation of a single hot metal desulfurizing facility for submarine transport ladles serving both the BOF Shop and the No. 4 Open Hearth Shop instead of separate facilities at each location. The Mag-Coke plunging method for hot metal desulfurization was selected because exploratory trials at Research had shown that:

- The reaction is rapid, requiring only 5 to 10 minutes for completion.
- The equipment is relatively inexpensive and easy to use.
- The method readily provides sulfur removal to levels of less than .010% S.

By the end of the trial program a process was developed which produced steel to a sulfur content of .005% ±.001. The main features of this process are:

- Desulfurization of the hot metal down to a range of .003 to .008% S by plunging Mag-Coke into the submarine ladles.
- Using only taphole submarines for maximum skimming of the hot metal treatment slag.
- Recycling low sulfur (.010% S max.) scrap to make up at least half of the scrap charge.
- Charging extra fluxes. In the BOF, the burnt lime was increased to 227 lb/ton compared with our normal addition of 168 lb/ton. Fluorspar usage was increased to 12 lb per 100 lb burnt lime. Fluxes needed in the open hearth were about 50% less than in the BOF.

Since the completion of the project, these desulfurization practices have been used to produce low sulfur heats as required at the Sparrows Point Plant.

Hot Metal Desulfurization

Mag-Coke

Mag-Coke is foundry coke impregnated with at least 43% magnesium by weight. When immersed in hot metal, the magnesium boils out of the coke and dissolves in the iron. It then combines with the sulfur as magnesium sulfide which floats to the surface. Mag-Coke is a patented product of the American Cast Iron Pipe Company (ACIPCO), Birmingham, Alabama, and is manufactured by dipping preheated coke in molten magnesium (2). Although crushed sizing is available, lumps of 2 to 5 lbs each were used to obtain a slower, more controlled release of magnesium. This sizing had a bulk density of 54 lb per cu ft.

Mag-Coke Plunging Equipment

The Mag-Coke plunging apparatus used for desulfurizing hot metal in submarine ladles is illustrated in Figure 1. Designed for handling by a crane, the apparatus consisted of a four-way lifting chain, steel ballast weighing about 10 tons, a steel shaft with flange connected to a graphite stem, and a graphite bell. The steel ballast was needed to compensate for the buoyancy of the Mag-Coke, the graphite bell, and the stem. The ballast, in combination with the four-way chain, prevented tilting. The bell was connected to the stem with the graphite pins through an expansion joint of a design developed during the project. A typical charge of 250 lbs of Mag-Coke was placed in a sheet metal can and lifted into the graphite bell with a forklift truck. The can was fastened in place with eight-inch spikes driven in through the ventholes in the side of the bell. When immersed in hot metal, the can melted, exposing the Mag-Coke. After eight minutes of treatment, the plunging assembly was removed, and the spent Mag-Coke floated to the bath surface.

An early problem was the rapid failure of the plunging bells resulting from progressive cracking due to differential thermal expansion and contraction in the transition zone between the thin tubular sides of the bell and the massive stem. An expansion joint design (Figure 2) that prevented cracking was developed and bell life increased from an average of 7 dips (range 6-10) to an average of 21 dips (range 14-31). Easy bell replacement is another advantage of this design.

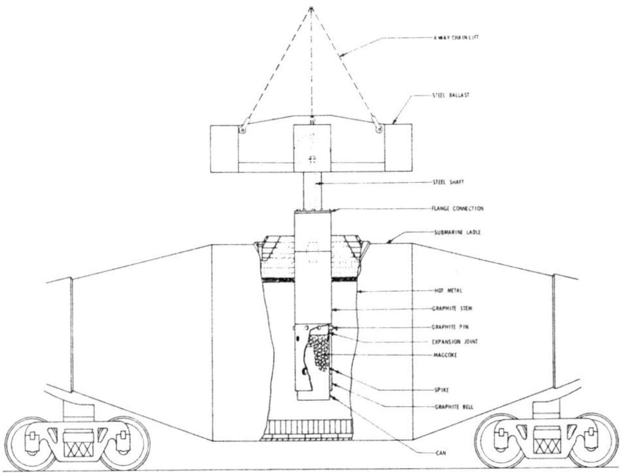

Fig. 1 - Mag-coke plunging assembly

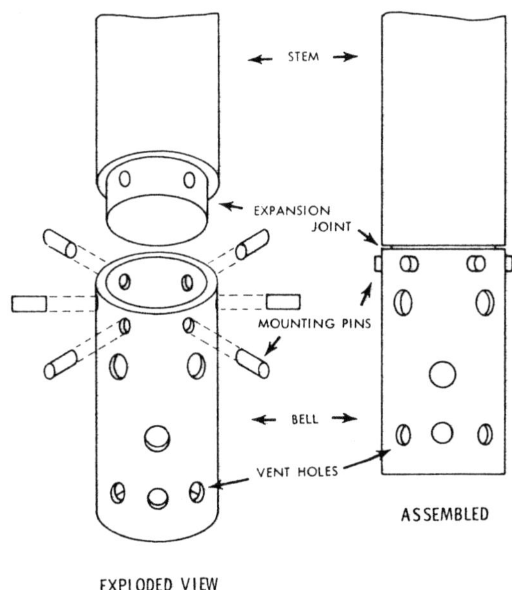

Fig. 2 - Expansion joint and bell

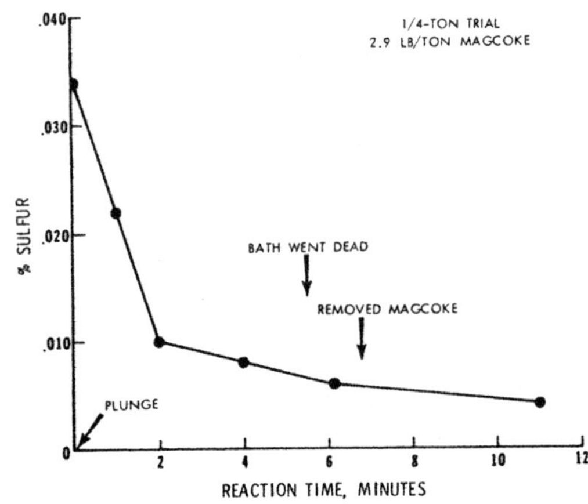

Fig. 3 - Fast reaction of mag-coke

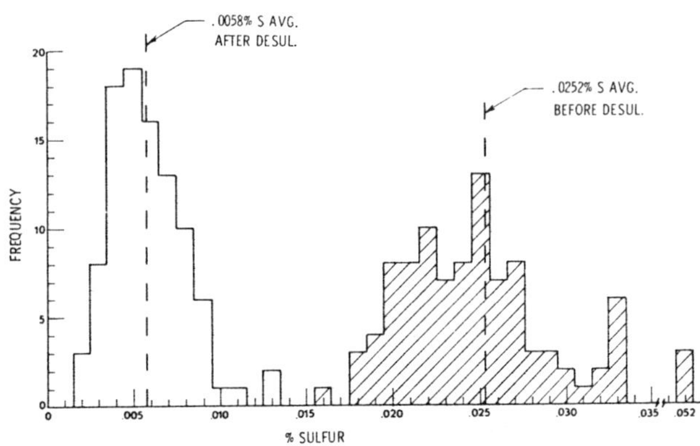

Fig. 4 - Sulfur in hot metal before and after mag-coke treatment

Hot Metal Desulfurization Results

The ability of Mag-Coke to remove sulfur rapidly to low levels is illustrated in Figure 3 by data from a 1/4-ton exploratory trial. The sulfur content, initially .034% S, decreased to .010% S in two minutes and to .006% S in six minutes. A further decrease to .004% S was noted after eleven minutes.

At Sparrows Point, 97 submarine ladles, each containing approximately 100 tons of hot metal, were desulfurized with Mag-Coke for use in low sulfur BOF heats during the trials. The frequency distribution of sulfur contents before and after Mag-Coke treatment is shown in Figure 4. On the average the sulfur content decreased from .0252% S down to .0058% S. The trials included three submarines of hot metal with a much higher than normal sulfur content (.052% S). These were successfully desulfurized to .003, .009, and .009% S with two 250-lb Mag-Coke plunges each.

The theoretical quantity of Mag-Coke required depends not only on the stoichiometric amount of magnesium needed to form its sulfide (.76 lb per lb S) but also on the solubility of magnesium in the hot metal. This solubility depends inversely on the sulfur content of the hot metal, as shown in Figure 5. For example, based on the solid line summarizing our results, the magnesium solubility is .005% Mg at .020% S but increases to .025% Mg at .005% S. The dashed curve from Speer and Parlee (3) shows less magnesium solubility; however, their results were for a temperature of 2300 F as compared to a hot metal temperature range of 2400 to 2600 F in the present investigation. There was a tendency in our results for the magnesium content to decrease upon holding, especially in the 1/4-ton tests. The important point is that the soluble magnesium does not remove sulfur and remains dissolved in the hot metal as a residual. This must be added to the stoichiometric amount required for forming sulfide to determine the total theoretical magnesium that should be pro-

Table I. Average Hot Metal Slag Analyses at Various Locations

	% S	% Fe	% CaO	% MgO	% SiO_2	% Al_2O_3
Blast Furnace Runner 1971	1.3	0.3	36.0	13.0	34.0	14.0
Subs Before Desulfurization	0.3	8.6	16.3	5.7	47.1	5.9
Subs After Desulfurization	1.1	16.4	11.3	10.1	42.7	4.6

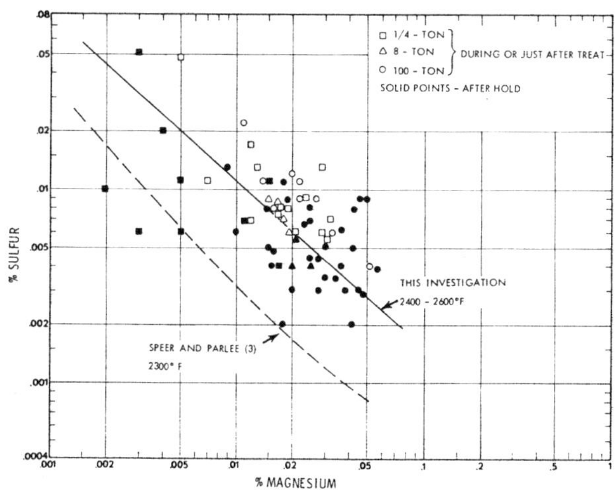

Fig. 5 - Mutual solubility of sulfur and magnesium in hot metal

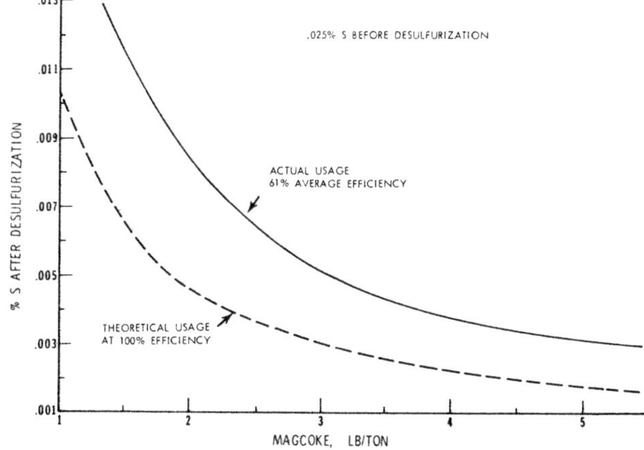

Fig. 6 - Effect of mag-coke useage on percent S after desulfurization

vided. The residual is more significant at low final sulfur levels.

On the basis of total theoretical magnesium, the average magnesium efficiency during our trials was found to be 61% when the initial sulfur ranged from .016 to .033% S.

$$\text{Mg Efficiency (\%)} = \frac{\text{Total Theoretical Mg}}{\text{Actual Mg Added}} \times 100$$

The corresponding Mag-Coke usage is shown in Figure 6 as a function of %S after desulfurization (solid curve). Typically, a final sulfur content of .006% S was obtained using 2.7 lb/ton Mag-Coke, as compared with a theoretical usage of 1.7 lb/ton depicted by the dashed curve at 100% magnesium efficiency for an initial sulfur of .025% S. The magnesium efficiency did not vary significantly with Mag-Coke immersion times ranging from 6 to 11 minutes. An immersion time of eight minutes was generally used.

Temperature loss during a Mag-Coke treatment averaged 25 F in the case of submarine ladles that had well-soaked linings from steady use. However, the temperature loss increased to 54 F either when a fresh submarine with a relatively cold lining was first brought on stream or when a submarine was used intermittently.

Deslagging Hot Metal

The average hot-metal slag analyses at various locations are summarized in Table I. Important points to note are:

- The chemistry of the slag in the submarines on arrival at the desulfurization station was quite different from that of the as-cast blast furnace slag. It was significantly lower in S, CaO, MgO, and Al_2O_3 but richer in SiO_2 and Fe.

- During desulfurization, the slag sulfur content increased from 0.3 to 1.1% S, the magnesium expressed as MgO increased significantly, and the iron content increased significantly.

Based on these results, it was concluded that:

- The basicity of the slag in the submarines is much less than that of the as-cast slag because of dilution by silica from the runner and/or from the acid lining of the submarine ladle.

- Since a significant amount of the removed sulfur floats out and enters the hot metal slag during and after treatment, this slag should be skimmed.

Several submarines had refractory dams with tapholes installed in their mouths similar to the tapholes described by Pause and Strathdee (4). To evaluate the effectiveness of the tapholes, we measured the depth of the slag in the transfer ladle after it was filled from the submarine ladles. The average transfer ladle slag depth was found to be three inches after pouring with taphole submarines but was five inches after pouring from regular submarines of hot metal that has been Mag-Coke treated. Although regular submarines exhibited a degree of self skimming action, the amount skimmed was less and the action more erratic than with taphole submarines. Overall, the tapholes significantly increased the degree of skimming.

Low Sulfur Steelmaking

Procedure

Our objective was to develop a practice for consistently producing .005% sulfur steel. Sulfur material balance calculations showed that with normal BOF practices about .009% S would be obtained in steel from hot metal desulfurized down to .006% S. The calculations showed that the desired steel sulfur content of .005% S would be attainable if sulfur from the scrap was decreased about 50% and if the slag-metal sulfur partition was increased to about 8:1 from the normal level of about 4:1. Accordingly, the following changes in BOF practice were investigated:

- Replacing part of the scrap with low-sulfur (.010% S max) scrap butts recycled from previous low-sulfur heats.

- Using extra lime and spar to increase sulfur transfer to the BOF slag.

The heats were closely observed to note the effects of these changes on operating conditions and to obtain slag samples. A total of 41 low-sulfur trials were made in the BOF and three heats were made in No. 4 Open Hearth.

BOF Results

The successful achievement of .005% S steel is summarized chronologically in Figure 7. The results occurred in two distinct periods:

1. An initial period during which taphole submarines were seldom used, only normal sulfur scrap was charged, and varying amounts of lime, spar and ore were added.

2. A final period of best practice during which only taphole submarines were used to maximize slag skimming, approximately one half of the scrap charge consisted of low sulfur scrap (.010% S max), and about 227 lb/ton burnt lime and 27 lb/ton spar were used (12 lb spar per 100 lb lime). In normal heats about 168 lb/ton lime and 5 lb/ton spar are used.

Although the steel sulfur varied erratically from .005% to .010% S in the first period, a consistent sulfur level of .005% S ±.001% S was obtained in the final best practice period of 13 consecutive heats.

The best low sulfur procedure was suitable for producing many different grades of steel with a ladle sulfur of .005% S ±.001. These grades included heats with:

- Final carbon levels ranging from .05 to 70% C.

- Various deoxidation practices, including rimmed, controlled rim, semi-killed, and killed grades.

Variables That Influence Final Steel Sulfur in the BOF

Regression analysis of the Mag-Coke heats was used to aid in determining the variables which influence the final steel sulfur. A library program, the Stepwise Linear Regression Program (SLIRP), was used. The resulting influences are summarized schematically in Figure 8.

Final steel sulfur was the principal dependent variable. Four variables exerted statistically significant influence (at least 95% confidence). They are, in order of their statistical significance, summarized below:

Variable	% Variance Explained	F Test*	Significance*
Hot Metal Sulfur	47	49.2	99%
Skim Code	23	13.9	99%
Fraction Low Sulfur Scrap	4	6.6	95%
S_{slag}/S_{steel}	3	6.0	95%

Total (four variables) 77% Multiple Correlation Coefficient = .878

*Minimum F at 95% = 4.08
Minimum F at 99% = 7.31 (40 heats)

The following equation resulted from the analysis considering the four variables:

Sulfur in Steel = .00939 + .419 (Hot Metal Sulfur) - .00906 (Skim Code) -

$.718 \times 10^{-5}$ (Low Sulfur Scrap, lb/ton Hot Metal) - 0.0027 (S_{slag}/S_{steel})

±.0030 (at 95% Confidence).

1 Standard Error = +.0015% S (67% Confidence),

2 Standard Errors = ±.003% S (95%).

Hot metal sulfur is the % sulfur contained in the hot metal. The skim code is an arbitrary value assigned to describe the degree of skimming of slag on the hot metal as follows:

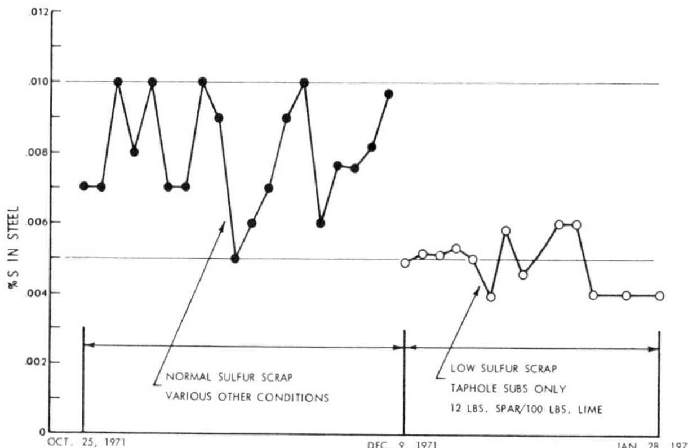

Fig. 7 - Chronological chart of percent S in low-sulfur steel heats

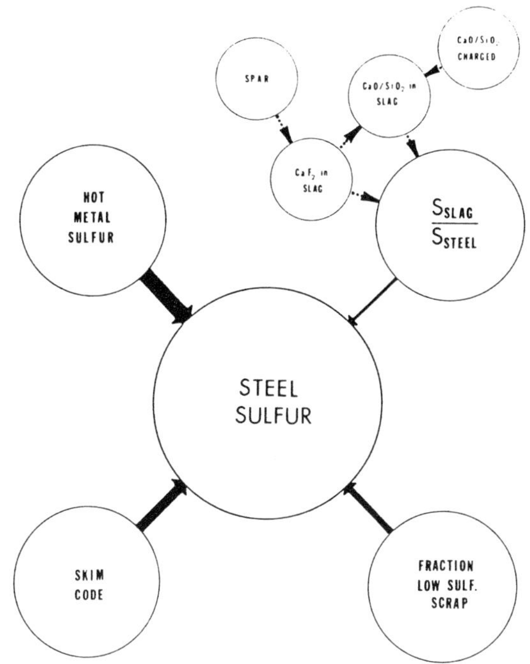

Fig. 8 - Variables that influenced final steel sulfur. All variables shown here are judged significant. Arrow thickness represents the relative statistical significance of each variable. Strength of the physical effect is given by regression equation.

Skim Code	Description of Hot Metal Slag Skimming
1	No tapholes submarine, variable slag retention.
2	Partly taphole submarines, i.e., if two submarines were poured, only one had a taphole.
3	All submarines had tapholes.
4	All submarines had tapholes plus additional transfer-ladle skimming

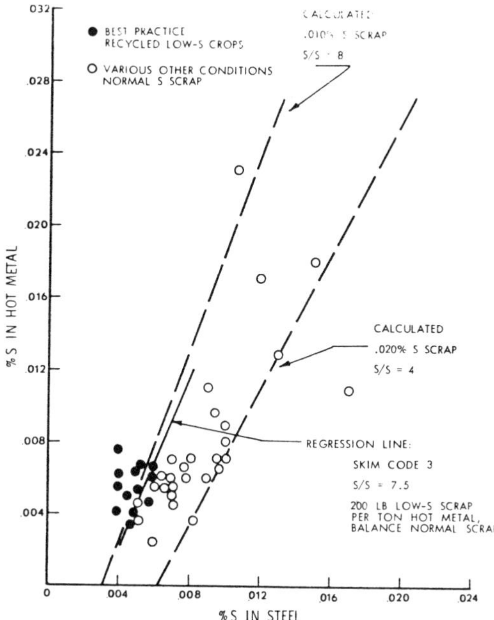

Fig. 9 - Percent S in steel versus percent S in hot metal

either by raking or through the slag freezing and forming a shell which did not go into the furnace.

The fraction of scrap is expressed in pounds of scrap per ton of hot metal. Finally, the S_{slag}/S_{steel} is the chemical percent ratio of sulfur in slag to sulfur in steel.

Assuming that all the variables in the above equation are statistically significant, according to the equation the following changes will raise the steel sulfur about .001% in the range of .003 to .011% sulfur:

- Raising the hot metal sulfur by .003% raises the steel sulfur by .0013%.

- Lowering the Skim Code by 1 unit raises the steel sulfur by .0010%.

- Decreasing low sulfur scrap from 140 lb/ton of hot metal (33% average scrap charge) to zero increases steel sulfur by .0010%.

- Increasing S_{slag}/S_{steel} from 4.0 to 7.5 or from 7.5 to 11.0 lowers steel sulfur by .0010%.

The equation predicted final steel sulfur to within ±.003% 95% of the time. Although this seems very high, the practical value of the relationship is evident in the best practice period in Figure 7, where these variables were controlled to produce steel sulfurs at or near .005% S.

A plot of the % S in steel as a function of the most significant independent variable, hot metal sulfur, is given in Figure 9. Also shown are two dashed lines calculated by sulfur material balances, one for normal amounts of fluxes and scrap sulfur content (.020% S) and the second for low-sulfur scrap (.010% S) and extra fluxes. The actual data generally fall within the practical operating band defined by the two dashed curves. All the best practice heats made with low-sulfur scrap (solid circles) lie on the low steel sulfur side of the band. Also plotted in Figure 9 is the regression line of the equation presented above. The conditions of the plotted line are all taphole submarines (Skim Code 3), 200 lb low sulfur scrap per ton of hot metal, and a sulfur partition ratio of 7.5:1. These empirical results agree very closely and lie slightly to the right of the calculated curve for low-sulfur scrap and an 8.0:1 partition ration.

Based on these results, we would expect the steel sulfur to vary as follows with hot metal sulfur and practice:

% S in Hot Metal:	.005	.010
% S in Steel - Normal Scrap and Fluxes:	.009	.012
% S in Steel - Low S Scrap and Extra Fluxes:	.005	.007

The difference between the calculated CaO/SiO_2 charged and the CaO/SiO_2 actually found by analyses of the turndown slags depends on the amount of spar charged. When less than 10 lb spar per 100 lb lime were used, the CaO/SiO_2 charged exceeded the actual by about 0.7. But when more than 10 lb spar per 100 lb lime were added, the difference between charged and actual CaO/SiO_2 decreased to zero, showing that much more lime was fluxed. These results reinforce the regression correlations in Figure 8 which show that the partition ratio and lime-silica ratio are significantly influenced by the CaF_2 level in the slag. Accordingly, for the best practice we decided on an aim addition of 12 lb spar per 100 lb lime.

No. 4 Open Hearth Shop Results

Details of the three low-sulfur 425-ton open hearth heats made from desulfurized hot metal skimmed with taphole submarines are summarized in Table II. The charge materials for each heat differed with respect to scrap sulfur content and ferromanganese added. Normal sulfur home scrap was used in the first two heats, but in the last heat most of the scrap charge consisted of extra-low sulfur scrap crops. Ferromanganese usage varied from a total of 27,000 lb (64.3 lb/ton) in the first heat to none in the last heat. A steel sulfur content of .005% S was obtained both with the high ferromanganese normal sulfur scrap charge and with the no-ferromanganese, low sulfur scrap charge. Most of the manganese was oxidized into the slag, which contained 18.5% MnO as compared with 4.2% MnO when no ferromanganese was used.

Table II. 425-Ton Open Hearth Heats

Trial No.	1	2	3
Hot Metal:			
Average \$ Si	.94	.92	1.01
Average % S	.011	.005	.008
Weight, lb	669,000	667,000	643,000
Scrap:			
Low-Sulfur (.010% S max), lb	0	0	262,000
Total lb	295,000	291,000	327,000
Other Charge:			
Ferromanganese, lb	22,000	0	0
Limestone, lb	20,900	17,100	19,500
Burnt Lime, lb	22,450	21,100	23,250
Furnace Additions:			
Ferromanganese, lb	5,000	5,000	0
Burnt Lime, lb	5,600	0	11,200
Spar, lb	2,000	5,000	5,000
Ore, lb	14,000	0	7,000
Slag: CaO/SiO_2	4.3	6.5	4.1
FeO, %	20.8	28.6	28.2
MnO, %	18.5	7.6	4.2
CaF_2, %	2.5	5.6	3.7
% S Steel, Average Ladle	.005	.007	.005

Other departures from the normal practice used in the open hearth were:

- Early initiation of roof lance oxygen blowing to increase refining temperatures as much as 100 F.

- Increase of the limestone charge to 17,100/ 20,900 lb versus 10,300 lb normally.

- Increase of the total charged or fed burnt lime to 21,100/34,450 lb versus 21,000 lb normally.

- Increase of the spar to 5000 lb versus 1000 lb normally.

Although more fluxes than normal were added, the amounts used for the low sulfur heats made in the open hearth were only about half as much per ingot ton as those required in the BOF low sulfur practice. Another advantage of the open hearth procedure over the BOF was that a much higher proportion of the charge consisted of scrap.

Conclusions

In summary, a hot metal desulfurization practice and steelmaking controls were established to produce steel with a sulfur content of .005% ±.001% (ladle teem stream) for 13 consecutive heats in the BOF at Sparrows Point. The best practice for the 220-ton BOF included:

- Desulfurization of the hot metal to a range of .003 to .008% S with Mag-Coke.

- Using only taphole submarines for maximum skimming of the hot metal treatment slag.

- Recycling low sulfur (.010% S max) scrap to make up at least half of the scrap charge.

- Charging 227 lb/ton lime and 27 lb/ton spar as compared with normal additions of 168 lb/ton lime and 5 lb/ton spar.

Additional findings were:

- Steel with .005% S can be produced in our 425-ton open hearth furnaces with an approach similar to that developed for the BOF.

- Since desulfurization is complete before tapping, a low sulfur content can be obtained in rimmed, controlled rim, semi-killed, and killed grades of steel.

- The average hot metal sulfur was .0252% S before and .0058% S after the Mag-Coke treatment. Mag-Coke usage was 2.7 lb per ton of hot metal. The average efficiency of magnesium utilization (including residual magnesium) was 61%. Hot metal with an as-cast sulfur up to .052% S was successfully desulfurized and used to make .005% S steel.

- A new Mag-Coke plunging bell design and expansion joint was developed which increased life from 7 to 21 dips.

- Taphole submarines significantly improved skimming, decreasing the average depth of hot metal slag in the transfer ladle from 5 to 13 inches.

Disadvantages of the overall process are: (a) skulling problems in the submarines and maintenance of tapholes, (b) less scrap charged to the BOF because of extra fluxes and lower hot metal temperature.

References

1. Roe, G.J., Slimmon, P.R. and Melloy, G.F., "Sulfur - Some Effects on Steel Processing and Steel Properties," AIME National Open Hearth Proceedings, V. 55, 1972, pp. 186-202

2. Snow, W.E., "Desulfurization of Blast Furnace Metal with MAG-COKE (Magnesium Impregnated Coke)," AIME National Open Hearth Meeting, Southwestern Section, February 8, 1972

3. Speer, M.C. and Parlee, A.D., "Dissolution and Desulfurization Reactions of Magnesium Vapor in Liquid Iron Alloys," Ductile Iron Society Annual Meeting, Bulletin No. 49, 1970, pp. 28-33

4. Pause, U. and Strathdee, B.A., "Tapholes in Hot Metal Torpedo Cars," AIME National Open Hearth Proc., V. 54, 1971, pp. 14-22

THE LIME-MAG PROCESS FOR DESULFURIZATION OF HOT METAL

P.J. Koros, R.G. Petrushka and R.G. Kerlin

Jones & Laughlin Steel Corporation
Pittsburgh, Pennsylvania 15263

This paper originally appeared in the Iron & Steelmaker, Vol. 4, No. 6, 1977, p. 34.

ABSTRACT

A new process for desulfurization of hot metal, based on injection of pure magnesium powder and fine lime, has been brought to large scale commercial use. Advantages over calcium carbide and other desulfurizers in general use are: better cost performance, near insensitivity to blast furnace slag in the subs, improved reproducibility, safety and wide flexibility for use in varied steel plant conditions. The Lime-Mag process is now in use at the production scale facilities at J&L Steel's Aliquippa, Pa. and Cleveland, Ohio Works.

INTRODUCTION

J&L Steel, with a product mix oriented toward flat rolled, HSLA, tubular and bar and rod grades, and with coke sulfurs at or above 1%, started to rely on hot metal desulfurization treatments in 1964. Since then, we have been through a series of steps to upgrade processes and capacity, with the result that now at Aliquippa and Cleveland Works production scale hot metal injection stations are in service, Figure 1. A totally new process, Lime-Mag, developed at J&L's Cleveland Works[1,2] is used. Several hundred subs are treated monthly in each plant; typically 20 to 30% of the hot metal is desulfurized. Subs are designated for treatment either for "salvage" of otherwise unuseable iron -- from >.040S to <.020S -- or for preparation of extra low sulfur hot metal (<.010S) for .010 and .015S max. BOF heats. Very often the "salvaged" subs become the "low sulfur" feed stock to the BOF. The impact of the availability of the desulfurization stations is illustrated in Figure 2, which shows the typical pattern for selection and disposition of the treated iron.

Our tests of the various large scale desulfurization processes and the development and rationale for the Lime-Mag process are presented in this paper.

BACKGROUND

Starting in 1964, increasing use was made of external desulfurization of hot metal at J&L's Cleveland Works. By the end of 1974, more than 25% of the BOF heats were charged with hot metal desulfurized to some extent. In the early years of this program[3], hot metal desulfurization was carried out simply by throwing soda ash on the bottom of the ladle utilized to charge hot metal into the steelmaking furnaces. This technique, used throughout the world, is reasonably effective and reproducible as long as moderately high sulfur levels in the hot metal ($\approx$.020/.030S) are acceptable. However, the steady tightening of steel sulfur specifications and the intolerable environmental effect of increased usage of soda ash, led to implementation of Mag-Coke desulfurization in 1972, first at the Cleveland Works and later at Aliquippa. Mag-Coke technology has been described in the literature[4]. By 1974, J&L was the largest consumer of Mag-Coke in the United States and second in the world. Mag-Coke served us well; however, the high cost and slow batch nature of this process made us realize that a better solution to the problem must be found. Experience had already been developed both in and outside the U.S. with desulfurization of hot metal by injection of various reagents, based on calcium carbide[5-8] lime[9,10] and in the USSR, on magnesium[11].

Trial Program

With this in mind, an experimental injection station was built at J&L's Cleveland Works[1], Figure 3, to develop expertise in the use of injection technology and to provide the opportunity for evaluation of the materials available commercially. Cost effectiveness, safety, and reproducibility of results were the key parameters, particularly for the low sulfur product mix at Cleveland and the relatively heavy contamination of the hot metal system by blast furnace slag. Of course, one aim was an immediate increase in desulfurization capacity.

Fig. 1 - J & L Steel - Aliquippa Works hot metal desulfurization station

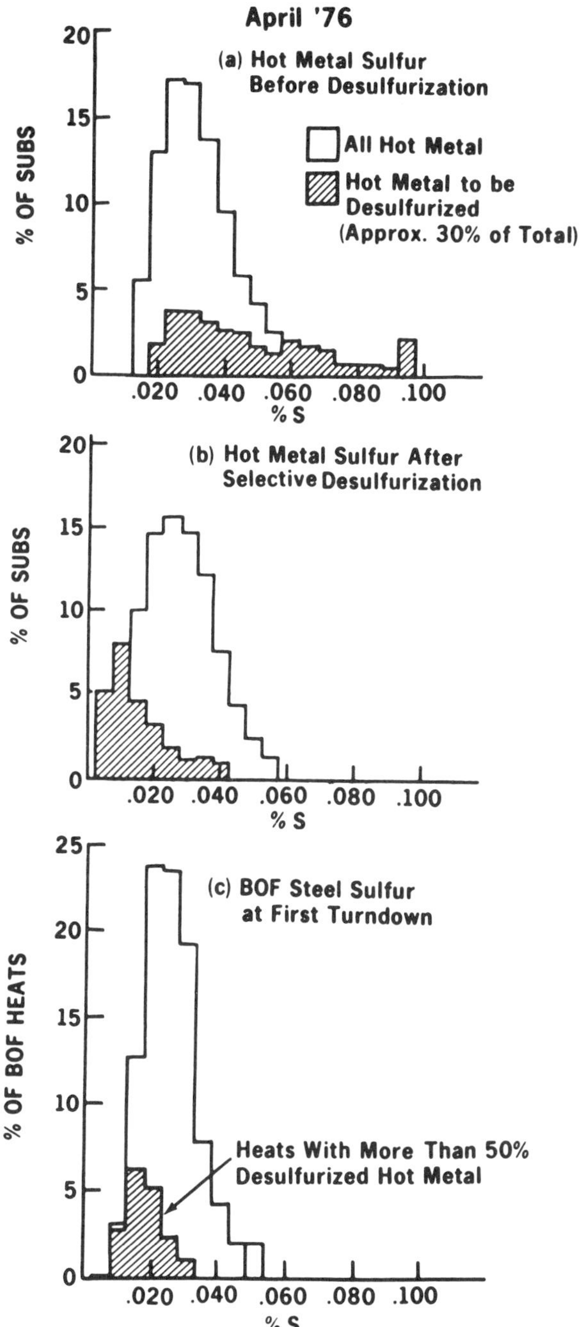

Fig. 2 - Effect of external desulfurization of sulfur in hot metal in April, 1976 - Aliquippa Works

In the program in Cleveland, nearly all materials available in the U.S. for injection were tested, Table I. As reported earlier [1], our results with calcium carbide, both commercial grade and doped with a gas releasing component, as well as trials involving use of lime/natural gas and magnesium-aluminum alloy powders, provided results within the range expected based on reports in the literature. An analysis of the conclusions is presented in Table II; in our view, the commercially available injectable reagents had shown deficiencies:

Fig. 3 - J & L Steel - Cleveland Works experimental desulfurization station

Lime injection, although the least costly and the safest, even at a 2:1 replacement against carbide leads to incremental slag volume in the subs which is intolerable.

P.J. Koros is Director - Process Metallurgy; R.G. Petrushka is Supervisor - Desulfurization and Dekishing, Steel Producing Department, Cleveland Works and R.G. Kerlin is General Supervisor, Ironmaking Department, Aliquippa Works.

Carbide injection, although mechanically easy, led to safety problems in wet handling of the slags left in the subs. Furthermore, the "efficiency" of carbide was found very sensitive to the presence of blast furnace slag in the subs: 25% reduction in efficiency and a 50% increase in standard error of end of treatment sub sulfur content [1]. A result of this sensitivity to blast furnace slag contamination is that attainment of hot metal sulfur levels $<.010$ with carbide necessitates extreme care in slag control.

Mg-Al alloy was found very costly because: (1) copious magnesium fumes are released, particularly at low sulfur contents, this being indicative of large scale waste of magnesium; (2) the readily injectable 50 to 65% Mg metallic alloy powder is priced equal

TABLE I

J&L STEEL HOT METAL DESULFURIZATION TRIALS REAGENTS AND EXPERIMENTAL CONDITIONS FOR INJECTION TRIALS (1)

Reagents	Injection Conditions	Sub Conditions	Hot Metal
Calcium Carbide	Transport Gas	Slag Contamination	S in .020 - .100
Commercial Grade	Nitrogen	Normal (30-100 lb. slag/NTHM)	out .077 - .003
Doped (20%)	Argon	"Clean" (10-30 lb. slag/NTHM)	
	Natural Gas		Si 0.7 - 1.80
Lime - Pure		Deslagging	Mn .05/1.0
Doped (5-20%)	Fluidizing Gas	Dry Drain Out	Ti .05/.20
	Nitrogen	Wet Drain Out	T 2540 - 2760°F
Mg-Al (Alloy)	Argon		
50% Mg		Iron Load (NTHM	
65% Mg	Gas/Solids Ratio	108 - 258	
75% Mg	.031 - .24 cf./lb.		
		Fill Range	
Mag-Lime (20% Mg)	Rate (lb./min.)	60% - 100%	
Lime & Mag (Variable Ratio)	68 - 250 CaC_2 Lime		
& Al	8 - 30 Mg-Al Alloy		
& Mg Al (Alloy)			

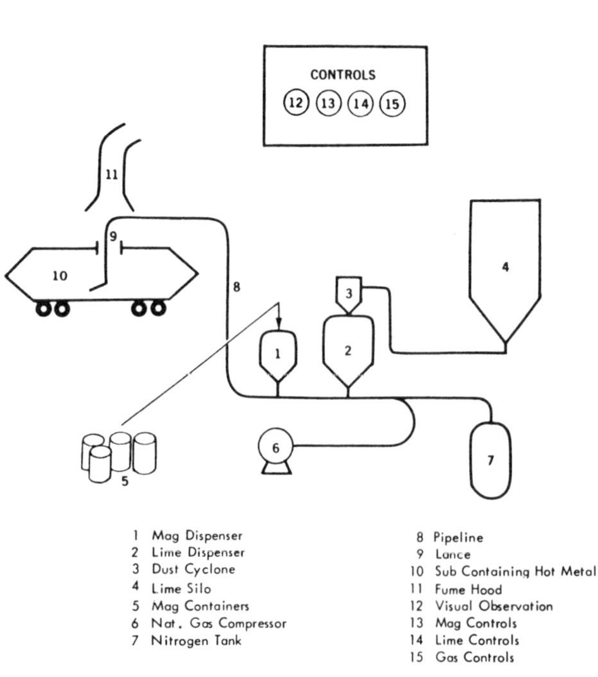

Fig. 4 - Schematic of lime-mag desulfurization process*

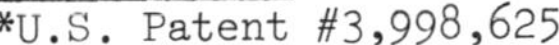

*U.S. Patent #3,998,625

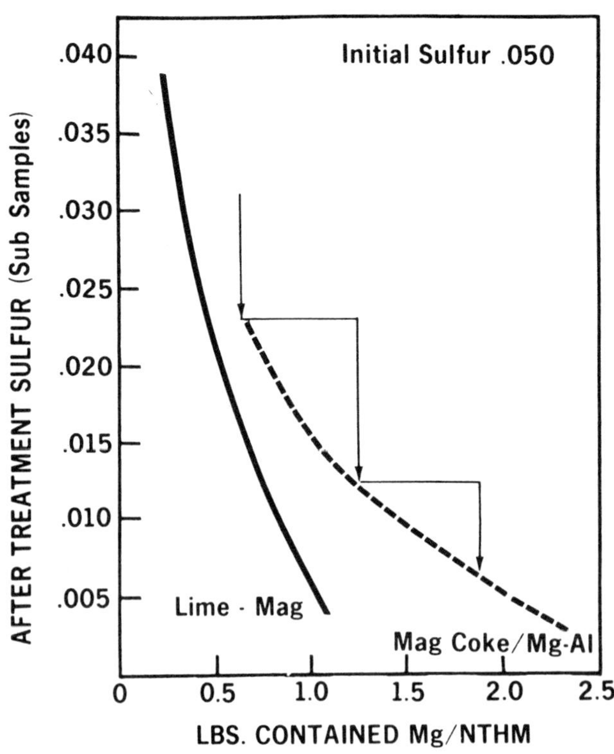

Fig. 5 - Comparison of magnesium consumption for desulfurization of hot metal by the commercially utilized magnesium process

TABLE II

CONCLUSIONS FROM PROGRAM TO EVALUATE DESULFURIZATION PROCESSES (1)

CALCIUM BASED DESULFURIZERS

Material	Cost Capital / Treatment	Advantages	Disadvantages
Carbide	Highest / Medium	Easy to Inject Simple Lance Extensive Experience Fast	Safety-acetylene generation with powder Difficult Fume Handling Sensitive to B.F. Slag Difficult Slag Disposal -- acetylene Some Cannot Fill Subs Attacks Sub Lining No Secondary Desulfurization
Lime	Medium / Lowest	Easy to Inject Safest Easy Slag & Fume Disposal	Large Slag Generation Sensitive to B.F. Slag No Secondary Desulfurization Not Practical <.020S Slow

MAGNESIUM BASED DESULFURIZERS

Material	Cost Capital / Treatment	Advantages	Disadvantages
Mag-Coke	Least / High	Simple Secondary Desulfurization Reproducible Insensitive to B.F. Slag Intermittent Use	Time-Batches Fumes Bell/Stem Breakage Cannot Control Magnesium Release Rate
Mg-Al	Low / Highest	Low Volume - Fast Intermittent Use Insensitive to B.F. Slag Secondary Desulfurization	Lance Vibration and Failures Fumes Safety with Powder
Mag-Lime (Pre-Mix)	Medium / High	Secondary Desulfurization Insensitive to B.F. Slag	Inhomogeneous Mix Causes Lance Vibration Difficult Disposal of Unused Material Limited Supply of Pre-Mixed Powder Difficult to Control Mg Release Rate

TABLE III

LIME-MAG PROCESS - MAGNESIUM CONSUMPTION

Initial (Sub Tests)	Final Sulfur Sub	Final Sulfur Transfer Ladle	Lbs.Mg / NTHM
.070	.025	.022	.82
.045	.020	.017	.50
.030	.005	.003	.60

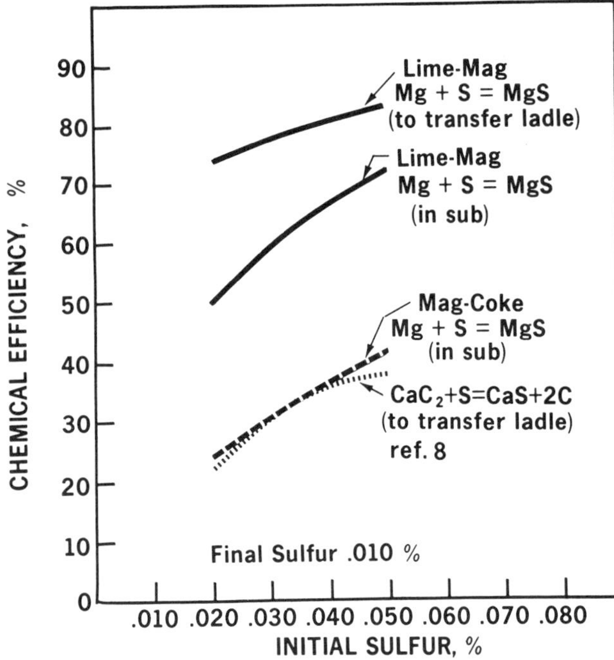

Fig. 6 - Comparison of chemical efficiencies in desulfurization of hot metal by injection and mag-coke

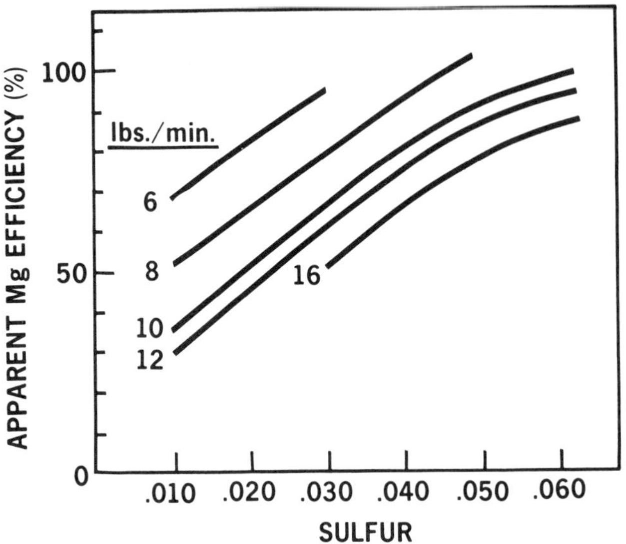

Fig. 7 - Influence of the rate of injection of magnesium on desulfurization "Efficiency" Calculated by assuming all sulfur removed by Mg without credit for sulfur removal by lime but including secondary desulfurization by Mg dissolved in the hot metal.

to pure Mg per pound of powder (e.g., 30 to 50% higher per unit of Mg); and, (3) lance cost is high and failure made frequent due to the vibration, breakage and pluggage caused by the action of the vaporizing magnesium.

Mag-Lime, pre-mixed at a nominal 20% Mg, was costly (per contained unit of Mg). The difficulties in attaining homogenity in the powder led to obvious local differences in concentration of Mg with the result that lance vibration and surges in Mg vapor release were intolerable. Accurate flow control was nearly impossible.

THE LIME-MAG PROCESS

It became clear that magnesium offers several advantages as a desulfurizer -- nearly total insensitivity to presence of blast furnace slag, no residual entrapment of reactive reagent in the slag, saturation of the iron with Mg (which prevents S reversal from the slag), no release of toxic or dangerous fumes, etc. -- so that further effort was warranted toward its implementation as a practical, economic desulfurizer. Looking at our problems with injection of the pre-mixed material, the (now) obvious next step was separate feeding of lime and magnesium powders with mixing in the transport line on the way to the sub. A schematic representation of what we call the Lime-Mag Process is shown in Figure 4. Typical levels of magnesium consumption are provided in Figure 5 and Table III.

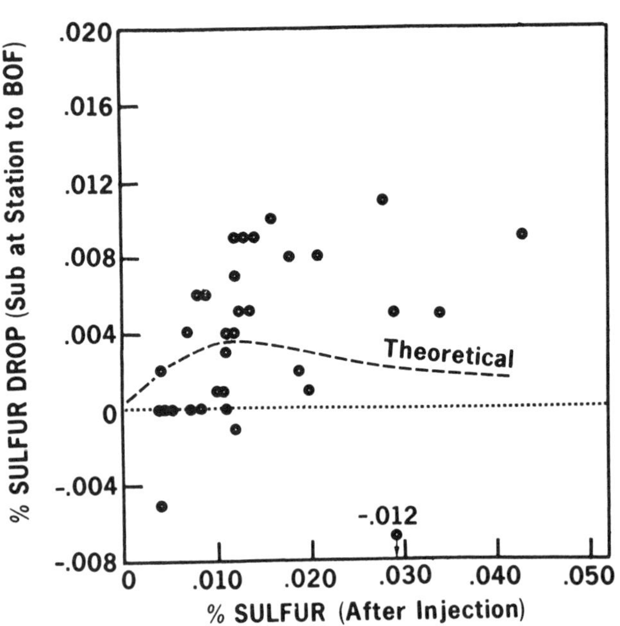

Fig. 8 - Secondary desulfurization obtained after injection treatment in hot metal saturated with magnesium.
Theoretical corresponds to desulfurization expected in cooling from 2500°F to 2400°F: $\%Mg \times \%S = 7.0 \times 10^{-5}$ (2400°F) and 11.5×10^{-5} (2500°F)

TABLE IV

PREDICTABILITY OF END OF TREAMENT SULFUR LEVEL
J&L CLEVELAND WORKS TRIALS(1)

Reagent	Sub Condition Rating	Slag Depth	Difference Predicted-Actual	No. Of Subs
Calcium Based				
Carbide* - 75% CaC_2	Normal	2" - 36"	.0073S	139
	Clean	2" - 3"	.0050S	25
- Doped	Normal	4" - 18"	.0040S	7
Lime/Natural Gas	Clean	1" - 12"	.0093S	32
Magnesium Based				
Mag-Coke	Normal	2" - 36"	.0045S	89
Mg-Al (Alloy)	Normal	2" - 36"	.0081S	23
Lime-Mag	Normal	2" - 36"	.0044S	117

*Literature figure for doped carbide, "clean" subs, is .006S(5).

TABLE V

ABILITY TO ATTAIN LOW SULFUR BOF TURNDOWNS
J&L - CLEVELAND WORKS - AIM .015S MAX.

Success of Attempts	Carbide	Mag-Coke	Lime-Mag	Untreated
At Ordered .015S Max.	44.7%	51.9%	82.6%	65.1%
At Attained <.010S	1.6%	15.7%	21.7%	9.3%
No. of Heats Monitored	123	185	69	43

Conditions

Hot Metal: >75% charge <.020S after treatment; average of .022S (sub samples)
Scrap: Exclude No. 2 bundles, no special selection of low sulfur home scrap
Raker: Attempt max. removal
Slag: Aim 3.5 V.R.

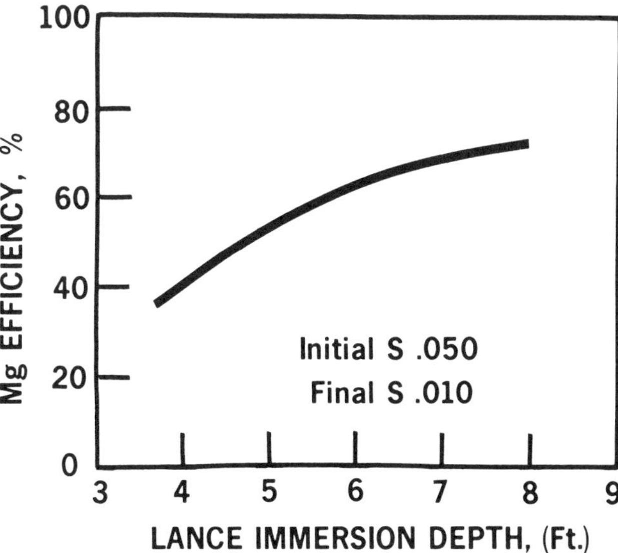

Fig. 9 - Effect of depth of hot metal on chemical efficiency during injection of magnesium

Efficiency

The dramatic difference between Lime-Mag's, Mag-Coke's and Mg-Al's magnesium consumption is illustrated in Figure 5. The apparent greater chemical efficiency, Figure 6, is attributed to a combination of several factors:

Mg bubble size is kept small by the pulverized lime acting as a dispersant. This is in contrast to Mag-Coke and Mg-Al alloy for which there is no means to prevent formation of the large, rapidly rising, bubbles described by Guthrie(12).

Mg Flow Control - With separate, independent control of lime and magnesium rates and with homogenity of the blend assured by the fluidized transport, stability of powder flow in the transport line is assured by holding constant the rate of lime injection and adjusting the magnesium flow rate to minimize fume evolution. Gas/solids ratio in the transport line is approximately .10 cf/lb., this assures minimum entrainment (i.e., loss by "carry through") of lime and magnesium powders. The philosophy is to use the desulfurizer as the stirring agent. The ability of the iron to "accept" magnesium decreases as the sulfur content is lowered; for optimum utilization, the rate of magnesium delivery must be lowered as the sulfur is lowered, Figure 7.

Lime does participate, to a slight extent, in the reaction, thereby complementing the magnesium effect. This is made clear by the attainment of magnesium "efficiencies" near 100% as shown in Figure 7.

Magnesium dissolution in the iron is an important advantage because it leads to "secondary" desulfurization, that is, a drop of sulfur which occurs while the hot metal cools in transit from the injection station to the BOF shop, Figure 8.

Injection Depth - A concern on the use of a vapor (magnesium) for desulfurization of hot metal is that there may be sensitivity with respect to the depth at which the reagent is released in the hot metal. As indicated in Figure 9, although the chemical efficiency does depend to some extent on the depth of lance immersion, in reality, the pool of hot metal available in a typical torpedo car (8 to 12 ft.) is such that unless the sub is considerably underfilled (<75% of capacity) the desulfurization results obtained are almost insensitive to the degree of filling. This was clearly the case in our initial trials at Cleveland, where sub filling generally was held in the 75-100% range. At Aliquippa, the subs treated in the 50% fill range have shown a loss in efficiency although desired end points are achieved by use of the same total magnesium quantity as if the sub were full.

Reproducibility - It was learned at the pilot operation in Cleveland that with proper use of the process

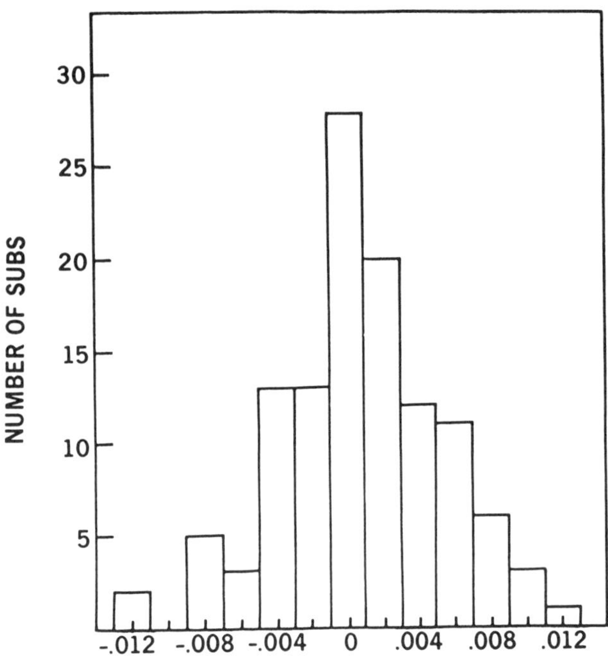

Fig. 10 - Lime-mag process control capability. Cleveland Works station data, 117 consecutive sampled subs.

Fig. 11 - J & L Steel - Aliquippa Works desulfurization station lime silo

TABLE VI

ENERGY USAGE COMPARISON FOR MAGNESIUM AND CALCIUM CARBIDE

	Materials			For Equivalent Desulfurization (.050S ➞ .010S)	
	Magnesium[14]	Calcium Carbide		Mg: .80 Lb./NTHM	Carbide: 8.9 Lb./NTHM of 70/30 Blend[5]
Year		Commercial Grade[15] (80% CaC_2)	70/30 Blend (56% CaC_2)		
Current - 1975/6	165,000 BTU/Lb.	27,500 BTU/Lb.	19,300 BTU/Lb.	132,000 BTU/NTHM	171,000 BTU/NTHM
Forecast - 1980	125,000	27,500	19,300	100,000	171,000

TABLE VII

MAGNESIUM AND CALCIUM CARBIDE AVAILABILITY IN THE UNITED STATES

		Capacity*	Consumption*	% Consumed By Ferrous Industry
Magnesium[16]	1976	270	185	13%
	1980	415	275	40%
Carbide[17]	1976	720 **	580 **	
	1980	720 **	580 **	

*Millions of Lbs./Yr.

**Unabated, idle but operable, capacity is 408 MM lbs.

Fig. 12 - J & L Steel - Cleveland Works desulfurization station, injection lance after blow

there is nearly total insensitivity of the reaction between magnesium and sulfur to the presence of blast furnace slag. This makes results more reproducible than with calcium carbide, Table V. The fact that the materials are introduced continuously, with start and stop controlled by the operator (as against a batch type process like Mag-Coke) provides the ability to attain end point sulfurs within a standard error of .0044%S, Figure 10.

This accuracy, plus the effect of the magnesium dissolved in the iron, have the result that when magnesium powder is used correctly, the need to "overshoot" the desired sulfur level is diminished as the operator can trust his control charts and aid the BOF in low sulfur turndowns, Table V.

Injection Materials

Pulverized lime and magnesium powder are used. Lime, sized >90% under 325 mesh, is received from several sources. The fine grain is required to maximize magnesium dispersion and to promote smooth powder flow to the sub. Lime is received in blower equipped trucks and transferred into the lime storage vessels. At each station, there is 50 NT lime holding capacity, Figure 11. Although currently magnesium is received in 250 lb. cans,

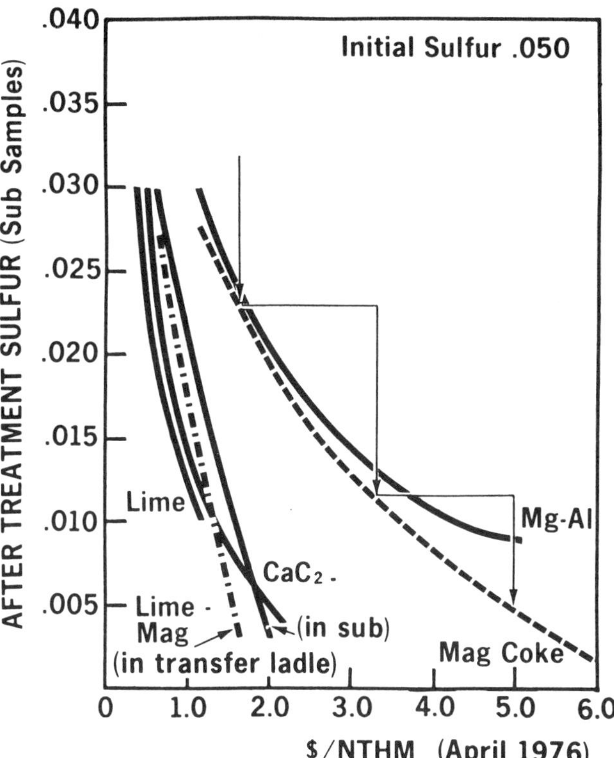

Fig. 13 - Comparison ($/NTHM) for hot metal desulfurization by several processes

within a few months we can expect to receive the powder in pressurized 4500 lb. boxes.

Temperature Loss

Lime-Mag injection causes only a small temperature loss measured at 18°F (160 NT subs) vs. 13°F for carbide (250 NT subs)(5). Although the difficulty in attaining a representative temperature reading prior to the stirring caused by injection sheds some doubt on these figures, the thermochemistry(13) of magnesium injection -- vs. that for calcium carbide doped with a carbonate gas release agent -- suggests that temperature loss for Lime-Mag should be about 3°F lower for a .050S to .010S treatment. In effect, the temperature loss is that due to the extra track time for the sub.

Lance Performance

A light injection lance, Figure 12, is used at both stations. This is made possible by the minute size of magnesium bubbles formed in the liquid iron and release of the magnesium at a distance from the shaft to prevent direct upward channeling of Mg vapor. The vertical entry system is less costly to build and the lances need be less rigid than the more conventional straight pipe -- angled entry. Lance life at Cleveland is of the order of 5-7 treatments; by patching errosion and light cracks, the record now stands at 23 at Aliquippa. Failures often involve mechanical damage by contact with the top of the sub, or with excessive sub bottom build-up. Lance cost is of the order of $0.05 per NTHM.

Cost

The Lime-Mag Process has very competitive cost per unit of sulfur, Figure 13. The reason for this is reliance on two commercially available products, pure magnesium powder -- cheaper per pound of magnesium than its alloys -- and pulverized lime. These two materials are available in what turns out to be a highly competitive marketplace.

Cost growth rate for magnesium and carbide are seen as very closely related to the rise in energy costs and the potential for process efficiency improvements by the producers. As shown in Table VI, although the energy requirement -- electrict and thermal -- for production of calcium carbide is lower per lb. than for magnesium,[14,15] for equivalent sulfur removal, say .050S down to .010S in the hot metal, magnesium has and is expected to continue to have, a significant advantage. Therefore, the sharp further escalation in energy costs expected by these two industries[14] should have a lower inpact on desulfurization cost via magnesium based systems.

Reagent Supply

A key consideration in the selection of a desulfurizing reagent is its availabilability, particularly in light of the tightness in magnesium supply through the 1972-74 period and the shutdown of a significant portion of the U.S. carbide producing capacity due to technological changes in the chemical industry.

Magnesium supply, at least through 1980, is forecast[16] to exceed demand by a relatively large margin, Table IV. Furthermore, even assuming that between a third and half of all hot metal produced in the U.S. is desulfurized, ferrous industry usage of magnesium is expected[16] to be far below that of the aluminum industry. For carbide[17], excess in capacity over demand appears to be less and support of extensive growth in hot metal desulfurization may require construction of new facilities.

Safety

Contrary to the problems with residual CaC_2 left in the slag on top of the sub, desulfurization with Lime-Mag presents minor safety problems. Repeatedly[1], experiments were conducted to determine whether magnesium is trapped by the slag in the sub -- the result always has been a categorical "no". The reason for this is that the slag contained in the subs is at a temperature which is above the vaporization temperature of pure magnesium. The safety precuations in handling magnesium powder are of the same nature as those involving calcium carbide. A major plus on the magnesium side is that when station fires occur, magnesium powders burn and are extinguishable by the operators. With calcium carbide, particularly in the relatively high humidities prevalent in the northeastern United States, spills result in release of acetylene, which, in turn can create an explosive atmosphere; the odor around the carbide injection stations tells of this potential safety problem.

CONCLUSIONS

1. The Lime-Mag desulfurization process has been brought successfully to the large tonnage scale.

2. Magnesium has definite advantages over its main competitor -- calcium carbide -- in safety, slag handling, independence from interference by blast furnace slag, etc.

3. Lime-Mag desulfurization cost is competitive with carbide in the typical .060-.020S range; below .015S, Lime-Mag is definitely less costly and more reproducible.

OPERATING EXPERIENCE AT YOUNGSTOWN STEEL WITH INJECTED SALT COATED MAGNESIUM GRANULES FOR EXTERNAL DESULFURIZATION OF HOT METAL

Allan M. Smillie and Richard A. Huber

Brier Hill Metallurgical Dept.
Tubular Division
Jones & Laughlin Steel Corporation
Pittsburgh, PA 15230

Close control of the sulfur content of hot metal entering the open hearth is essential to achieve optimum productivity. Pursuit of this goal led to the development of a low cost production scale hot metal desulfurization system based on the injection of salt coated magnesium granules. External desulfurization of about 90% of the hot metal supply has enabled the Brier Hill Steelworks to produce all seamless grades to a maximum sulfur specification of 0.020% and steels with 0.010% sulfur have been tapped with normal refining times.

INTRODUCTION

The Brier Hill Works of the Youngstown Sheet and Tube Company produces steel rounds for the manufacture of standard and line pipe and oil country goods.

Prior to the introduction of oxygen roof lances in late 1977, the open hearth shop consisted of nine 185 ton conventional furnaces. A review of operating data showed that the refining time had increased by 1.6 hrs. per heat over a seven year period. This was almost wholly the result of an increase in furnace sulfur load as evidenced by an increase in preliminary sulfur from 0.036% to 0.049% during the same period.

STEEL PLANT SULFUR LOAD

As a first step towards solving the sulfur problem and regaining the lost productivity, a research group conducted a detailed evaluation of the individual sulfur sources using a mass balance approach. The results of this study are summarized in Table I and discussed in detail below.

TABLE I Typical Furnace Sulfur Load Prevailing At The Beginning Of The Investigation

Composite Mass Balance Results
Prelim Sulfur Range - 0.045 to 0.050%

Material	Quantity (Lbs.)	% S	Wt. Sulfur (Lbs.)
Hot Metal (As Cast)	230,000	0.035	81
Hot Metal (In Transit Reversion)		0.008	18
B.F. Carryover Slag	6,600	0.80	53
Scrap	180,000	0.030	54
Sulfur Pick-up from 1% S Fuel Oil	-	-	24
			230

Hot Metal

A 55% charge of hot metal with a typical as-cast sulfur of 0.035% accounted for 35% of the total furnace sulfur load.

In Transit Sulfur Reversion

Detailed investigations at both the Indiana Harbor BOF Shop and the Brier Hill Works in Youngstown revealed that the sulfur content of hot metal received by the steel plant was about 0.008% higher on average than the cast analysis. This corresponded with about 8% of the furnace sulfur load. Samples of carryover slag taken from the transfer ladle revealed a dramatic decrease in both sulfur content and base/acid ratio. Typical slag analyses are shown in Table II. Similar slag chemistry changes have been reported by Duquette (1). Further work led to the conclusion that in-transit reversion was the transfer of sulfur from blast furnace carryover slag to hot metal due to the combined effects of compositional modification of the slag and a reduction in slag temperature (2). This change resulted from oxidation of metalloids, ingestion of runner sand and vermiculite insulation, plus solution of ladle refractory. A typical material balance depicting the relative contributions from each source is shown in Figure 1 (3).

TABLE II Average Blast Furnace And Transfer Ladle Slag Analyses For Samples Taken During Preliminary Investigation

	Blast Furnace Slag[1]		Transfer Ladle Slag[2]		Calculated Diluent[3]	
	Mean %	Std. Dev.	Mean %	Std. Dev.	Mean %	Std. Dev.
CaO	38.8	2.0	27.7	3.4	0	0
SiO_2	35.1	2.0	48.2	3.4	81.5	3.6
MgO	14.8	0.8	10.4	1.4	0	1.9
Al_2O_3	9.0	0.7	9.5	0.8	10.8	2.3
MnO	0.56	0.38	2.15	0.91	6.1	1.8
TiO_2	0.32	0.06	0.71	0.18	1.6	0.5
S	2.09	0.29	0.92	0.38		
B/A	1.22	0.10	0.66	0.11		

---

1. Average composition of slag samples from 100 casts at No. 4 blast furnace.

2. Average composition of 25 slag samples taken from the hot metal transfer ladle at Brier Hill.

3. Based on a constant weight of CaO.

Figure 1:
A typical material balance illustrating the mechanics of in-transit slag to metal sulfur reversion.

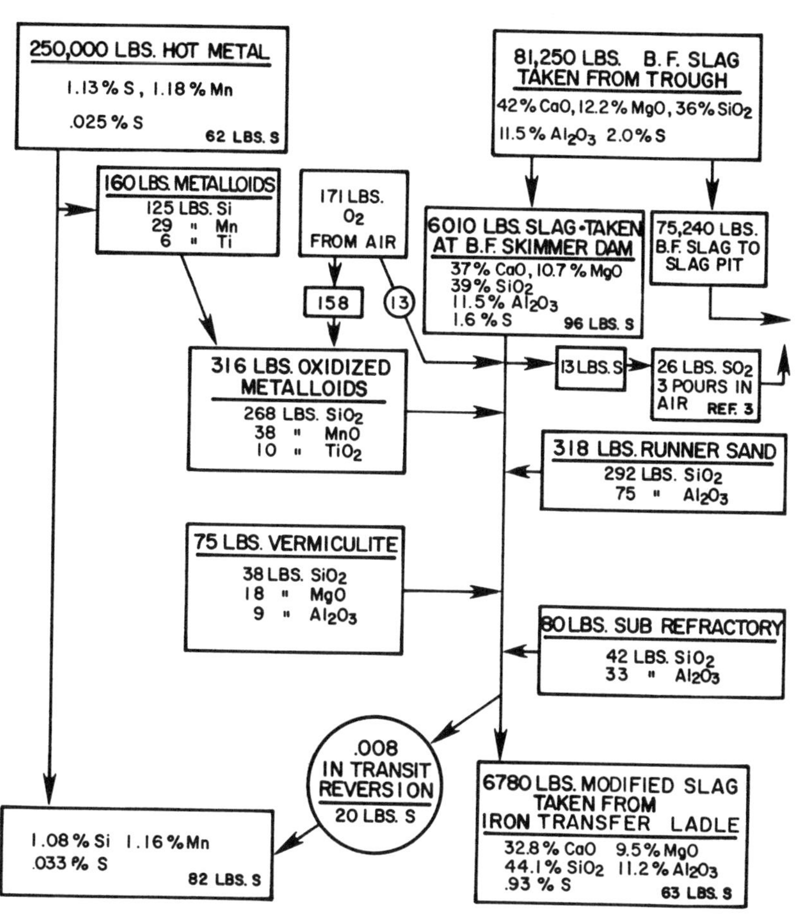

Figure 2:
General layout of the Youngstown Steel injection station.

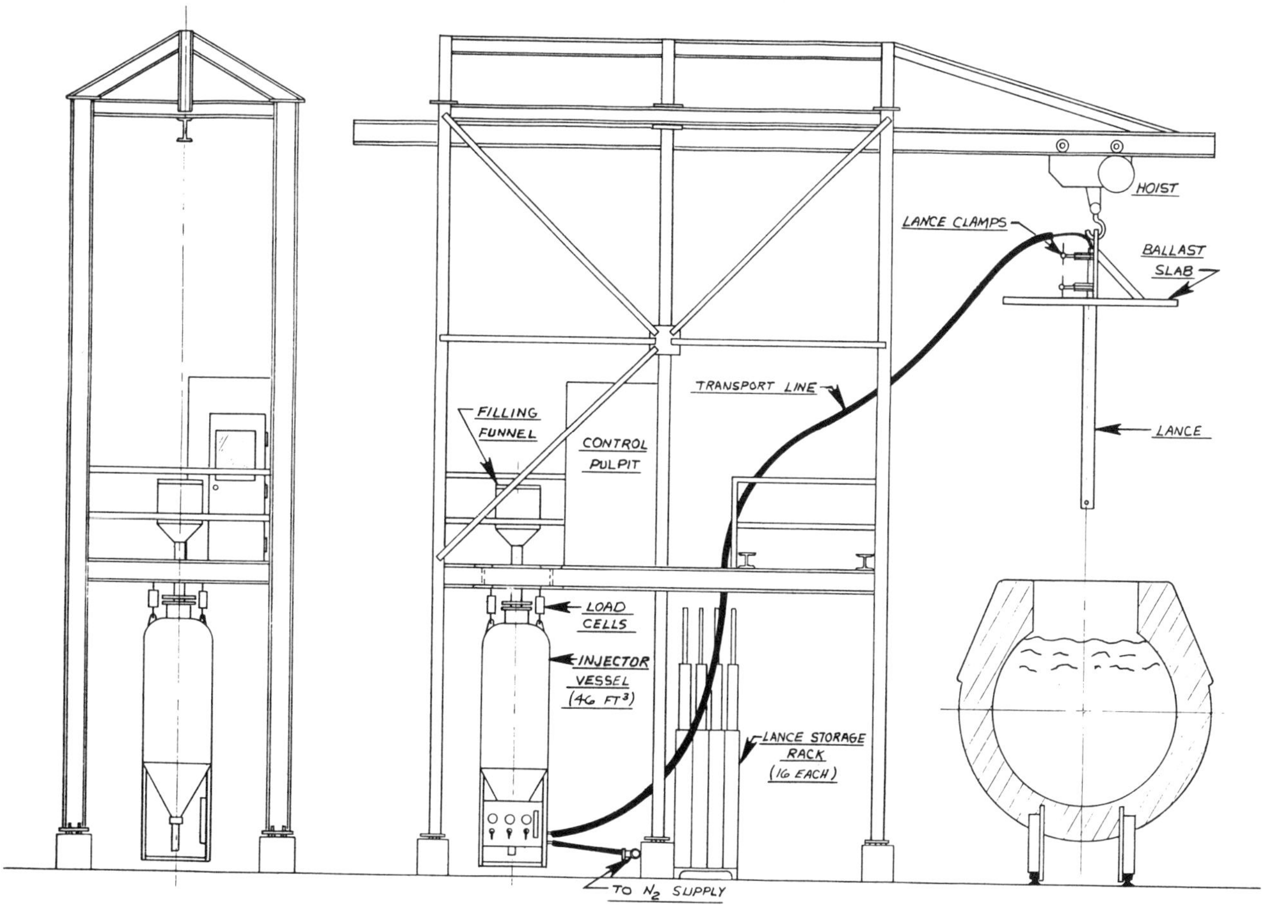

Carryover Slag

Even though much of the sulfur content of the original blast furnace slag had been depleted by reversion, this factor still represented 23% of the furnace sulfur load.

Scrap

Scrap sulfur accounted for 23.5% of the furnace sulfur load and was essentially a function of steel ladle sulfur since the shop consumed mainly home scrap.

Fuel Oil

A detailed sulfur mass balance indicated that the high sulfur fuel oil (1.0% S max.) contributed 10.5% of the sulfur load. A switch to a low sulfur fuel oil (0.6% S max.) reduced this contribution to about 4.5%. This is consistent with the findings of Dewing, Herty, and Edwards (4,5,6).

The hot metal and the accompanying carryover slag accounted for between 65 and 70% of the total furnace sulfur load. Clearly, the solution to the sulfur problem and the associated loss in productivity lay in finding a practical and economically sound system for external hot metal desulfurization coupled with an effective slag skimming system. The logistics of the existing operation dictated that all desulfurization be done in the submarine ladle just ahead of the steel plant, while skimming would have to be done in the O.H. transfer ladle.

SELECTION OF EXTERNAL DESULFURIZATION SYSTEM

Preliminary Trials with Mag-Coke

Because of its demonstrated effectiveness and low capital cost, Mag-Coke was chosen to evaluate the impact of low sulfur hot metal on the steel plant (1,7,8). No after treatment slag skimming was included in this trial, although several methods to control carryover slag were tried at the blast furnace with various degrees of success. By the third month of operation, a peak of 73% of the hot metal supplied was desulfurized. Samples taken from 105 submarine ladles revealed that the as-received sulfur content had been reduced from 0.044% to 0.018%, with an accompanying .013% decrease in first preliminary sulfur and a 50 minute reduction in refining time.

The combined disadvantages of the system; primarily the inability to treat narrow mouth or skulled submarine ladles, and the relatively high operating cost of any of the proprietary plunging reagents, pointed to the need for a flexible, inexpensive injection system to replace plunging.

Injection of Magnesium-Aluminum Powder

Since injection systems using magnesium-aluminum powder were being used successfully by others, required substantially less capital expenditure than either calcium carbide or magnesium-lime based systems, and could be installed rapidly, an experimental injection facility designed to treat submarine ladles was constructed (9,10). Every effort was made to hold down costs and the result was a very simple yet practical injection system.

Basically, the system consisted of a 46 cubic foot pressure vessel capable of holding about 2000 lbs. of reagent. This injector was equipped with controls for appropriate regulation of reagent and carrier gas, and was suspended on three load cells so that reagent inventory and flow rate could be monitored on a digital readout. Carrier gas flow was monitored by an in-line electronic flowmeter. This instrument alerted the operator to the onset of plugging at the lance tip so that timely corrective action could be taken. Reagent was transmitted to the refractory injection lance through a pipe connected to a length of flexible sand blast hose. The lance was clamped to a steel ballast slab which covered the mouth of the submarine ladle and minimized ejection of hot metal. Originally, the lance mounting was rigid, but experience proved that a spring mounted system improved lance life. The ballast slab was raised and lowered by a simple monorail hoist. This hoist was also used to load reagent and to change spent lances. Control of the operation was conducted from an elevated control pulpit which housed the injector controls, the hoist controls and the reagent weight monitoring unit. The entire system was supported by a 10 ft. wide by 20 ft. long by 36 ft. high structural steel framework. The basic configuration of the system is shown in Figure 2. Details of the control system are shown in Figure 3.

Although this injection facility was originally intended to be a pilot plant, the system operated smoothly and it soon became a regular production unit, operated by one hourly employee who also dekished submarine ladles between treatments. Consequently, routine injection of all hot metal supplied to the Brier Hill Open Hearth Shop began in September 1976.

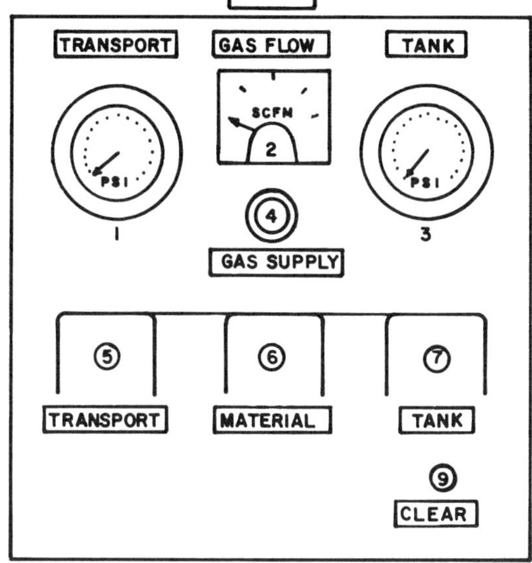

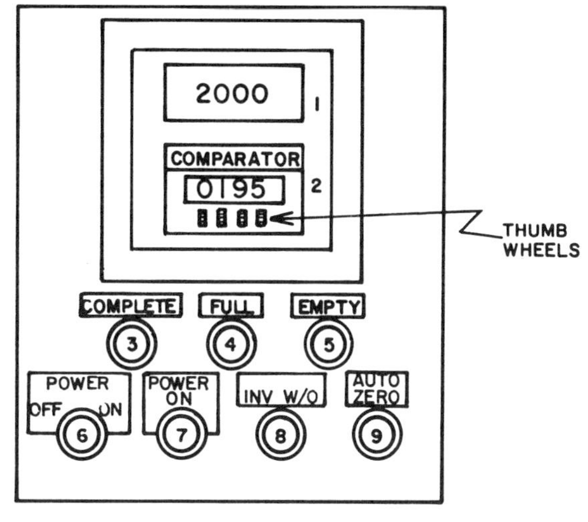

REMOTE INJECTOR CONTROLS

1. TRANSPORT PRESSURE GUAGE
2. FLOWMETER READOUT
3. TANK PRESSURE GUAGE
4. N_2 SUPPLY PRESSURE INDICATOR
5. TRANSPORT ACTIVATOR BUTTON
6. MATERIAL " "
7. TANK " "
8. LOW FLOW INDICATOR LITE ENERGIZES AT FLOWS LESS THAN 40 SCFM
9. ORIFICE CLEANER

WEIGH CONTROLLER

1. DIGITAL WT. READOUT IN POUNDS FOR INVENTORY AND WEIGH OUT
2. DESIRED WEIGH OUT DIALED IN WITH THUMBWHEELS
3. LITE ENERGIZES WHEN WEIGHOUT = COMPARATOR
4. " " WHEN INVENTORY EXCEEDS 2000 POUNDS
5. " " WHEN INVENTORY IS LESS THAN 600 POUNDS
6. POWER SWITCH
7. LITE ENERGIZES WHEN POWER IS ON
8. MODE SELECTOR SWITCH
9. AUTOMATIC ZERO FOR WEIGHOUT MODE

Figure 3:
Injection station pulpit control panel.

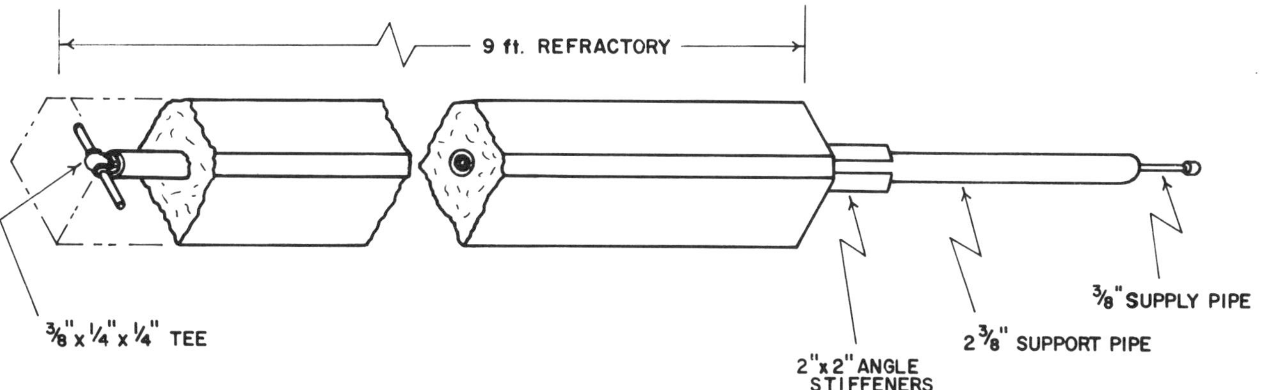

Figure 4:
Injection lance construction.

Three areas of concern were encountered in the first few months of routine injection operations.

1. Lance Life

The lances initially supplied with the injector comprised a 2" diameter solid steel rod for support surrounded by a 2" thick layer of castable refractory. The 3/8" diameter reagent supply pipe ran down the side of the support rod and straight out the bottom of the lance. These lances generally failed after 3-4 injections. Youngstown's Technical Services Department, working with several suppliers, designed an improved lance currently averaging 20 injections and occasionally lasting over 40 injections. The current lance design is illustrated in Figure 4. The T-shaped design of the lance outlets simplifies the removal of plugs and permits the operator to check visually to ensure that all obstructions have been removed.

2. Reaction Violence

This system was inherently more vigorous than other injection techniques because large quantities of magnesium vapor were generated per unit time. Excessive violence was controlled by reducing the injection rate; however, it was necessary to protect the mouth area of the submarine ladles with a relatively inexpensive gunning refractory to prevent periodic splashes of hot metal from full submarine ladles from contacting the steel shell.

3. Reagent Cost and Chemical Efficiency

The injection system was started up using a commercially available 65% magnesium - 35% aluminum alloy powder. Aluminum was included in the alloy to reduce plugging, but dramatically increased the cost of contained magnesium. Almost immediately, 75% magnesium - 25% aluminum alloy powder was tested and found satisfactory. Prompted by this success, further experimentation with 98% magnesium granules was conducted. Excessive plugging with the pure magnesium forced the return to the 75% magnesium - 25% aluminum alloy.

During this same general time period, two mechanical changes occurred which significantly improved the chemical efficiency of the system. The transport piping was altered to reduce all unnecessary flow restrictions and the exit end of the lance was changed from a single hole, bottom discharge to a two hole, side discharge.

The combined effects of these changes were a 100% increase in the transport efficiency (lbs. reagent/SCF gas carrier) and a 44% increase in magnesium desulfurization efficiency at mean trial conditions.

Magnesium and sulfur determinations made on 120 hot metal samples taken immediately after desulfurization yielded the following relationship:

$$\% \underline{Mg} \times \% \underline{S} = 1.5 \times 10^{-4} \quad (2400\text{-}2600^oF)$$

Like Duquette, no temperature dependence was discernible in the range 2400°F to 2600°F (1). As may be clearly seen from the solubility curve in Figure 5, desulfurization much below 0.010% S results in excessive magnesium solubility losses. Currently, a programmed injection is used with a target endpoint sulfur of 0.015 $\pm$ .005% S. The initial hot metal sulfur content is determined by a sample taken by the desulfurizer operator from each submarine ladle. The pin portion of the sample is run by the operator on a rapid combustion sulfur analyzer and the result is used to determine the quantity of reagent to be injected. The disc portion of the sample is sent to the main chemical laboratory for x-ray determination of sulfur, silicon and manganese.

Trials of various pre-blended mixtures of pure magnesium and other solids have been conducted in an effort to increase the chemical efficiency while maintaining freedom from plugging. Although the general desulfurization efficiency of each mixture was similar to that of 75% Mg-25% Al powder, the variability was much greater. A variety of handling and operating problems were also encountered with the premixes which made them unsatisfactory for daily use. No suitable replacement for 75% Mg-25% Al powder was found until the introduction of salt coated magnesium granules.

INJECTION OF SALT COATED MAGNESIUM GRANULES

Early in 1977, representatives of the Dow Chemical Company unveiled samples of a proprietary magnesium product soon to be made available commercially. This product took the form of metallic magnesium granules with a salt skin. The chemical and physical specifications of the salt coated magnesium granules are listed in Table III. These granules were generally spherical and appeared to have excellent flow properties. Unlike commercially ground Mg-Al powder, the salt coated granules were sized -10 +100 mesh and contained only a trace of fines, predominantly from the salt coating. To highlight the attractive safety features

TABLE III Salt Coated Magnesium Granules

Chemical Spec.

Metallic Magnesium: 88 - 92%

Chlorides of Na, Ca, Mg, K: Balance

Approximate Composition of Salt Coating:

54% NaCl, 20% $MgCl_2$, 13% $CaCl_2$, 13% KCl

Physical Spec.

Granulation, U.S. Standard Sieve:

Thru #10	100%
Retained on #100	100%
Thru #100	Trace

of the new granules, the Dow Chemical Company conducted tests in which batches of granules were heated in air above the melting point of pure magnesium (1202°F) without spontaneous ignition occurring (11).

In addition to the safety related aspects, the new granules were very attractive from an economic standpoint since they were significantly less expensive than 75% Mg-25% Al powder on a per pound of contained magnesium basis. A similar material had already been produced in the U.S.S.R. (12,13). Voronova has described in detail the advantages claimed for granular magnesium and has indicated that it is now the preferred reagent for external desulfurization of hot metal in the U.S.S.R. (14,15). The successful application of a similar product in the Soviet Union, when combined with the anticipated savings in reagent cost and safety considerations, provided a strong incentive to explore the compatibility of salt coated magnesium granules with the existing injection system at the Brier Hill Works.

An initial trial was carried out in May, 1977, using a 500 lb. batch of granules containing a nominal magnesium content of 85%. Based on the limited data obtained during this trial, it appeared that there was little difference between 75% Mg-25% Al powder and the salt coated granules with respect to chemical desulfurization efficiency, but it was observed that the granules possessed important advantages in the following areas:

a) safety with respect to fire and explosion hazards
b) reduction in lance plugging tendency
c) less material handling due to the higher Mg content.

No significant problems were uncovered during this trial, and it was concluded that the granules were compatible with the existing injection equipment.

Early in 1978, Dow Chemical Co. notified Youngstown that it was in a position to supply granules commercially in sufficient quantity to cover the normal magnesium consumption of Brier Hill Works. Before making a commitment to use the granules in place of Mg-Al powder, a second trial was carried out with batches of granules under operating conditions. The successful conclusion of this trial led to the changeover from 75% Mg-25% Al powder to salt coated magnesium granules, and the Brier Hill Works has consumed granules at the rate of about 20 NT/month since June, 1978.

Desulfurization Efficiency

The primary purpose of the second trial was to assess the chemical efficiency of salt coated magnesium granules and compare it with historical data on 75% Mg-25% Al powder and Mag-Coke. Table IV contains typical operating data from the second trial. Since submarine ladles with nominal capacities ranging from 100-150 NTHM were used, and since these ladles were filled to approximately 90% of capacity, the actual weight of hot metal treated varied from 95-135 NT. Consequently, the depth of immersion of the lance outlets varied from 4-6 ft. but was generally around 5 feet. While no attempt was made to quantify the effect of lance immersion depth on chemical efficiency, it was recognized that the relatively shallow immersion depth had a deleterious effect on desulfurization efficiency (16). This view was supported by the obvious increase in white MgO fume evolved during the injection of partially filled submarine ladles.

Although a relatively slow injection rate was desirable to improve desulfurization efficiency and reduce reaction violence, particularly in view of the shallow lance immersion depth, long injection times causing delays in the supply of hot metal to the steelmaking shop had to be avoided. A granules injection rate of about 15 lbs./minute was established as a convenient compromise and the time taken to desulfurize a submarine ladle was generally less than 10 minutes.

During the second trial, the temperature of the hot metal was measured and a sample was taken prior to injection.

TABLE IV <u>Injection of Salt Coated Magnesium Granules</u>
<u>Typical Operating Data From The Second Trial</u>

Parameter	Unit	Average	Range
Weight of Hot Metal Treated	N.T.	112	95-135
Treatment Time	mins.	9.0	3.5-13.25
Magnesium Injection Rate	lbs./min.	13.1	9.9-16.7
Magnesium Consumption	lbs./NTHM	1.02	0.73-1.31
Initial Hot Metal Temperature	^{o}F	2421	2360-2540
Initial Hot Metal Sulfur	%	0.040	0.024-0.064
Final Hot Metal Sulfur	%	0.010	0.002-0.021

Immediately after injection, another sample of hot metal was taken. The hot metal samples provided a disc for x-ray analysis of silicon and manganese plus a pin for Leco analysis of sulfur. This same procedure had been followed during earlier work with 75% Mg-25% Al powder and with Mag-Coke. Analyses of the data pertaining to the various reagents indicated that the desulfurization efficiency of each reagent could be described adequately in terms of an equation linking the sulfur content and the temperature of the hot metal prior to desulfurization. Typical results obtained with these desulfurization reagents are summarized in Table V. These results are presented graphically as a function of initial sulfur content in Figure 6.

Although the chemical efficiencies of the reagents were all strongly dependent on the initial sulfur content of the hot metal, the chemical efficiency of the salt coated magnesium granules was also observed to be more temperature sensitive (Figure 7). Over the range of hot metal temperatures encountered during the second trial, the average desulfurization efficiency of the salt coated magnesium granules was higher than that achieved with either Mag-Coke or 75% Mg-25% Al powder. The slower magnesium injection rate associated with the salt coated magnesium granules was believed to be at least partly responsible for the observed increase in desulfurization efficiency.

Lance Plugging

Whereas attempts to inject pure magnesium at flowrates in the range 10-20 lbs./minute were unsuccessful due to lance plugging problems, practically no plugging problems were encountered with salt coated granules despite their relatively high magnesium content. In fact, the minimum hot metal temperature restriction of 2350°F which was applied during injection of 75% Mg-25% Al powder has been replaced by a lower limit of 2300°F for current operations with granules.

Fume Formation

The fumes evolved during the injection of salt coated magnesium granules come from two distinct sources. Firstly, there is greyish-white fume containing a mixture of kish and magnesium oxide, the latter produced by unreacted (with sulfur) magnesium vapor which reaches the surface of the liquid iron and combines with atmospheric oxygen. Secondly, there is white fume released upon vaporization of the 10% salt coating on the granules. These combined fumes form a cloud of white, opaque smoke.

The results of limited sampling studies conducted during injection of salt coated magnesium granules suggest that approximately 80% of the injected salt is captured by the residual blast furnace slag present in the submarine ladle and only 20% of the salt is released as fume. The composition of the salt fume is essentially the same as the composition of the salt coating on the original magnesium granules. Thus, it is advisable to protect those parts of the pollution control system which come into contact with the salt fume by painting them with an epoxy based paint or other corrosion resistant coating.

The bulk of the injected salt is absorbed by the residual blast furnace slag which is eventually skimmed off the transfer ladle or dumped out of the submarine ladle at the dekishing station. The actual quantity of salt absorbed by the slag is small, amounting to 0.33% by weight of slag for a typical case in

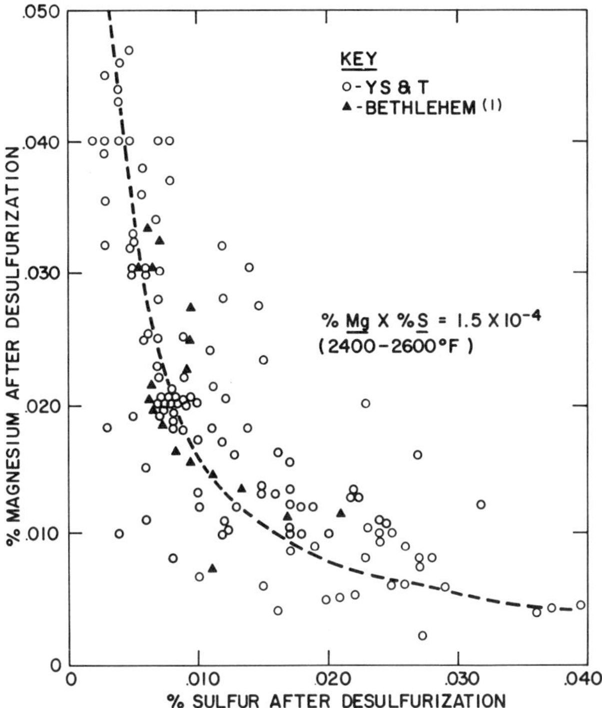

Figure 5:
The mutual solubility of magnesium and sulfur in hot metal just after desulfurization.

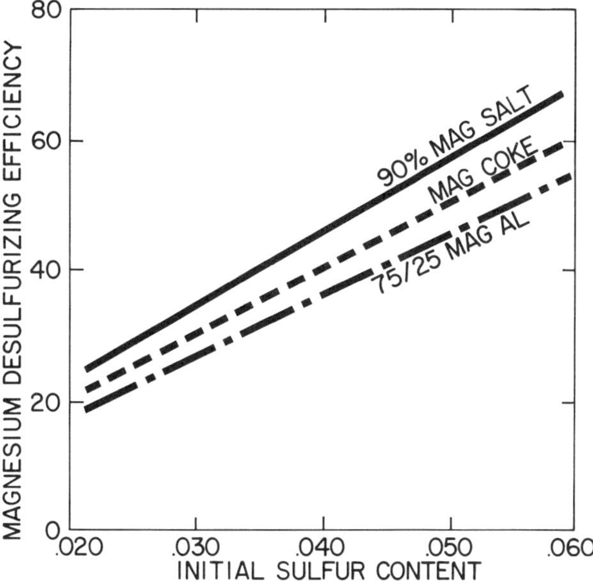

Figure 6:
Comparison of the desulfurizing efficiency of various magnesium based reagents at 2450°F as a function of initial hot metal sulfur content.

which the consumption of salt coated magnesium granules is 1 lb./NTHM (0.9 lb. Mg/NTHM) and the corresponding slag quantity is 30 lb./NTHM. However, if the mixture of cold iron and high sulfur slag from the skimming and dekishing operations is recovered in order to recycle a portion of the cold iron, caution should be exercised to ensure that the magnetic separation effectively discards the bulk of the high sulfur slag containing the salt. Also, this high sulfur slag should not be allowed to contaminate steelplant slags which are destined for recycle to the blast furnace. Routine monitoring of the alkali content of blast furnace slags has shown no significant increase in alkali levels despite the changeover from 75% Mg-25% Al powder to salt coated magnesium granules.

Material Handling

Although the salt coating on the magnesium granules contains 20% $MgCl_2$ and 13% $CaCl_2$, two noticeably hygroscopic salts, no difficulty has been experienced with moisture pick-up from air. The salt coated granules are delivered in drums containing 440 lbs. net and the granules are poured through air into the injector. In this way, the granules are only exposed to air briefly. With a view to lessening the hygroscopic nature of the salt coating, Dow Chemical have supplied a two component salt coated magnesium granule with a 50% NaCl-50% KCl coating. Preliminary testing of this product began in October, 1978.

OPERATING ADVANTAGES OF SALT COATED MAGNESIUM GRANULES

Safety is of paramount importance when handling reagent in any routine desulfurization operation. The salt coating on the magnesium plus the relatively coarse particle size of the granules combine to give a magnesium based product which can be handled safely under a wide variety of operating conditions. Unlike the acetylene generation problem associated with the use of calcium carbide, or the danger of explosion when handling finely divided magnesium or magnesium-aluminum alloys, salt coated magnesium granules are, for most practical purposes, inert.

Operating cost is attractive when compared with other desulfurization processes (Table VI). The total cost of consumables (reagent, lances, nitrogen, etc.) is less for salt coated granules than for other magnesium based processes, but somewhat greater than that for calcium carbide based systems with high reagent efficiency (16,17). However, when total operating costs including

TABLE V Comparison Of Desulfurization Reagents

Multiple Regression Equations

Reagent	Magnesium Desulfurization Efficiency (%)
Salt Coated Mg Granules:	1108 (Initial % S) -0.052 (H.M. Temp. °F) +128
75% Mg-25% Al Powder:	911 (Initial % S) -0.034 (H.M. Temp. °F) + 84
Mag-Coke (270 lb. Can):	1005 (Initial % S) -0.030 (H.M. Temp. °F) + 74

Basis

Reagent	Data Sets	Desulfurization Efficiency (%) Mean	Std. Dev.	Hot Metal Temp. (°F) Mean	Std. Dev.	Initial Sulfur (%) Mean	Std. Dev.	Final Sulfur (%) Mean	Std. Dev.	Avg. Magnesium Injection Rate (lbs. Mg/min.)	Avg. Depth of Immersion (ft.)
Salt Coated Mg Granules	30	45.9	13.6	2421	47	.040	.010	.010	.007	13	5
75% Mg-25% Al Powder	47	32.8	14.1	2481	47	.038	.014	.010	.007	18	5
270 lbs. Mag-Coke	124	39.3	13.3	2492	49	.039	.012	.014	.005	-	-

Note: Magnesium desulfurization efficiency (%) $= \frac{\text{calc. wt. of Mg converted to MgS}}{\text{weight of Mg injected}}$

$$= \frac{1516 \times \text{NTHM} \times \Delta\%S}{\text{lbs. Mg injected}}$$

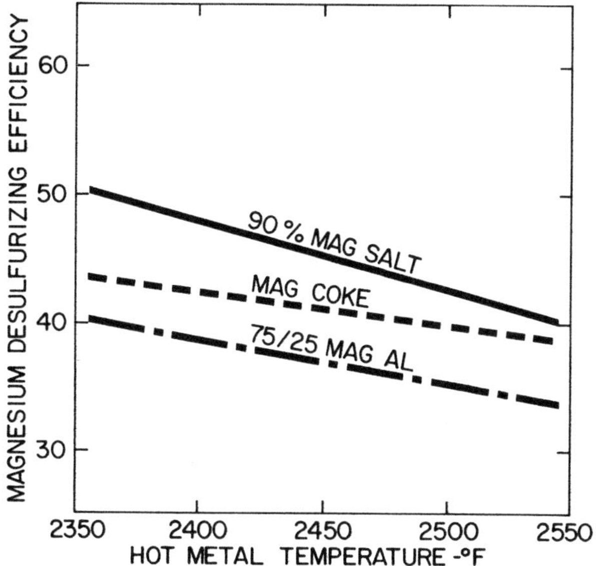

Figure 7:

Comparison of the desulfurizing efficiency of various magnesium based reagents as a function of hot metal temperature.

labor and maintenance are considered, a simple system based on salt coated magnesium granules is hard to beat.

The inherent simplicity of the injection system required for salt coated granules is a major plus since it confers several important advantages, namely:

a) extremely low capital cost for relatively high throughput installation,

b) facilities can be constructed and brought on stream rapidly,

c) maintenance costs are almost negligible,

d) manpower requirement is held to an absolute minimum,

e) space requirements for reagent storage are small.

TABLE VI Cost of Hot Metal Desulfurization In Submarine Cars With 90% Magnesium Salt Coated Granules

Basis:

Initial Sulfur: .050% Final Sulfur: .015%
Magnesium consumption: 1.0 lb./NTHM
Injection depth: 5 ft.
Reagent cost (delivered):
$1.30/lb. contained Mg
Desulfurization of 1500 NTHM/day

Item	$/NTHM
Reagent	$1.30
Labor(1) and Maintenance(2)	0.27
Lances	0.12
Gunning refractory(3)	0.09
Other consumables (4)	0.06
Total	$1.84

Notes: (January, 1979)

(1) One operator/turn dekishes at least 50% of the submarine ladles in addition to desulfurization duties.

(2) Overhead is included in labor and maintenance cost.

(3) Protective refractory coating applied to submarine ladle shell.

(4) Other consumables comprise nitrogen (dominant cost) plus minor equipment items.

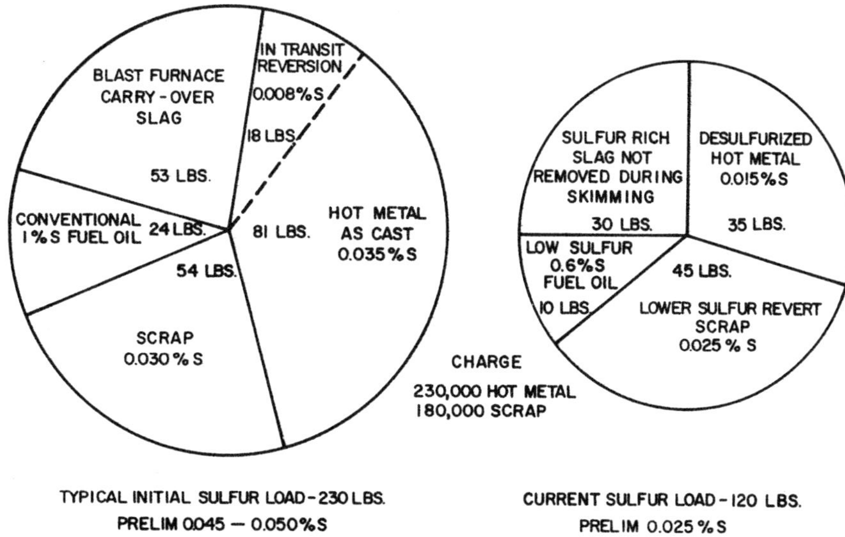

Figure 8:
Comparison of typical furnace sulfur loads before and after the sulfur control program.

TABLE VII Estimated Reagent Quantities For External Desulfurization of 10,000 NTHM/Day

Basis: Initial Sulfur: .050% Final Sulfur: .015%

Process	Reagent Consumption (lbs./NTHM)	Daily Reagent Requirement (NT)
Salt Coated Mg	1.15 lbs. granules	5.6
75% Mg-25% Al	1.4 lbs. alloy powder	7.0
ATH (18)	5.3 lb. CaC_2 + 0.6 lb. $CaC0_3$	29.5
Nippon (Kaiser) (17)	4.2 lb. CaC_2 + 1.8 lbs. $CaC0_3$	30.0
Lime-Mag (16)	0.7 lb. Mg + 10 lbs. B.L.	53.5
Burnt Lime/ Natural Gas (19)	20 lbs. Ca0	100.0

Material handling is reduced to a minimum when compared with other desulfurization processes (Table VII), (16,17,18,19).

Slag volume is essentially unchanged and, therefore, no additional slag skimming equipment is required other than the customary transfer ladle slag rake or ladle mouth slag dam. Also, no additional load is placed on slag disposal equipment.

Slag chemistry shows little change apart from a significant increase in sulfur content due to entrained magnesium sulfide and a minor increase in alkali content.

If a significant improvement in magnesium desulfurization efficiency can be achieved by injecting salt coated magnesium granules into relatively deep hot metal transfer ladles rather than into shallow submarine ladles, it follows that the reduction in reagent consumption would significantly lower operating costs making the salt coated magnesium granules approach very cost competitive. Further, the relatively high magnesium content of the salt coated granules promotes rapid desulfurization which is essential in a high productivity BOF shop. Consequently, injection of salt coated magnesium granules into the hot metal transfer ladle may prove to be an ideal solution for the BOF shop faced with the problem of making a wide variety of sulfur specifications.

IMPACT OF SULFUR CONTROL PROGRAM

The successful implementation of the sulfur control program has dramatically affected both steel sulfur levels and productivity. The combined effects of external desulfurization of almost all hot metal followed by slag skimming, use of low sulfur fuel oil, and a noticeable reduction in the sulfur level of recycled scrap have practically halved the original furnace sulfur load (Figure 8). The monthly average preliminary sulfur level at melt-in is currently 0.022%-0.026% and the corresponding steel ladle sulfur averages 0.016-0.018%. Furthermore, the average consumption of burnt lime during refining has been reduced from about 30 lbs./NT originally to less than 20 lbs./NT currently. It is not unusual to have weeks in which over half the steel produced has 0.015% S or less and there are numerous heats made with final sulfur levels below 0.010%.

The drastic reduction in furnace sulfur load was directly responsible for about a 30% gain in productivity observed with the conventional open hearth furnaces. With the advent of oxygen roof lances, full benefit of the low sulfur load was realized. Despite a decrease in maximum steel sulfur specification from 0.025% to 0.020%, the combination of oxygen injection and charge sulfur control has produced a substantial increase in furnace productivity.

Conclusions

1. The safety and relatively low operating cost associated with the use of salt coated magnesium granules for external desulfurization of hot metal have been demonstrated.

2. Based on experience gained during routine desulfurization of about 1500 NTHM/day at the Brier Hill Works, construction costs of a similar facility

capable of treating about 5000 NTHM/day should not exceed $250,000. In a large plant handling 10,000 NTHM/day, three separate injection stations would be required; two on-line and one on standby. The projected capital cost of three submarine ladle injection stations plus a central pollution control facility should not exceed $1 million.

3. The simple equipment required and the rapid desulfurization achieved suggest that injection of salt coated magnesium granules into relatively deep hot metal transfer ladles is a viable solution for BOF shops faced with the problem of consistently producing steels with sulfur contents as low as .010%. Coupled with scrap control, this approach eliminates the need for bulk additives for steel ladle desulfurization.

References

1. Duquette, W. H., Griffing, N. R., Miller, T. W., "Low Sulfur Steel (.005%) from Mag Coke Desulfurized Hot Metal", TMS-AIME Open Hearth Proceedings, Vol. 56, 1973, pp. 79-90.

2. Muan, A., and Osborn, E. F., Phase Equilibria Among Oxides in Steelmaking, Addison-Wesley, 1965, pp. 151.

3. Richardson, F. D. and Fincham, C. J. B., "Sulfur in Silicate and Aluminate Slags", Journal of the Iron and Steel Institute, Vol. 178, September 1954, pp. 12.

4. Dewing, E. W. and Richardson, F. D. "Thermodynamics of Mixtures of Ferrous Sulphide and Oxide", Journal of the Iron and Steel Institute, Vol. 194, April 1960, pp. 449.

5. Herty, C. H. Jr., Christopher, C. C. F, Freeman, H., and Sanderson, J. F., "The Control of Iron Oxide in the Basic Open Hearth Process", Deoxidation of Steel (Herty Memorial Volume), AIME, 1957, pp. 563.

6. Edwards, C. A., "The Relative Merits of Low-and-High Sulfur Oil in Open Hearth Steelmaking", Journal of the Iron and Steel Institute, Vol. 178, October 1955, pp. 141.

7. Gilby, S. W., "External Desulfurization of Hot Metal Using Magnesium", TMS-AIME Ironmaking Proceedings, Vol. 32, 1973, pp. 133-142.

8. Thorp, R. B., and Dastur, P. N., "Desulfurization of Hot Metal with Mag-Coke", TMS-AIME Open Hearth Proceedings, Vol. 56, 1973, pp. 75-78.

9. Wood, J. K., Schoeberle, G. E., and Pugh, R. W., "External Desulfurization of Hot Metal - A Pneumatic Injection Technique", Symposium on External Desulfurization of Hot Metal, McMaster University, Hamilton, Canada, 1975, 11, pp. 1-11.

10. Koros, P. J., and Petrushka, R. G., "Materials for Hot Metal Injection", Ibid., 7, pp. 1-23.

11. Private communication with Lloyd Lockwood, Dow Chemical Company, Midland, Michigan.

12. U.S. Patent No. 3, 881, 913.

13. U.S. Patent No. 3, 969, 104.

14. Voronova, N.A., Pliskanovskii, S. T., Shevchenko, A. F., Laurent'ev, M. L., and I. Ya. Emel' yanov, "Desulfurization of Pig Iron by Injecting Magnesium into Hot Metal Ladles", Steel in the U.S.S.R., Vol. 4, 1974, pp. 261-265.

15. Voronova, N. A., "External Desulfurization of Hot Metal by Magnesium Injection, "Proceedings of the 34th Annual Meeting of the International Magnesium Association, Columbus, Ohio, 1977, pp. 44-52.

16. Koros, P. J., Petrushka, R. G., and Kerlin, R. G., "The Lime-Mag Process for the Desulfurization of Hot Metal", I&SM, Vol. 4, No. 6, 1977, pp. 34-40.

17. "External Hot-Metal Desulfurization - an Old Idea Finds New Favor with Big Steel", 33 Metal Producing, July, 1978, pp. 33-37.

18. Haastert, H. P., Meichsner, W., Rellermeyer, H., and Peters, K-H., "Operational Aspects of the Injection Process for Desulfurization of Hot Metal, "Iron and Steel Engineer, October, 1975, pp. 71-77.

19. Pückoff, U. and Kister, H., "Desulfurization of Hot Metal at Hoesch-Estel", Symposium on External Desulfurization of Hot Metal, McMaster University, Hamilton, Canada, 1975, 8, pp. 1-20.

DESULFURIZATION OF HOT METAL WITH BURNT LIME

Tatsuo Ohya, Fumio Kodama,
Hisashi Matsunaga, Minoru Motoyoshi,
and Masaya Higashiguchi

Nippon Steel Corporation
Hirohata Works
Himeji, JAPAN

This paper originally appeared in the Steelmaking Proceedings, Vol. 60, 1977, p. 345.

Introduction

Burnt lime, limestone, soda ash, caustic soda, calcium cyanamide, etc. have been known as external desulfurizing agents of hot metal. In recent years, however, calcium carbide is mainly used for desulfurization process in due consideration of the factors such as industrial desulfurizing power, environmental protection and errosion of refractories caused.

Unfortunately, the manufacture of calcium carbide involves consumption of much electricity, and the price has risen considerably since the oil crisis, which in turn has caused an increase in the cost of desulfurization. This is the reason why the advent of a less expensive desulfurizer with an excellent desulfurizing power has been anticipated.

Lime has been known for many years as an inexpensive desulfurizer. Because of its slow reaction rate, however, it has hardly been used alone as an external desulfurizer of hot metal on an industrial scale. This slow rate of desulfurization reaction is considered to be attributable, on the one hand, to the solid reaction involved, and, on the other, to the necessity to simultaneously reduce oxygen contained in lime with carbon and silicon contained in hot metal.

Hirohata Works of Nippon Steel Corporation has found that the addition of proper amounts of carbon and fluorspar considerably improves the desulfurization rate of CaO, and the Works is using CaO as the external desulfurizer of hot metal on an industrial scale. An outline of the actual application of CaO to the desulfurizing process is given below.

Desulfurization Process

In NSC Hirohata Works, hot metal is carried in 70-ton hot metal ladles from the blast furnaces to the steelmaking plants. External desulfurization is accomplished by an impeller-stirring 100-ton KR reactor (1) shown in Fig. 1 and an N_2 bubble-stirring continuous ladle desul-

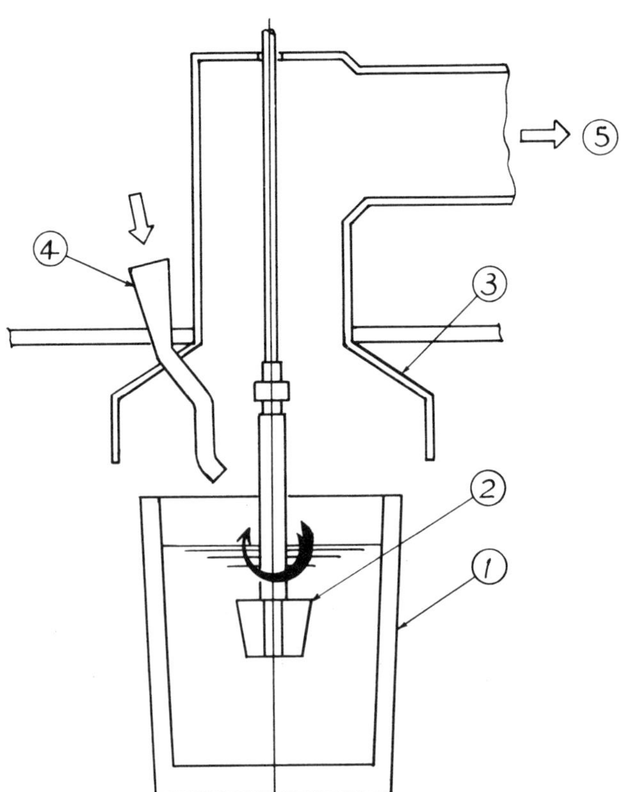

Fig. 1 - KR reactor (1. hot metal ladle, 2. impeller, 3. dust collecting hood, 4. chute for desulfurizing agent, 5. dust catcher)

furization equipment (2) located between the blast furnaces and the steelmaking plants. The former has a desulfurizing capacity of 34,000 t/month, and the latter 160,000 t/month.

Both equipments are of open ladle desulfurization types. Desulfurizing agent is added from above in the atmosphere, and the desulfurization process is accelerated either by impeller-stirring or by bubble-stirring. The treatment time is 10 minutes for the KR reactor and 15 minutes for the continuous ladle desulfurization equipment. In Hirohata Works, CaC_2 has in the past been used as desulfurizing agent. For the experiments of the improved desulfurizing power of burnt lime, the KR reactor was chiefly used.

The experiments were conducted at temperatures ranging from 1320 to 1420°C, and sulfur contents of hot metal ranging from 0.025 to 0.050%.

Desulfurization with Burnt Lime

The desulfurization of hot metal by CaO is considered to take place in accordance with either of the following two molecular formulas:

$$2(CaO)+\underline{Si}+2\underline{S} \text{ —— } 2(CaS)+(SiO_2) \quad ..(1)$$

$$(CaO)+\underline{C}+\underline{S} \text{ —— } (CaS)+CO \quad .....(2)$$

In the case of carbon-saturated iron, Si has a larger reducing power than C, provided that the Si content is more than 0.05% at a temperature of 1300°C (3). Therefore, the reaction of formula (1) above is usually considered to contribute more to desulfurization.

The equilibrium of desulfurization of hot metal by CaO has been thermodynamically calculated by H. Sawamura (4), B. Trentini et al(5), N. Tsuchiya et al(6), S. Eketorp (7) etc. Their calculations vary with the preconditions involved such as thermodynamic data, reaction formulae used, characteristics of the reaction products present, their concepts of reaction activity, chemical compositions of the hot metal concerned, etc. Their calculations show, however, that the sulphur equilibrium contents of hot metal desulfurized by CaO are very low at 0.2 to 10 ppm, which means that CaO has a sufficient power of desulfurization.

It is to be noted, however, that a 10-minute desulfurization treatment by the KR reactor with a consumption of 4 kg/t of CaO ranging in grain size from 0.2 to 1 mm produced a 65% desulfurization ratio and a 4% standard deviation. Therefore, the desulfurization at this slow reaction rate could not be used on an industrial scale.

As can be deduced from the aforesaid reaction formulas (1) and (2), the following steps should be taken in order to improve the reaction rate of desulfurization by CaO to an industrial scale.

1. To maintain a reducing atmosphere at the reaction interface.
2. To increase the reaction interface area
3. To increase the transport rate of the reaction products
4. To increase the transport rate of the sulphur contained in hot metal

The question of item 4 above may be solved if a desulfurization equipment is provided, and the approaches to industrial improvements for items 2 to 4 above were studied.

Optimum Composition of Lime-Fluorspar-Carbon Desulfurizing Agent

In order to provide a reducing atmosphere according to the idea described

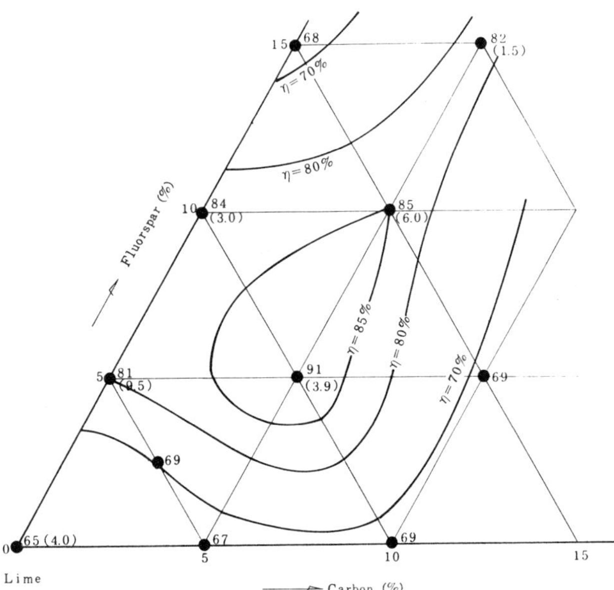

Fig. 2 - Relation between composition of lime-fluorspar-carbon desulfurizer and desulfurization ratio (figures mean average and standard deviation)

above, investigation was conducted in connection with the conditions of desulfurization of hot metal by CaO by adding coke which is inexpensive and easy to handle, and fluorspar which is inexpensive and effectively works, with a small quantity, to increase the transport rate of reaction products. The experiment was made with a consumption of 4 kg/t of desulfurizing agent and a treatment time of 10 minutes. Fig. 2 shows the relation between the ratio of fluorspar and coke added and the desulfurization ratio η. In this case desulfurization ratio is written as

$$\eta = \frac{[S]i - [S]f}{[S]i} \times 100\% \quad ......(3)$$

where [S]i and [S]f show sulphur contents before and after desulfurization treatment, respectively.

As is clear from the above, the desulfurization ratio was highest with a desulfurizing agent consisting of 90% burnt lime, 5% fluorspar and 5% coke, and it was found that a 10-minute KR treatment with 4 kg/t of this desulfurizing agent produces a desulfurization ratio of 90%.

Effect of Lime Grain Size

Desulfurization by lime is of a solid reaction. The specific reaction surface and reaction rate are considered to increase with decreasing lime grain size. Fig. 3 shows the desulfurization ratios against different grain sizes of burnt lime. The consumption related to Fig. 3 is 4 kg/t of desulfurizing agent consisting of 90% burnt lime, 5% fluorspar and 5% carbon, and the treatment time is 10 minutes. In this case, the desulfurization ratio increases with decreasing lime grain size.

In contrast with this, however, the desulfurization ratio decreased if the grain size is excessively small. This is because part of such small-size burnt lime is swallowed up into the dust catcher when it is added, thereby reducing the effective amount of the burnt lime added, and also because such smallsize lime grains cohere during the desulfurizing process, thereby reducing the specific reaction surface area. Accordingly, a grain size range of 0.2 to 1 mm is recommended for use in open ladle desulfurization.

Effect of Fluorspar Quality

There are various grades of fluorspar available for industrial use. They differ in content of gangue such as Al_2O_3 and SiO_2 and also in price. The effect of the quality of fluorspar upon desulfurization was examined under the same testing conditions as described above. Table I shows the chemical compositions of high purity fluorspar and low purity fluorspar. Table II shows the ratios of

Table I. Chemical Composition of Fluorspar (%)

	CaF_2	Al_2O_3	SiO_2	MgO
High purity fluorspar	90	2	7	tr.
Low purity fluorspar	70	12	16	1

Table II. Effect of Fluorspar Quality on Desulfurization Ratio

	Nos. of experiment	Average Sulphur Content		Desulfurization ratio
		before	after	
High purity fluorspar	7	0.0342 %	0.0087 %	75.8 %
Low purity fluorspar	49	0.0340 %	0.0031 %	91.0 %

Table III. Effect of Carbon Characteristics on Desulfurization Ratio

Kind of Carbon	Size (mm)	Average Sulphur Content		Desulfurization ratio
		before	after	
Electrode carbon	0.1-0.5	0.0350 %	0.0090 %	76.2 %
Oil coke	0.2- 3	0.0340 %	0.0031 %	91.0 %

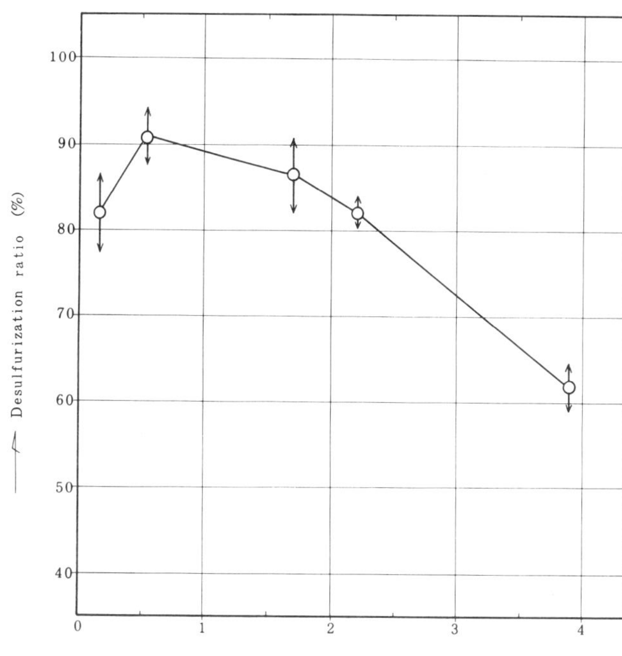

Fig. 3 - Relation between burnt lime size of 90% lime-5% fluorspar-5% carbon desulfurizer and desulfurization ratio (arrow lines mean standard deviation)

desulfurization using these grades of fluorspar, and it indicates that the use of low quality fluorspar results in a higher desulfurization ratio, which is probably because the presence of proper amounts of SiO_2 and Al_2O_3 lowers the melting point of and decreases the viscosity of reaction product, thereby accelerating the rate of separation of the reaction product from the reaction interface.

Effect of Carbon Characteristics on Desulfurization

It is assumed that the rate of reaction of desulfurizing agent with the atmosphere over the open ladle varies with solid carbon, and that the same amount of desulfurizing agent added produces varied reduction atmospheres and desulfurization ratios. For this reason, a comparison of oil coke and electrode carbon was made under the same testing conditions. The results obtained are shown in Table III, which indicates that oil coke gives a better desulfurization ratio. This is probably because oil coke whose surface is porous forms a reducing atmosphere more readily.

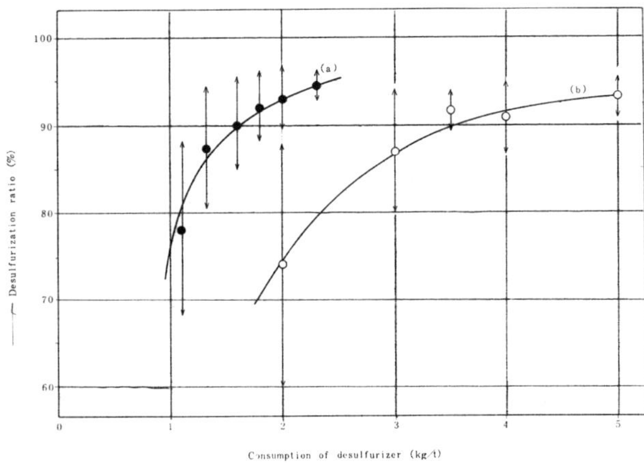

Fig. 4 - Relation between desulfurizer consumption and desulfurization ratio (a) CaC_2 (b) 90% lime-5% fluorspar-5% carbon desulfurizer (arrow lines mean standard deviation)

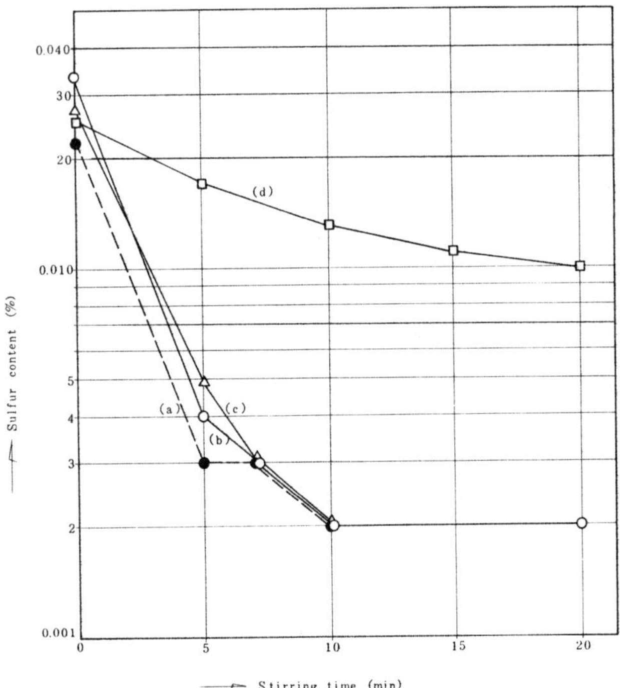

Fig. 5 - Stirring time and sulphur content of hot metal (a) CaC_2 1.5 kg/t, (b) (c) (d) 90% lime-5% fluorspar-5% carbon desulfurizer 4, 3, 1.5 kg/t respectively

Desulfurizing Agent Consumption, Desulfurization Ratio and Desulfurization Reaction Rate

It was made clear by the experiments conducted so far that a 90% desulfurization ratio can be obtained by a 10-minute KR treatment with 4 kg/t of CaO desulfurizing agent consisting of 90% burnt lime 0.2 to 1 mm in size, 5% low purity fluorspar containing SiO_2 and Al_2O_3 and 5% activated oil coke. Fig. 4 shows the relation between CaO desulfurizing agent consumption and desulfurization ratio. The KR treatment time is 10 minutes. The data on CaC_2 used are also given in Fig. 4 for comparison. As is clear from this, the larger the consumption of desulfurizing agent, the higher the desulfurization ratio becomes. The use of twice as much CaO desulfurizing agent as CaC_2 produces the same desulfurization ratio.

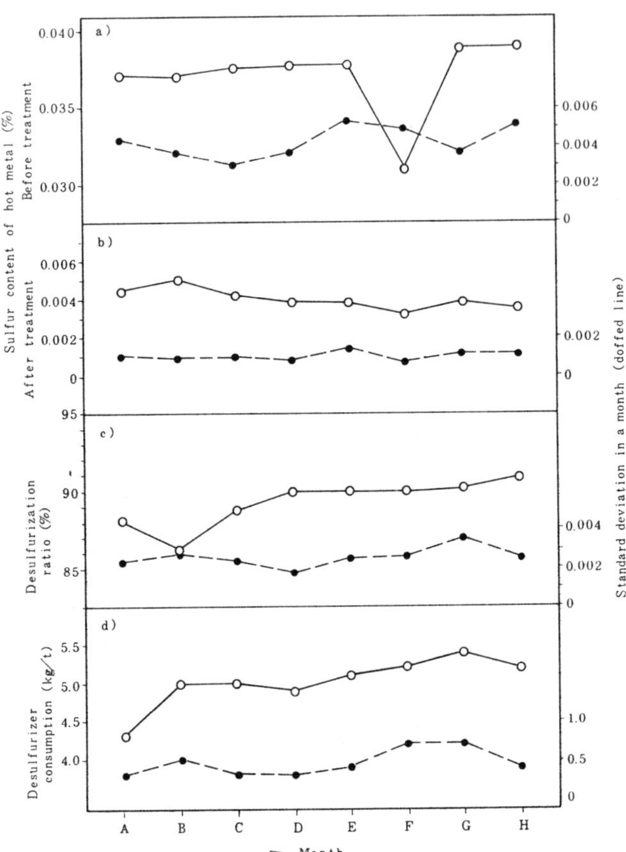

Fig. 6 - Monthly operational data of external desulfurization with burnt lime (a) initial sulphur content of hot metal (b) final sulphur content after 10 min. treatment (c) desulfurization ratio (d) desulfurizer consumption

Considering that the price of CaO desulfurizing agent is less than one fourth that of CaC_2, the desulfurizing cost using CaO is less than half the desulfurizing cost using CaC_2.

Fig. 5 shows variations in sulphur content of hot metal with the change of the consumption of lime-fluorspar-carbon desulfurizing agent. The index of CaC_2 is also plotted in Fig. 5. It shows that the apparent desulfurization ratio improves with increasing consumption of desulfurizing agent. With a consumption of 2 kg/t, no single sulphur can be obtained even if the stirring time is longer.

Application of Lime-Fluorspar-Carbon Desulfurizing Agent

In Hirohata Works, the actual application of the lime-fluorspar-carbon desulfurizing agent to hot metal desulfurization was commenced in 1976. Fig. 6 shows the initial sulphur content of hot metal, final sulphur content after 10-minute treatment, consumption of lime-fluorspar-carbon desulfurizing agent, monthly average desulfurization ratio and standard deviation. Some variations are seen according to the hot metal conditions of the month, but the 8-month average indicates that a 0.037% initial sulphur content of hot metal is reduced to 0.0039% with a 5 kg/t consumption of desulfurizing agent, which represents an approximate 90% desulfurization ratio.

Discussion of Desulfurization with Lime-Fluorspar-Carbon Desulfurizing Agent

The desulfurization of hot metal by CaO is of a solid-liquid reaction, and the important factor in this process is the separation of reaction product CaS. Fig. 7 gives the image patterns of samples taken from the slag after desulfurization reaction and examined by an electron probe microanalyzer. These image patterns are those near the interface of iron particle. It is seen that sulphur is not concentrated on the interface, but co-exists in dots with concentrated Si and Al. It is considered from this that reaction product CaS, together with CaF_2, SiO_2 and Al_2O_3 contained in the low purity fluorspar, and also with SiO_2 that forms as reaction product of molecular formula (1), forms slag of a low melting point and moves in liquid state inside the lime grains through gaps and cracks made during the calcinating process of burnt lime. Considering the diffusion of solids 0.2 to 1 mm in size, it may be said that an industrially acceptable rate of desulfurization was attained by the use of burnt lime of fairly large grain size.

The variations in the sulphur content of hot metal with time as indicated in Fig. 5 do not comply with exponential function, and the reaction rate depends on the consumption of desulfurizing agent. If this is because of the rate-determining stage of the sulphur contained in hot metal, it should be explained by the boundary film principle and may be written as follows on the assumption that the equilibrium sulphur content is very small.

$$-\frac{d[S]}{dt} = -k[S] \qquad (4)$$

$$[S] = [S]i \exp\{-kt\} \qquad (5)$$

where $[S]i$ = initial sulphur content

$[S]$ = sulphur content in time t

t = time

$$k = \frac{F}{V} \times \frac{D}{\delta} = \text{const.}$$

The above equation shows that the

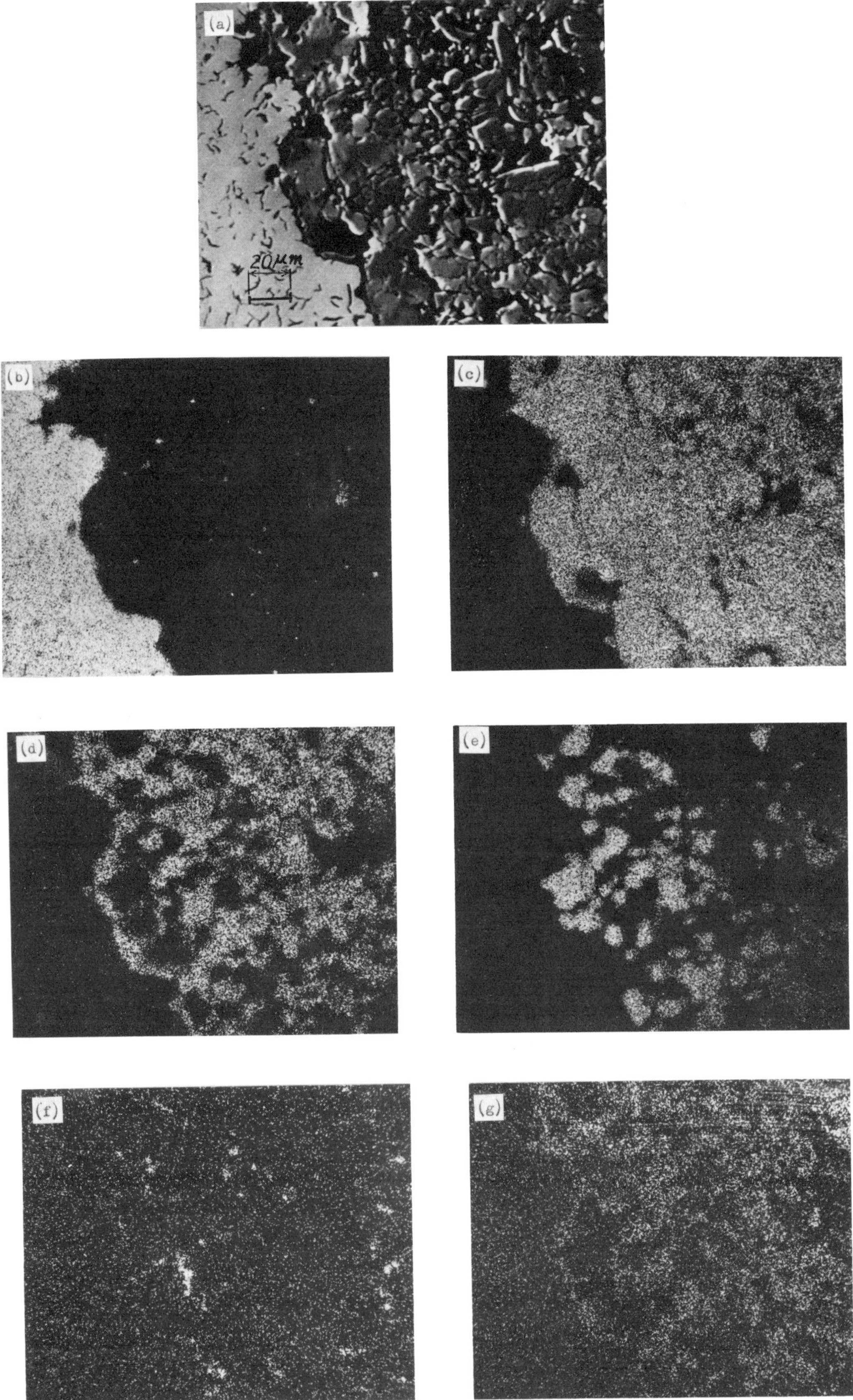

Fig. 7 - Image patterns of slag after treatment by electron probe microanalyser
(a) BSE (b) Fe (c) Ca (d) Si (e) S (f) Al (g) O

reaction rate does not depend on the consumption of desulfurizing agent. Therefore, the variations in the sulphur content of hot metal are considered to be due to the reaction being in the rate-determining step, provided, however, that the chemical reaction rate is very fast.

In respect of the dependence of the desulfurization rate upon the consumption of desulfurizing agent, the following interpretation may be given. The CaS formed on the surface of solid burnt lime grains is dissolved in the slag and permeate the grains through the porosities or cracks. However, the amount of this permeation is limited, and despite the presence of CaO, the burnt lime grains practically lose their desulfurizing power because they are covered with the slag consisting of CaO, Al_2O_3, SiO_2, CaF_2 and CaS.

For this reason, for the desulfurization to continue, it is necessary for fresh burnt lime grains to appear on the reaction interface. The frequency of replacing of burnt lime, that has virtually completed reaction, with fresh burnt lime increases with increasing number of lime grains, provided that the stirring rate is constant.

Accordingly, it may be assumed that the larger the number of burnt lime grains added, that is, the larger the consumption of desulfurizing agent, the higher the apparent rate of reaction becomes. Computed from the desulfurization critical value given in Fig. 5 and the material balance of sulphur, it is considered that as 15% of lime grains transform into CaS through desulfurization reaction, the desulfurizing power of the lime grain is practically lost.

Conclusion

Despite its low price, burnt lime has not been used for desulfurizing hot metal on an industrial basis in the past because of its slow rate of reaction. NSC Hirohata Works has found that an addition of 5% each of low purity fluorspar and oil coke sufficiently increases its rate of reaction, and the works is regularly using burnt lime at present as the external desulfurizing agent of hot metal in open ladle. Shifting from CaC_2 to the lime-fluorspar-carbon desulfurizing agent has reduced the desulfurizing cost to less than a half. Besides, this method of desulfurization has served to improve the labor environment in disposing of desulfurized slag associated with offensive order and to facilitate the recovery of metal from the slag.

References

1. Kanbara, K., Nisugi, S., Shiraishi, O. and Hatakeyama, T.: "On the desulfurization with impeller stirring," Tetsu-to Hagane, Vol. 58, No. 4, March 1972, p. 34.

2. Kumai, K., Kodama, F., Ohori, Y. and Higashiguchi, M.: "On the development of a new external desulfurization method of hot metal," Tetsu-to-Hagane, Vol. 60, No. 11, Sept. 1974, p. 100.

3. Kalling, B., Danielsson, C. and Dragge, O.: "Desulfurization of pig iron with pulverized lime," Journal of Metals, Vol. 3, Sept. 1951, p. 732-738.

4. Sawamura, H.: Riron Tetsuyakingaku, Maruzen, Tokyo 1955 p.321-324.

5. Trentini, B., Wahl, L. and Allard, M.: "An efficient method of desulfurizing liquid pig iron," Journal of Metals, Vol. 9, Sept. 1957, p. 1133-1139.

6. Tsuchiya, N., Ooi, H., Ejima, A. and Sanbongi, K.: "Limit of desulfurization in blast furnace and external

desulfurization by sintered $CaO-CaF_2$ pellets," paper presented at Symposium on External Desulfurization of Hot Metal, McMaster Univ., May 1975.

7. Eketorp, S.: "Desulfuration par la chaux solide," Revue de Métallurgie, Vol. 52, Sept. 1955, p. 718-724.

LADLE - FURNACE PROCESS IN JAPAN

H. Ushiyama
General Manager
Technological Administration Department

G. Yuasa
Manager
Shibukawa Works

T. Yajima
Manager
Technological Administration Department

Daido Steel Company
Nagoya, Japan

This paper originally appeared in Ironmaking and Steelmaking, 1978, Volume No. 3, Page 121.

Abstract: To transfer the reducing period in the arc furnace to the ladle, the ladle-furnace process was developed in 1971 by Daido Steel, then Japan Special Steel. Argon stirring, submerged-arc heating, and the vacuum system are integrated to form the LF process, which is used in four steelworks in Japan as 30-150t units. Slag metallurgy is applied in LF for the refining of steels and sometimes of slags, and the role of vacuum is limited to hydrogen removal. In this paper, six years' experience in Daido together with that in NSC Yawata, JCF Tobata, and Mitsubishi Tokyo are reported on equipment, metallurgical processes and advantages, and quality improvements. Not only exact steel chemistry and temperature control, but also fewer inclusions resulting from the removal of such impurities as oxygen and sulphur can be obtained from the process leading to improvements in the mechanical properties of the products. Also, both higher recovery of alloying elements and productivity increase are enough to offset the operating costs which amounted to £4·3/t in the 20t unit at Daido in 1975.

Ladle metallurgy, here called secondary steelmaking technology, began with degassing processes such as RH, DH, and ladle degassing and has made great progress since the introduction of a heating device to the ladle. The ASEA-SKF process, developed in 1964, was a pioneer in this field. Argon stirring was introduced by the Finkl-Mohr process in combination with arc heating under reduced pressure. In the LF process described in this paper, slag refining technology was added to ladle metallurgy in combination with argon stirring and submerged-arc heating under atmospheric pressure.

The ladle-furnace process (LF) was first developed in April 1971 by the then Japan Special Steel Ohmori Works[1] (which merged with Daido Steel in September 1976) as a 20t unit and was equipped with a heating unit with three electrodes in combination with a stirring device by argon injection through a porous plug placed in the ladle bottom. If necessary, the ladle could be evacuated to remove hydrogen. Two years' experience in ladle bubbling and great progress in ESR technology in Ohmori led to the application of mainly slag-metal reactions in the presence of 2-8% slags for the refining of steels, and sometimes for the simultaneous refining of both steels and slags. In May 1973 the capacity was enlarged to 40t simply by making a new ladle and 30-40t heats combined from two EAFs could be processed for making large forging ingots. Coupled with electric arc furnaces, this unit played an important role in developing new refining practices, and its six years' life ended in March 1977 with the shutdown of Ohmori Works.

In July 1973, the second unit was installed for the EAF-LF duplex refining of austenitic stainless heats, in which the vacuum system was taken away and a deslagging device (tilting of the ladle cradle) was added. This unit was used for the next two years and processed about 1500 heats and was then transferred in January 1977 to Shibukawa Works of Daido Steel as a remodelled 30t unit.

During the past six years' operation in Ohmori, a total of 5700 heats were processed in various ways, depending on the steel quality, and it is now believed that, irrespective of steel quality the reducing period in arc furnace practice can be completely transferred to the ladle furnace and all the metallurgical reactions in the reducing period can be easily and accurately reproduced in every heat.

The process attracted keen interest in another steelworks in Japan, leading to its installation in three different steelworks. The first was December 1973 in Yawata Works of Nippon Steel Corporation (NSC) coupled with a 70t BOF for making cleaner, and therefore tougher, heavy plates; the second was in September 1975 in Tobata Works of Japan Castings and Forgings Corporation as a 150t unit for making mainly forging ingots; and the third was in August 1976 in Tokyo Works of Mitsubishi Steel Company, Ltd., to increase their 50t arc furnace productivity and ensure quality.

Except for one steelworks where the requirements were different in some respects, this process was examined in terms of the metallurgical possibilities in every steelworks, especially in NSC Yawata Works, from other theoretical and practical points of view.

In this paper, Daido's six years' experience in Ohmori, together with the results offered by courtesy of Nippon Steel Corporation, Japan Castings and Forgings Corporation, and Mitsubishi Steel Co. Ltd. are reported. Below, LF equipment, and especially its metallurgical and operational reliability, is described in relation to heating and stirring devices, ladle, and refractories. Four metallurgical processes are then summarized and some metallurgical advantages of each processes are mentioned. Finally, metallurgical reactions in the LF process, especially deoxidation and desulphurization (investigated in NSC Yawata) and their effects on the mechanical properties of the products are briefly reviewed. The shape control of inclusions by the addition of calcium in a 20t unit in Ohmori is also described. Average annual operating costs in Daido are mentioned in the last section.

LF EQUIPMENT

The LF equipment consists of four parts: a conventional ladle with a stirring device (by argon injection), a water-cooled lid, a heating unit, and a vacuum system. A water-cooled flange is attached to the top of the ladle shell to protect a rubber gasket seal mounted on the lower edge of the lid from heat during processing. The gap between the ladle flange and the lid must be tightly sealed to prevent the entry of air, and thus a good reducing atmosphere can be provided during processing. The lid can be connected to the vacuum duct for degassing if the ladle is vacuum tight. The lid and the ladle need not be so designed if RH or DH units are available. Three electrode openings in the lid are electrically insulated by inserting a refractory sleeve and covering with a small lid during vacuum treatment. Figure 1 shows an outline of the LF equipment in Daido Shibukawa and Fig. 2 shows the equipment in NSC Yawata where a DH unit is available.

Table 1 gives the outline of the heating unit and vacuum systems used in five steelworks. For each of them, a secondhand furnace transformer is used and the specific transformer rating varies widely from 40kVA/t in the 150t unit to 167kVA/t in the 30t unit. However, the highest heat efficiency (35%) is reported from the 150t unit which has the lowest kVA/t. The electrical system should be so designed as to permit high current flow at low tap voltage. The choice of the vacuum system depends on the method of processing; the vacuum is, in principle, applied only for hydrogen removal. The ladle degasser is, however, recommended here because the strong stirring action generated under reduced pressure can be applied in special cases for the acceleration of slag-metal reactions.

Table 2 gives the ladle and bath dimensions and the nominal capacity of each unit. To indicate the bath shape the ratio of bath depth to diameter H/D was chosen; this varies from 0·49 in Mitsubishi to 1·13 in Yawata. In the 20t unit in Ohmori, the H/D ratio in the 20t ladle can be varied between 0·49(13t) and 0·86(23t). Although the bath depth of the 13t heat is only 819 mm, even ELC stainless and superalloys can be processed easily

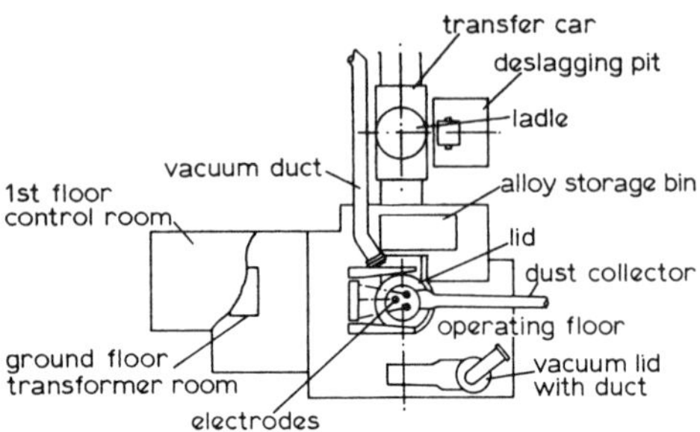

Fig. 1 - Plant layout of 30t LF unit in Daido Shibukawa

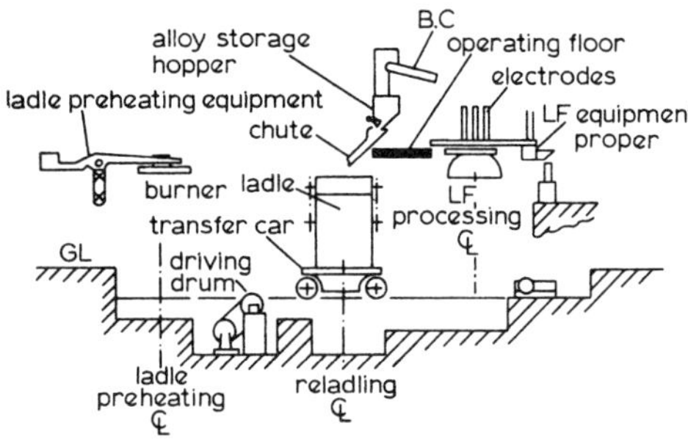

Fig. 2 - Outline of 60t LF unit in NSC Yawata[2]

Table 1 - LF equipment data

	Daido Ohmori	Shibukawa	NSC[2] Yawata	JCF Tobata	Mitsubishi Tokyo
Installation date	April 1971	Jan. 1977	Dec. 1973	Sept. 1975	Aug. 1976
Primary furnace	20t EAF	30t EAF	70t BOF	100t EAF	50t EAF
LF capacity					
t/heat	20	30	60	150	50
t/month	2000	2500	4500	4000	15000
Electrical system					
transformer, kVA	3000	5000	6500	6000	7500
secondary rated voltage, V	220/95	235/85	225/75	275/110	250/102
secondary rated current, A	10600	14400	23860	17000	17320
electrode dia., mm	254	254	356	356	305
pitch circle dia., mm	650	600	810	940	900
Vacuum system	Mechanical L degas	Steam ejector L degas	Steam ejector DH	Steam ejector L degas	Steam ejector L degas
Specific transfer rating, kVA/t	150	167	108	40	150

Table 2 - Ladle and bath dimensions

	Daido Ohmori	Shibukawa	NSC[2] Yawata	JCF Tobata	Mitsubishi Tokyo
LF capacity					
nominal, t	20	30	60	150	50
actual, t	13/23	18/33	60	100/150	45/50
Dimensions, mm					
shell dia.	2200	2400	2600	3900	2924
inside dia. *D*	1676	1948	2070	3164	2430
total height	2300	2500	3150	4330	3040
inside height	1995	2195	2740	4000	2770
bath depth *H* (at *X* t)	1260(20)	1402(30)	2340(60)	2754(150)	1348(45)
Depth/dia. at *X* t *H*/*D*	0·75	0·72	1·13	0·87	0·49

Fig. 3 - Wear of slag-line bricks after 9 heats: 20t ladle (Daido Ohmori)

Table 3 - Wear lining data

Works	Slag line	Side wall	Bottom
Daido, Ohmori	Synthetic Mg–dol Burnt and tar impregnated MgO 87% CaO 7% Tar 6%	High alumina Burnt or chemical bond Al_2O_3 87% SiO_2 9%	High alumina Burnt or chemical bond Al_2O_3 87% SiO_2 9%
NSC, Yawata	Synthetic Mg–dol Burnt and tar impregnated MgO 88% CaO 10%	High alumina Burnt Al_2O_3 86% SiO_2 9%	High alumina Burnt Al_2O_3 86% SiO_2 9%
Mitsubishi, Tokyo	Rebond Mg–Cr Burnt MgO 76% Cr_2O_3 11% Al_2O_3 6% Fe_2O_3 5%	High alumina Burnt Al_2O_3 87% SiO_2 10%	Chamotte Burnt Al_2O_3 35% SiO_2 60%

and rapidly. This wide applicability to heat size means that, for example, one 18t furnace can be used from 72 to 128% of its nominal capacity using one ladle, and thus the ability to fulfil order size is increased. The largest (150t) unit in JCF Tobata can process a minimum 50t heat simply by changing the ladle.

It can be seen from both tables that the distance between electrodes and the nearest side wall is comparatively short (only 380 mm in the 20t unit in Ohmori). Nevertheless, hot spots are not observed on the slag-line bricks because of submerged-arc heating.

In the following sections, the present state of ladle lining, heating and stirring efficiencies, and furnace atmosphere are described in detail. The operational experiences from four steelworks are described.

LADLE LINING

About 70% of the operation costs in Ohmori are incurred by refractories (wear and safety lining, the sliding-gate valve, and the porous plug). Table 3 gives the wear lining data in Daido, NSC Yawata, and Mitsubishi where standard 114mm thick bricks are used. In general, several courses around the slag line are lined with basic materials and the rest of the wear lining with 87% alumina bricks. The life of the wear lining varies widely, depending on the process applied and the use of the ladle, for example, continual, as in Mitsubishi, or intermittent, as in Daido Ohmori. The mean lives of the wear linings, excluding slag-line refractories are 48, 40 and 35 heats in NSC Yawata, Mitsubishi, and Daido Ohmori, respectively.

The lives of the slag-line refractories are shorter, in spite of using higher quality bricks: 20 heats for rebond Mg-Cr bricks in Mitsubishi and 15 heats for synthetic tar-impregnated Mg-dolomite bricks in Daido, and these lives determine the repair cycle. Wear of the slag-line refractories is mainly caused by chemical attack (corrosion) in Mitsubishi and thermal spalling in Daido Ohmori as shown in Fig. 3. Hot repair by gunning is not used in either steelworks. Slag-line refractories are usually used until the moment when the safety lining is partially exposed after processing a heat, and therefore the safety lining behind it must also be lined with basic bricks, usually standard MgO grade which should be replaced after every six campaigns (about 240 heats). The rest of the safety lining is lined with standard chamotte bricks. A breakdown of the wear lining has never occurred during operation, except twice in the earlier stage of development in Ohmori when chamotte bricks were used by mistake for the safety lining behind the slag line. Carbon in tar-impregnated bricks does not lead to carbon increase during processing, thus allowing the treatment of even stainless heats in a newly relined ladle. No special precautions are required for the sliding-gate valve, which can be used for three heats in Ohmori. In Mitsubishi, the sliding-gate valve, of high-alumina grade, can be used for six consecutive heats.

A maximum shell surface temperature of 350°C was recorded in the last heat of a campaign when the mean thickness of the wear lining was reduced to 50mm, and 200°C in the first heat. However, this has not caused any damage on the shell plate and the first ladle is still used after six years with no deformation.

POROUS PLUG

This is naturally the most important component, being a centre of the process, and is therefore mentioned separately. To give a maximum theoretical stirring energy to the bath one must use a good porous plug with high open porosity uniformly distributed in the plug which allows a large volume of gas to flow at the lowest pressure. In addition, the plug must have high strength to endure mechanical and thermal stresses at bath temperature, and high resistance to the slag attack which arises at the end of pouring.

Figure 4 shows the plug now used for a maximum 40t heat in Daido and some physical and chemical properties. The plug is inserted from outside the

composition Al_2O_3 96% TiO_2 2%
porosity 31%
refractoriness under load (Ti) (2·5kg/cm^2) 1700°C
crushing strength at 1400°C 140kg/cm^2
refractoriness SK 38+

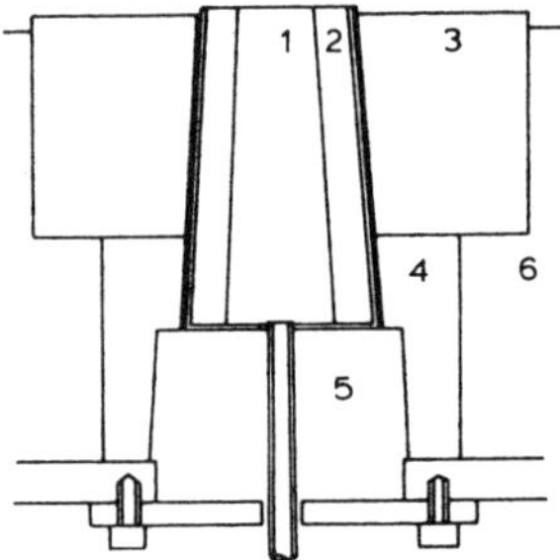

1 porous plug; 2 outer sleeve; 3 upper seating block; 4 lower seating block; 5 rammed MgO; 6 bottom lining

Fig. 4 - Schematic diagram of porous plug (Daido)

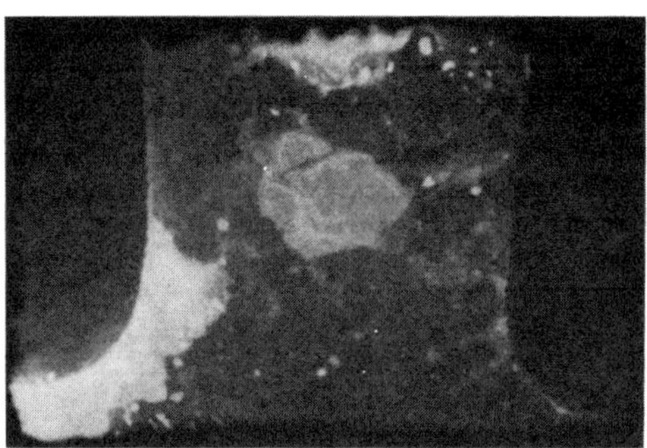

Fig. 5 - Submerged-arc heating in 20t ladle: steel quality AISI 316L (Daido Ohmori)

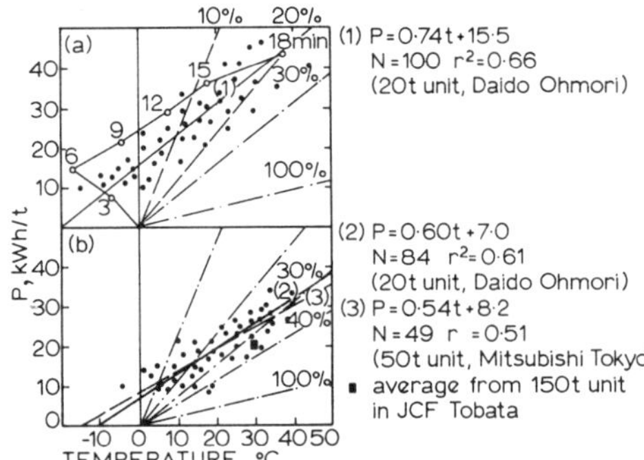

Fig. 6 - Energy efficiencies during heating a in preheated ladle and b in continually used ladle

ladle bottom and fixed tightly in a high-alumina seating block with a magnesia mortar. The open space at the bottom of the lower seating block is then filled with rammed MgO to form a safety lining. The outer sleeve, of high-alumina grade, forces the gas to flow only through the plug and not through the gap between the steel cladding and the sleeve, thus preventing local wear at the opening of the upper seating block which can be used for 15 consecutive heats.

Great care has been taken in maintaining the plug in Daido. After pouring every heat, the molten slag which penetrates the open pores is removed by reblowing argon through the plug until all the slag is flushed out of the ladle. If the boiling of slag, owing to the reblowing does not happen, or severe wear is observed by visual inspection, then the plug must be replaced. The mean life of a plug is 3·8 heats: in 1975 it was a maximum of 8 heats. The operating pressure, however tends to increase with successive use of the plug owing to the partial blockage of open pores by slag.

No serious damage to the plug during processing causing leakage of molten steel through the plug has occurred in the past 5700 heats in Daido, and the blockage of the plug is less than 0·5% at present when using only one plug. Even when a blockage occurs, LF operation can be continued by lancing argon every 10 min. through an opening in the lid.

ENERGY EFFICIENCIES

During heating, the arc is completely submerged in slag and the energy efficiency is therefore higher in the LF than in the arc furnace. This is clearly shown in Fig. 5 where a 13t ELC stainless melt is processed in a 20t ladle by supplying 2000kW power at 127V. Two electrodes, top centre (unseen) and below left, are energized and an open bubbling site can be seen at the centre of three electrodes. After about 10s, the site is also covered with the slag being expelled under the electrodes. By placing a porous plug at the centre of the ladle the stirring force is the strongest where the heat supply is the greatest, and the reverse applies in the periphery where slag attack on bricks is feared, through which rapid heat transfer and thus less slag-line wear can be expected. In the 20t ladle, the slag is usually superheated by 30° - 50°C without causing any wear in the slag line. In spite of dipping electrodes in the slag, carbon increase during heating negligible (about 1 ppm/min for ELC stainless).

Figure 6a and b shows temperature increase as a function of power consumption (kWh/t) in the 20t unit in (a) a preheated and (b) a continually used ladle, respectively. When the efficiency is calculated only the heat absorbed in ladle steel is taken into account on the basis of a specific heat of 0·23kWh/t/°C(0·20kcal/kg/°C). The power supply is kept constant at 2000 kW at 127V. In Fig. 6a a regression line is expressed as

$$P = 0{\cdot}74t + 15{\cdot}5 \quad (1)$$

where P is power consumption, kWh/t; t is temperature increase, °C; N = 100; and $r^2 = 0{\cdot}66$.

The temperature increase in a newly relined ladle, measured every 3 min. up to 18 min. is plotted in Fig. 6a, from which the first 24kWh/t power can be seen to be consumed only in heating up the lining. A high rate of temperature increase of

4·6°C/min can, however, be attained in the last 12 min.

Figure 6b shows the same relationship in which the ladle is used continually and the following regression line is obtained:

$$P = 0{\cdot}60t + 7{\cdot}0 \quad \ldots\ldots\ldots\ldots\ldots\ldots\ldots\ldots\ldots\ldots\ldots\ldots\ldots (2)$$

with N = 89 and $r^2 = 0{\cdot}61$.

The constants 15·5 and 7·0 in the respective equations (1) and (2) express the power loss in the first few minutes. Although the slopes of the two regression lines run parallel to the 30 and 40% efficiency lines in equations (1) and (2), respectively, the mean efficiency is low (15% in equation (1) and 25% in equation (2), owing to the initial losses).

A regression line obtained from a 50t unit in Mitsubishi, namely,

$$P = 0{\cdot}54t + 8{\cdot}2 \quad \ldots\ldots\ldots\ldots\ldots\ldots\ldots\ldots\ldots\ldots\ldots\ldots\ldots (3)$$

with N = 49 and $r^2 = 0{\cdot}50$, is added in Fig. 6b (as a broken line) when supplying 7000kW at 250V. The mean efficiency is the highest there (about 35%). Comparing those two lines in Fig. 6b, it can be clearly seen that the same relationship can be expected in both a 20 and 50t unit. The rate of temperature increase is faster in the 50t unit (3°C/min) owing to the greater power input rate, compared with 2°C/min in the 20t unit.

The filled square in Fig. 6b shows the average from 150t heats in JCF Tobata. Although the rate of temperature increase is slow (only 1°C/min) owing to the under-powered transformer, the high efficiency (35%) should be noted. A power consumption of 33·9kWh/t is reported from Mitsubishi and 74kWh/t from Daido, where electrode consumption is 0·35 and 0·65kg/t, respectively. The figure for electrode consumption is about one-hundredth that of power consumption.

STIRRING ENERGY

Argon bubbling is the sole energy source for stirring the bath in LF. To give the largest stirring energy from the smallest volume of argon, the gas should be injected at the lowest pressure and the smallest bubble size. Under the condition that the gas is injected at the same pressure as that of the ladle bottom and is heated there from room temperature T_1 to bath temperature T_2 and then rises up to the bath surface, the energy input can be expressed as[3]

$$E = \frac{Q}{22{\cdot}4} RT_2 \left(1 - \frac{T_1}{T_2} + \ln\frac{P_1}{P_2}\right) \ldots\ldots\ldots\ldots\ldots\ldots (4)$$

where

- E = Stirring energy, W
- Q = argon flowrate, Nm^3/s
- R = gas constant (8314 N/m/K/mol)
- T_1 and T_2 = argon and bath temperatures respectively, K
- P_1 and P_2 = bottom and surface pressures, respectively, N/m^2
- 1 atm = 1013mb = $101300N/m^2$.

If one assumes that T_1 = 298K, T_2 = 1873K (1600°C), and P_2 = 1 atm and one converts Q into q(l/min), the following equation can be obtained:

$$\begin{aligned} E &= 11{\cdot}58q\,[0{\cdot}841 + (2{\cdot}303 \log P_1 - 2{\cdot}303 \log P_2)] \\ &= 11{\cdot}58q\,[0{\cdot}841 + (2{\cdot}303 \log P_1 - 11{\cdot}58)] \quad \ldots(5) \end{aligned}$$

where 0·841 gives the energy by isobaric expansion and $2{\cdot}303 \log P_1 - 11{\cdot}58$ that by isothermal expansion. Figure 7 shows the stirring energy calculated from equation (5). In the 20t unit, the average flowrate of 13l/min during heating gives about 230W and a specific energy of $110W/m^3$ (bath). The authors have not observed chemistry and temperature segregation and therefore skull formation in the periphery of the ladle bottom, and believe that this stirring energy is enough to make a homogeneous bath.

In Mitsubishi Tokyo, where the smallest H/D (0·49) is used, 100 l/min argon giving $280W/m^3$ allows the refining of a 45t heat within 40 min. In Tobata, a 150t heat is stirred by 200 l/min argon giving the energy of 4480W and $215W/m^3$. For a 100t heat in the same ladle, 200 l/min argon gives 3900 W and $280W/m^3$.

If the furnace chamber is evacuated to reduce the pressure inside, the stirring energy becomes larger. Under the conditions of one atmospheric pressure and of 10 l/min argon, the stirring energy of 178W in a 1·44m deep bath increases suddenly to 883, 1085, and 1200W under 1/10, 1/100, and 1/1000 atm, respectively. The same amounts of energy can theoretically be produced by increasing the flowrate to 50-70 l/min under atmospheric pressure, as shown in Fig. 7. The broken lines in Fig. 7 indicate the energy generated under 1/100,000atm. The high stirring energy generated under reduced pressure is often used for the acceleration of slag-metal reactions together with hydrogen removal.

Figure 8 shows an example of chemistry control around the desired analysis in a 20t unit. A decrease of about 50% in the standard deviations of Mn, Cr, and Mo can be obtained. Scattering of C on the positive side of the desired analysis is due to the fact that an increase of about 0·01% during heating is not considered at alloying. Another example of chemistry and temperature control is given below.

INERT FURNACE ATMOSPHERE

The reducing, i.e. less oxidizing, furnace atmosphere must be provided during the reducing period to allow for the application of highly fluid and reactive reducing slags for the refining of steels. In the LF process this condition is ensured, first of all, by the complete sealing of the furnace chamber against the infiltration of air. In addition, the oxygen left in the furnace at the start of operation quickly reacts with carbon in the electrodes to form CO. Also, the large volume of CO formed by the reactions of electrode carbon with oxygen in such oxides as FeO, MnO, and Cr_2O_3 in slag maintains the reducing atmosphere during the heating period, and at the same time,

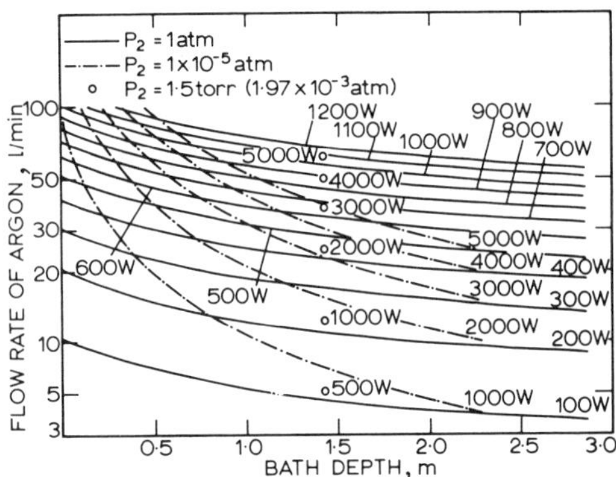

Fig. 7 - Stirring energy calculated from equation(5)[3]

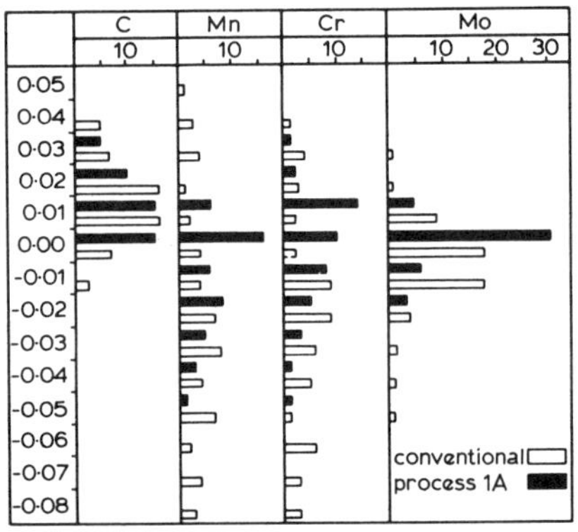

mean value and standard deviation of C, Mn, Cr, and Mo

	N	C mean	C S	Mn mean	Mn S	Cr mean	Cr S	Mo mean	Mo S
process 1A	45	·011	·010	--·011	·016	·003	·017	--·017	·007
conventional	54	·016	·013	--·023	·035	--·024	·037	--·004	·013

S, standard deviation, %

Fig. 8 - Chemistry control around desired analysis: constructional steels (Daido Ohmori)

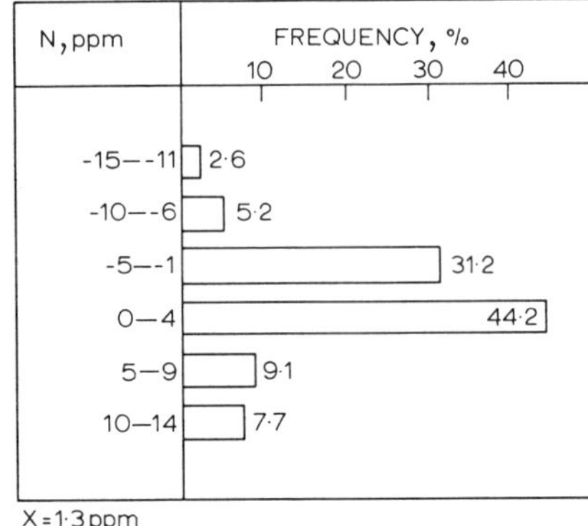

Fig. 9 - Difference in nitrogen before and after LF processing[4]

this reaction accelerates the formation of reducing slags containing less of these oxides. Free oxygen in the atmosphere is thus reduced down to 0·5% within the first 10 min of heating and kept at this level all the time during ladle refining. Most of the furnace atmosphere is replaced with CO during heating and with N_2 during bubbling. However, little nitrogen increase is observed, as shown in Fig. 9[4] in which changes in nitrogen content before and after LF processing of a 60t BOF heat containing low nitrogen (20-45ppm) are shown. This is probably due to submerged-arc heating and a thick slag cover which prevents nitrogen dissociation and absorbtion.

It should be mentioned here that even hydrogen removal down to 3 ppm can be expected under atmospheric pressure. This is due to the reduction of moisture in the air and can only be attained inside the tightly sealed furnace chamber. Detailed information is given below. From these facts, one can clearly understand how a good atmosphere is provided in the LF process.

METALLURGICAL PROCESSES

Several metallurgical processes now used in the LF operation are reported in this section together with their results on quality control and alloy savings. The reducing period generally starts with the removal of oxidizing slags from a furnace in the case of plain carbon or low-alloy steel melting practice, and with slag reduction by adding Fe-Si in high-alloy tool or stainless heats. The LF process can be used for both types of reducing period, and in this paper, the former is called process 1 and the latter process 2. Both processes can be finished with vacuum degassing if less hydrogen is required.

Process 1 is sub-divided into processes 1A and 1B, depending on the method of deoxidation: precipitation deoxidation is mainly used in process 1A and diffusion deoxidation in 1B. Vacuum carbon deoxidation is also included in the latter. Process 2 is also sub-divided into processes 2A and 2B. Process 2A is the so-called single-slag practice in which furnace oxidizing slag is first reduced during tapping to recover alloying elements from slags and then converted gradually to highly reducing slags in the course of refining in the LF. This process is used for the production of alloy tool steels and high-speed steels in which the amount of slag is usually around 8%. In process 2B, furnace oxidizing slags are reduced during tapping and then removed from the LF. The refining proceeds with fresh reducing slags being added and formed in the LF. This process can be regarded as a double-slag practice and is used for stainless heats. Figure 10 shows an outline of those four processes now applied in four steelworks in Japan.

PROCESS 1A

Furnace oxidizing slags must be removed by flushing in the EAF and by reladling in the BOF operation[2] to prevent oxide contamination and rephosphorization. Additional dephosphorization can be carried

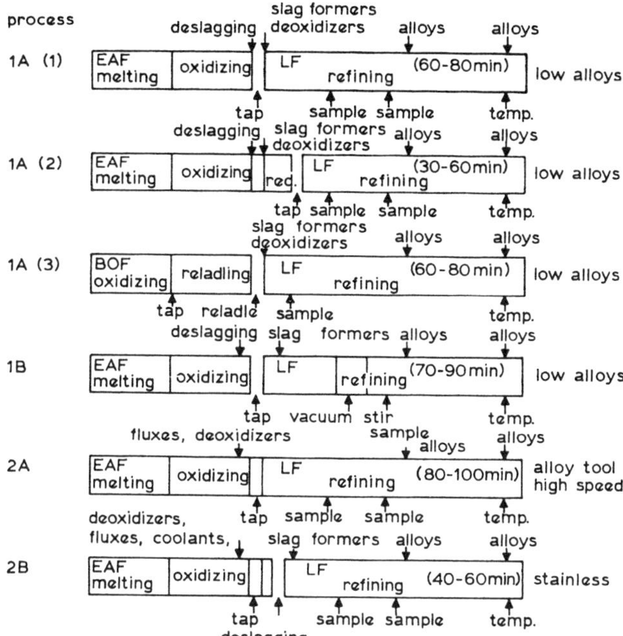

Fig. 10 - General pattern of LF processing

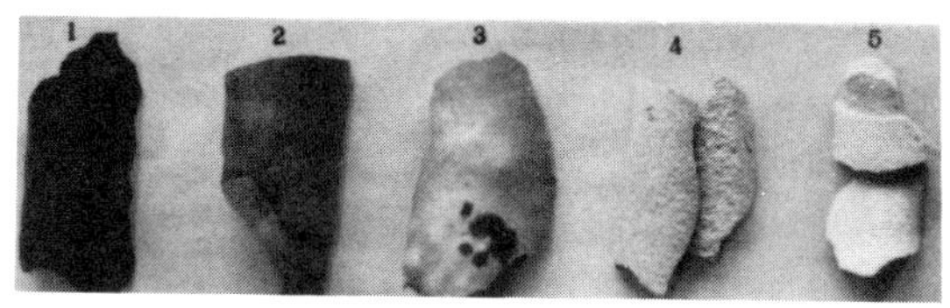

No.	Time, min	Slag chemistry, wt-% CaO	MgO	SiO_2	Al_2O_3	S	FeO	MnO	Cr_2O_3
1	18	36·0	14·1	28·7	10·6	0·078	1·20	0·74	0·70
2	22	34·4	15·7	28·1	11·4	0·075	0·56	0·54	0·50
3	29	41·5	13·5	26·7	11·1	0·095	0·30	0·31	0·20
4	44	43·8	13·1	26·3	10·6	0·099	0·33	0·20	0·20
5	74	44·4	13·1	25·3	12·0	0·157	0·24	0·10	0·10

Fig. 11 - Change in chemistry of slag by process 1A: steel quality H13 (Daido Ohmori)

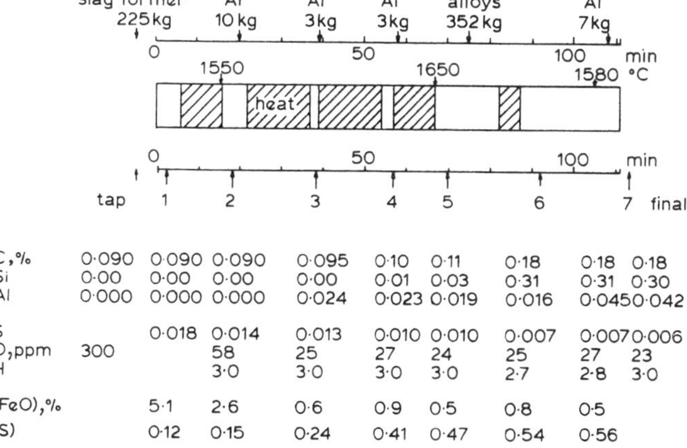

	tap	1	2	3	4	5	6	7	final
C,%	0·090	0·090	0·090	0·095	0·10	0·11	0·18	0·18	0·18
Si	0·00	0·00	0·00	0·00	0·01	0·03	0·31	0·31	0·30
Al	0·000	0·000	0·000	0·024	0·023	0·019	0·016	0·045	0·042
S		0·018	0·014	0·013	0·010	0·010	0·007	0·007	0·006
O,ppm	300		58	25	27	24	25	27	23
H			3·0	3·0	3·0	3·0	2·7	2·8	3·0
(FeO),%		5·1	2·6	0·6	0·9	0·5	0·8	0·5	
(S)		0·12	0·15	0·24	0·41	0·47	0·54	0·56	

Fig. 12 - Heat log of process 1A for open steel (Daido Ohmori)[1]

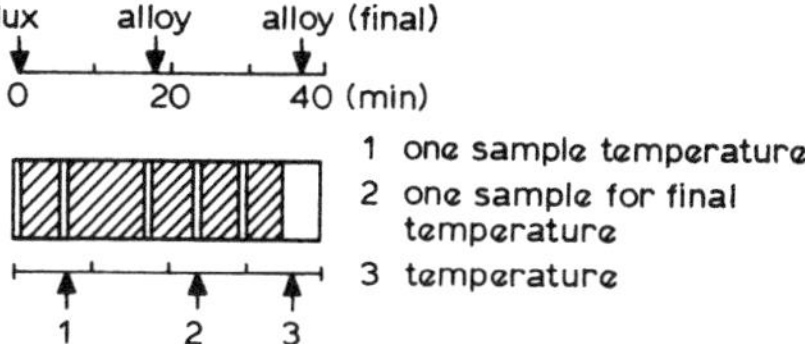

Fig. 13 - General processing patterns (Mitsubishi Tokyo)

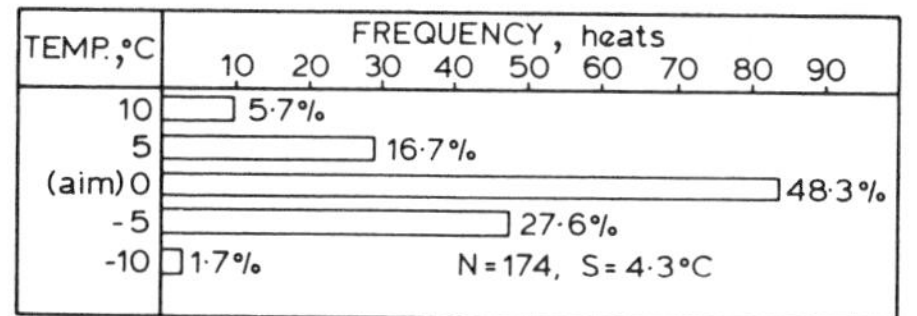

Fig. 14 - Pouring temperature distribution around desired value (Mitsubishi Tokyo)

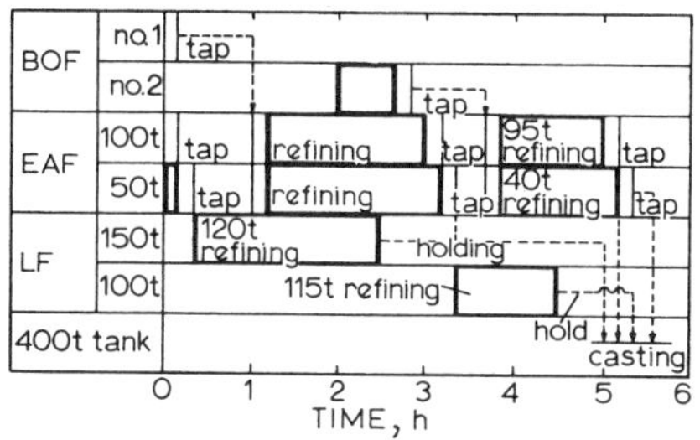

Fig. 15 - Time schedule for making 370t ingots by the triplex process (JFC Tobata)[5]

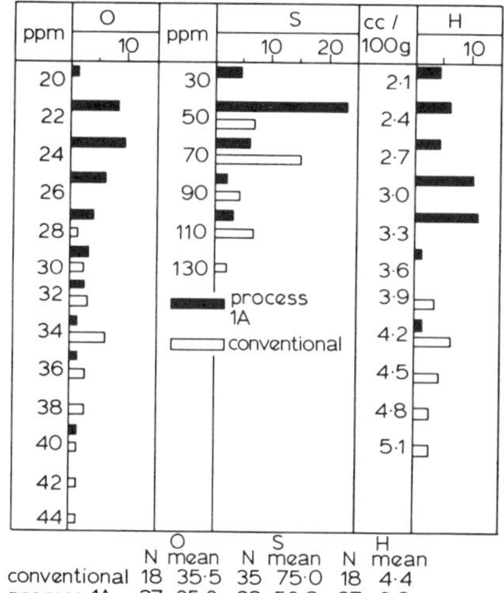

Fig. 16 - Comparison of oxygen, sulphur, and hydrogen contents (Daido Ohmori)

Table 4 - Chemistry control for making 370t ingot (JCF Tobata)[5], wt-%

No.	Process	Ladle weight t	Weight, t	Desired analysis C	Si	Mn	P	S	Cr
				0·38	0·25	0·75	..	..	0·30
1	EAF–LF	150 LF	118	0·55	0·24	0·72	0·003	0·002	0·31
2	BOF–EAF	100	99	0·40	0·26	0·73	0·005	0·002	0·32
3	BOF–EAF	100 LF	116	0·27	0·24	0·72	0·003	0·001	0·32
4	BOF–EAF	80	41	0·13	0·21	0·73	0·003	0·002	0·28
Total			374	0·38	0·24	0·72	0·004	0·002	0·31

out during tapping of the BOF heat into the transfer ladle. A reducing slag is formed either in the furnace or in the ladle in the case of the EAF operation. The former is mainly used in Daido and Mitsubishi. In the BOF operation, all the slag formers are added to the LF during reladling together with deoxidizers. About 40kg/t(4%) of slag are used for the refining of both EAF and BOF heats. The refining temperature is lower than that in the furnace, and is generally about 30°C higher than the specified pouring temperature in the 20t unit in Daido.

Figure 11 shows how the slag chemistry changes in the course of refining an alloy tool steel similar to H13 using furnace reducing slag. The first slag taken after 18 min in the LF still contains 1·2%FeO and looks dark. After refining for another 56 min while steel chemistry and temperature are adjusted, FeO decreases to 0·24% together with MnO and Cr_2O_3 to form pure white reducing slag. Desulphurization proceeds under this slightly basic (%CaO/%SiO_2=1·75) but highly reducing slag as clearly indicated from the sulphur increase in the slag.

Figure 12[1] shows a heat log from an early trial of a 16t case-hardening steel (JIS SCM 21) tapped in an open state, and 225kg of reducing slag formers are added during tapping. Within 38min in the LF, the initial 300ppm oxygen decreases to 58ppm and then to 25ppm, first by diffusion deoxidation under fresh slag additives and then by precipitation deoxidation 18 min after the addition of 10kg (0·625kg/t) Al and this oxygen level is kept constant all the time during refining. Later 6kgAl are added and are consumed merely to reduce the SiO_2 in the slag, producing 0·03% Si in the steel 30 min after the addition. FeO content follows the same path as that of oxygen in steel, from 5·1 to 0·6% at 38 min. The desulphurization rate is comparatively slow, from 180 to 60ppm in 110 min and about 1ppm/min under this highly basic slag. This is due partly to a weak stirring intensity of only 160W and 72W/m^3 and to a smaller slag volume of 13 kg/t, which is, however, enough to promote deoxidation. There is little increase of carbon and hydrogen from electrodes and atmosphere, respectively, in spite of the longer processing time in the LF.

Figure 13 shows the process now applied in a 50t unit in Mitsubishi Steel for constructional steels. During the period October-December 1976, only one month after start up, 586 heats were processed in this way and the whole refining was completed within 40 min, including 30 min heating. The exact steel chemistry and temperature control of a constructional steel are evident from the standard deviation of several elements(wt-%): C=0·007; Si=0·020; Mn=0·022; Cr=0·020; and Al=0·005, and from Figure 14 which shows the standard deviation of temperature of 4·3°C measured in the pouring stream. In general, the pouring temperature is kept constant during teeming, but drops slightly at the end. Homogeneous steel chemistry and temperature are thus ensured in the ladle having the smallest H/D of 0·49 and being stirred by 100 l/min argon. The oxygen level is controlled between 25 and 34ppm, with an average of 29ppm, and a standard deviation of 2·2ppm.

From JCF Tobata an interesting application of the LF for making 370t ingots is reported[5] using a triplex process, namely, two BOFs, two EAFs, (100 and 50t), and one LF (150t). Figure 15 shows the time schedule, starting with the tapping of the first two scrap-melt EAF heats totalling 120t steel into the LF where the refining is continued for 130 min to reach 1660°C and thereafter the first LF heat is transferred to the pit side and left there for 150 min until pouring. Meanwhile, the first BOF is charged and this is refined in two EAFs from which a 115t heat is tapped into the second ladle for the LF and refined there for 70 min. During this second LF refining, the two EAFs are refining their third heat (total 135t) from the second BOF, 95t in a large and 40t in a small furnace. Pouring is started 5h after the first EAF tapping with the first 120t LF heat followed by the 95t EAF, second 115t LF, and finally, the 40t EAF heat, and lasts for 45min. The temperature drop in the first LF heat during holding time is only 0·37°C/min for the first 100 min and thereafter 1·06°C/min by intensified stirring to reach the final limit of 1570°C. Table 4[5] gives the final steel chemistry from four ladle samples. A large variation of carbon content in each heat, to avoid segregation in the ingot, should be noted here.

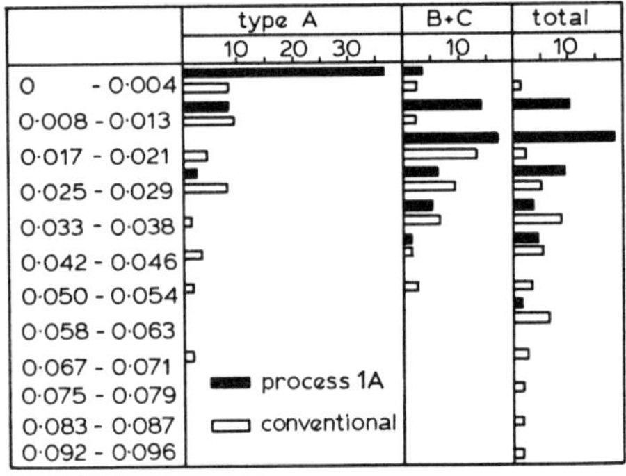

Fig. 17 - Comparison of inclusion contents (Daido Ohmori)

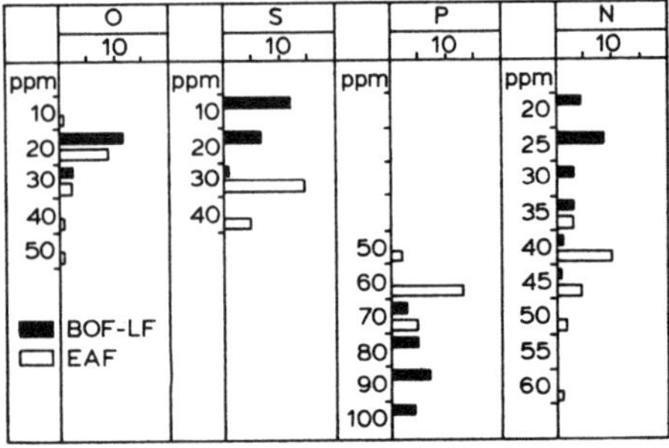

Fig. 18 - Comparison of ladle analysis of special steel plates (NSC Yawata)[2]

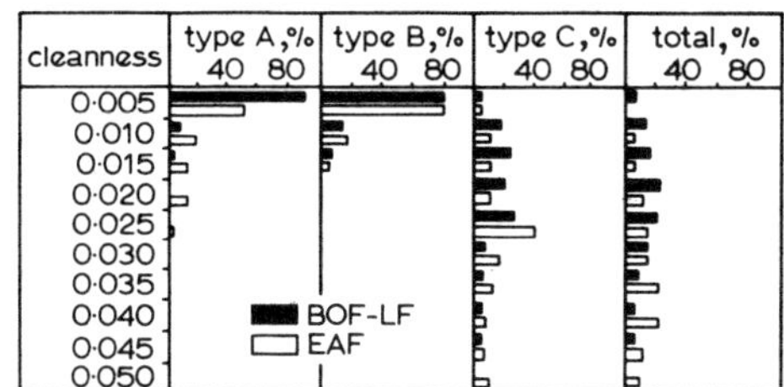

Fig. 19 - Comparison of cleanness of LF with EAF steels for special heavy plates[2]

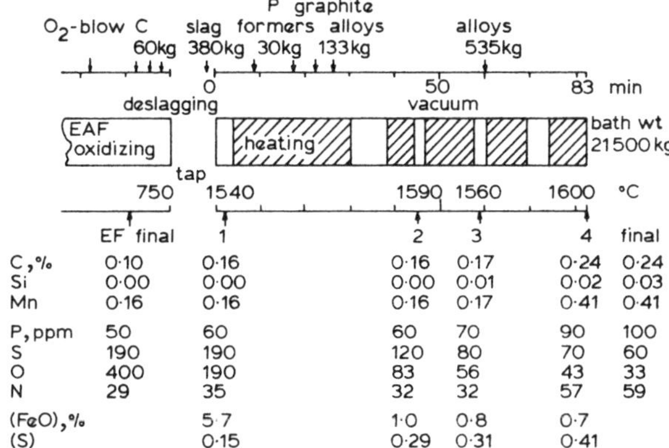

Fig. 20 - Heat log of process 1B (Daido Ohmori)

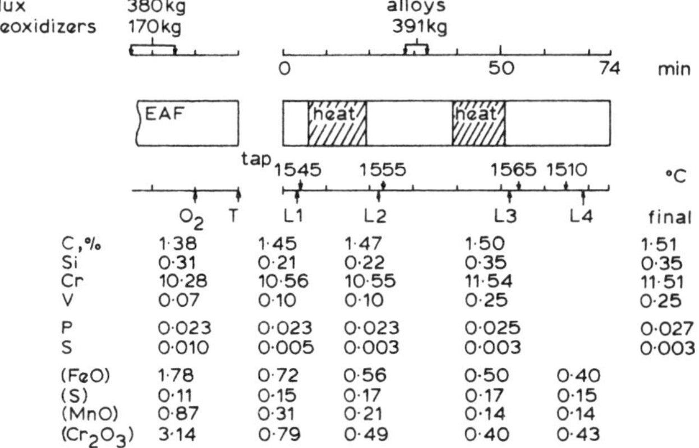

Fig. 21 - Heat log of process 2A for SKD 11 (similar to AISI D2) (Daido Ohmori)

Improvements in quality owing to the decrease of impurities and inclusions are shown in Fig. 16 and 17, respectively, (obtained from 20t heats of constructional steels in Daido Ohmori), in which results from process 1A and standard double-slag practice with ladle bubbling are also compared. A decrease of about 30% in the mean oxygen and sulphur content in LF results in a 50% decrease of the total inclusion content, and 75% of type A inclusions (sulphides and silicates) as shown in Fig. 17. Also, a hydrogen content about 34% lower (around 3ppm) can be obtained from LF heats.

The same results have been obtained from NSC Yawata,[2] as shown in Figs. 18 and 19 where impurities and inclusion contents, respectively, of high-tension steels made by 60t BOF-LF duplex process and by scrap-melt EAF heats, are summarized. In addition to lower oxygen and sulphur, and thus inclusion contents, a low nitrogen content (averaging 28ppm from the BOF) is ensured by the process.

PROCESS 1B

This process is designed to melt special-quality steels for which precipitation deoxidation is not allowed and vacuum carbon deoxidation is specified. To decrease the burden under vacuum one must strive to decrease oxygen in advance. This can be accomplished by applying the LF for diffusion deoxidation together with desulphurization, chemistry, and temperature control.

Figure 20 shows a heat log in a 20t LF in Daido Ohmori where oxygen decreases from an initial 190 ppm to 43ppm, mainly by diffusion deoxidation under 20kg/t basic reducing slag, first to 83ppm, then to 56ppm partly by intensified stirring and partly by vacuum carbon deoxidation under reduced pressure of 1/10atm for 11 min, and finally to 43ppm. The FeO content in the slag shows the same trend as oxygen in steel, from 5·7 to 0·7% by continual addition of pulverized graphite to the slag. Together with deoxidation, desulphurization proceeds under basic slag ($\%CaO/\%SiO_2=6\cdot0$) from 190ppm to a final 60ppm, and the rate is accelerated particularly by vacuum stirring to 3·6 ppm/min.

The melt is vacuum poured and a final oxygen content of 33ppm can be attained in the top of the ingot by vacuum carbon deoxidation. Table 5 shows the inclusion contents, according to the JIS and ASTM A methods in the disc of a rotor forging produced from the heat shown in Fig. 19. Most inclusions are converted into globular type (ASTM D type).

Table 5 - Oxygen, sulphur, and inclusion contents in rotor forging (Daido Ohmori)

	Impurities, ppm				Inclusion content JIS, %			ASTM *A*							
								A		B		C		D	
	P	S	O	N	A	B+C	total	T	H	T	H	T	H	T	H
Final	100	60	33	59	..	..	..	..	..	..	..	..	..	..	..
Rotor															
outer	100	60	34	51	..	0·004	0·004	..	..	..	..	..	..	1·5	0·5
1/2*r*	100	60	33	63	..	0·008	0·008	..	..	1·0	..	..	..	1·5	0·5
centre	110	70	37	69	..	0·008	0·008	0·5	..	1·5	1·5	..	..	1·5	0·5

Table 6[6] shows the behaviour of oxygen, sulpher, and inclusions in 150t heat in JFC Tobata. Oxygen is decreased first to 53 ppm in the LF under basic slags, then to 25·3 ppm by vacuum carbon deoxidation during vacuum casting, and then down to 19·5 ppm in the core after trepanning. Sulphur decreases finally to 45ppm in spite of the difficulties in desulphurizing the bath with such a low silicon content as 0·03–0·10%. Low oxygen and sulphur result in a lower inclusion content in the centre core where positive segregation of these two impurities and inclusions can be observed.

Table 6 - Oxygen, sulphur, and inclusion contents in rotor forgings (JCF Tobata)[6]

	Oxygen, ppm		Sulphur, ppm		Inclusion content (JIS), %		
	$\bar{X}$	Range	$\bar{X}$	Range	Type	$\bar{X}$	Range
LF final	53·3	45–63	45	20–70	A	0·013	0·004–0·020
Ingot top	25·3	18–34	..	..	B+C	0·003	0·000–0·012
Centre core	19·5	14–24	..	..	Total	0·016	0·008–0·025

N=13 heats; Si=0·03–0·10%.

Table 7 - Improvements in recovery of SKD 11 (Daido Ohmori), %

	N	Overall*	Cr	Mn	V
Conventional (EAF)	14	95·9	91·3	85·5	94·4
Duplex (EAF–LF)	8	98·7	96·4	94·0	98·5
Difference	..	2·8	5·1	9·5	4·1

*Fe–Si for slag reduction is taken into account.

Table 8 - Chemistry control of M2 by process 2A (Daido Ohmori)

Element	Desired analysis, wt-%	Mean value, % 10t EAF	18t EAF and LF	Standard deviation, % 10t EAF A	18t EAF and LF B	Reduction % B/A
C	0·89	0·904	0·893	0·029	0·018	62
Si	0·30	0·275	0·313	0·081	0·043	53
Mn	0·30	0·299	0·307	0·021	0·015	71
Cr	4·00	4·042	3·993	0·113	0·064	57
Mo	5·00	4·941	4·975	0·125	0·044	35
W	6·15	6·165	6·156	0·192	0·020	10
V	1·95	1·997	1·963	0·081	0·041	51

*N=30 heats.

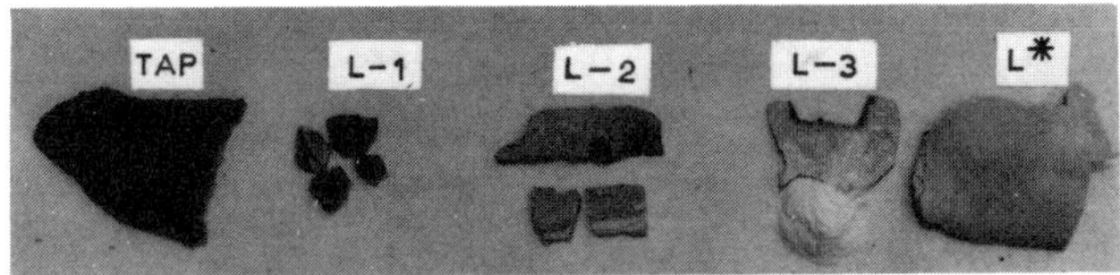

Slag analysis (Heat no. 51–1916, NRI)

Element	Tap	L–1	L–2	L–3	L*
CaO	36·49	45·81	54·50	44·47	40·84
CaF_2	9·42	8·31	8·97	9·64	12·66
MgO	5·39	5·93	6·92	7·30	7·28
MnO	0·87	0·31	0·21	0·14	0·14
Cr_2O_3	3·14	0·79	0·49	0·40	0·43
FeO	1·78	0·72	0·56	0·50	0·48
Al_2O_3	5·63	6·83	7·00	7·15	6·66
SiO_2	25·43	30·63	30·29	29·90	28·19
P_2O_5	0·039	0·027	0·014	0·012	0·015
S	0·111	0·149	0·170	0·165	0·154
MoO_3	0·13	0·04	0·03	0·04	0·03
Nb_2O_5	0·43	0·43	0·40	0·35	0·28
WO_3	0·04	0·05	0·05	0·05	0·04
V_2O_3	0·16	..	..	..	..

Fig. 22 - Change in chemistry of slag refined by process 2A: steel quality SKD 11 (Daido Ohmori)

PROCESS 2A

Since the installation of a second unit in Ohmori, this process has been widely applied as a duplex process (EAF-LF) for the melting of alloy tool steels (D1 and D2) and high-speed steels (M2, M35, and M100). Strong mixing at tapping is used for the reduction of Cr_2O_3, V_2O_5, MnO, and FeO in the slag, which is gradually converted into pure white reducing slag in the subsequent processing in the LF. Although the basicity of the slag is low (about 2) desulphurization proceeds under this highly reducing slag to a minimum sulphur content of 20ppm. This good desulphurization can be partly attributed to the high activity of sulphur in tool steels containing high carbon. Figure 21 shows an example of an alloy tool steel (JIS SKD 11, similar to AISI D2) practice in EAF

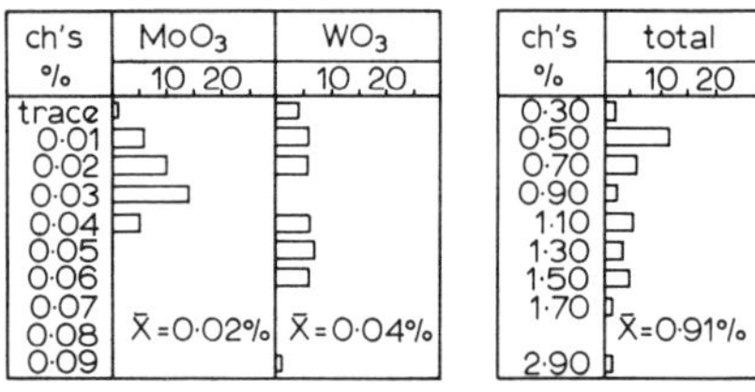

Fig. 23 - Six oxide contents in final LF slag by process 2A: steel quality M2 (Daido Ohmori)

Fig. 24 - Three oxide contents in LF slag after reladling of stainless heats by process 2B (Daido Ohmori)

and LF. Immediately after the meltdown of 17·5t scrap and alloys, slag reduction begins with the addition of Fe-Si fluxes to make the slag fluid. After 45 min from meltdown, steel and slag are together tapped into the LF and most of the reduction of oxides must be completed at this moment. Later, in the LF, the reduction of oxides proceeds with desulphurization, which can be observed from the change in slag chemistry of a total of 1400kg of slag shown in Fig. 22. One sample is enough to adjust steel chemistry as can be seen from the same values of the L1 and L2 samples. About 400kg of various alloys are then added and the final desired analysis is attained only 15 min after this addition (L3). It takes only 45 min in the EAF and 74 min in the LF, a total of 119 min. Table 7 shows the recovery improvements of this quality. A saving of Cr only is enough to offset the operation costs in the LF.

Figure 23 shows the contents of six reducible oxides in a final LF slag when processing high-speed steel (M2). The total contents of those oxides decrease to less than 1% on average. Higher content of V_2O_5, and thus FeO, resulting from the shortage of Fe-Si addition is due to larger slag volume exceeding 100kg/t by erosion of the furnace hearth. Care must be taken therefore with the shape of the furnace hearth before charging and any local wear must be repaired to make a spherical hearth.

The improvement in chemistry control is shown in Table 8 for steel M2, where chemistry deviation around the desired analysis is calculated for all the elements to be adjusted. A considerable decrease in the standard deviations of such heavy metals as W and Mo to 10 and 35%, respectively, should be noted. Before the application of LF, a 10t EAF was used for melting high-speed steels and Fe-W and Fe-Mo lumps were still observed on the hearth after tapping. In the LF process, these heavy alloys, which have higher melting points, are melted within 20 min owing to good stirring of the bath in spite of a lower bath temperature (about 1550°C).

PROCESS 2B

This process was used for two years until the installation of AOD in Ohmori and for more than 1500 heats of AISI 304 and 316, including 150 heats of ELC grades. The principle of refining is the same as process 2A, but the slag volume is so great after tapping into the LF that all the slag must be removed and fresh reducing slag must be added to the LF. The refining lasts for an average of 45 min under this slag with little carbon increase (1 ppm/min) and clean steel can be produced with exact steel chemistry and temperature. This process is not now used for austenitic grades, but may still be used for the refining of 13% Cr stainless for which the application of AOD is too expensive.

For AISI 304, an overall Cr recovery of 95·1% is obtained and the mean contents of Cr_2O_3, MnO, and FeO in slag after tapping are 2·2, 1·1, and 0·6%, respectively, as shown in Fig. 24. For heat-resistant steels for which a nitrogen addition is specified, nitrogen gas, instead of argon, can be injected both for alloying and stirring. A mean nitrogen yield of 30% can be obtained from a steel to which nitrogen has been added to 800ppm and a linear increase of N can be expected within this range.

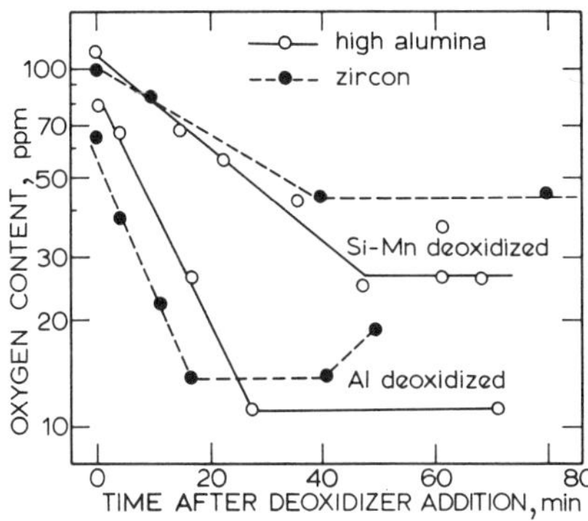

Fig. 25 - Effects of lining materials and deoxidizers on oxygen content (NSC Yawata)[4]

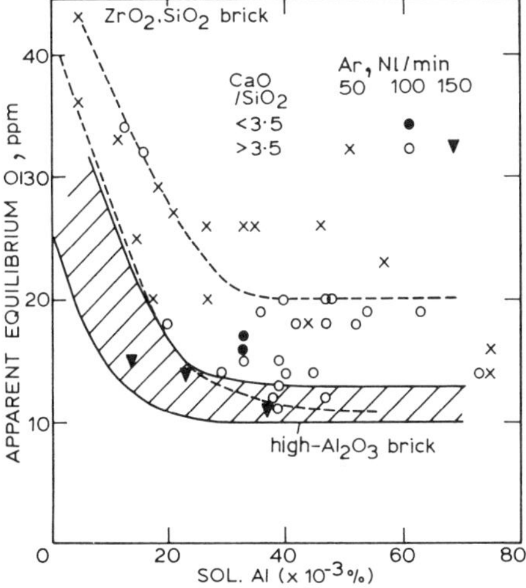

Fig. 26 - Relationship between apparent equilibrium oxygen and soluble aluminium[4,8]

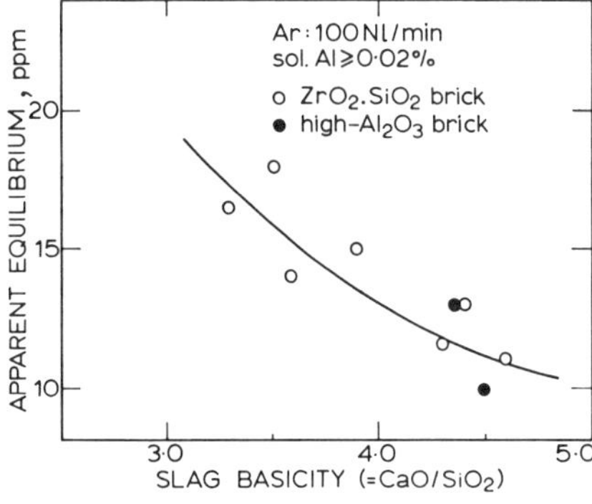

Fig. 27 - Effects of slag basicity on apparent equilibrium oxygen[4]

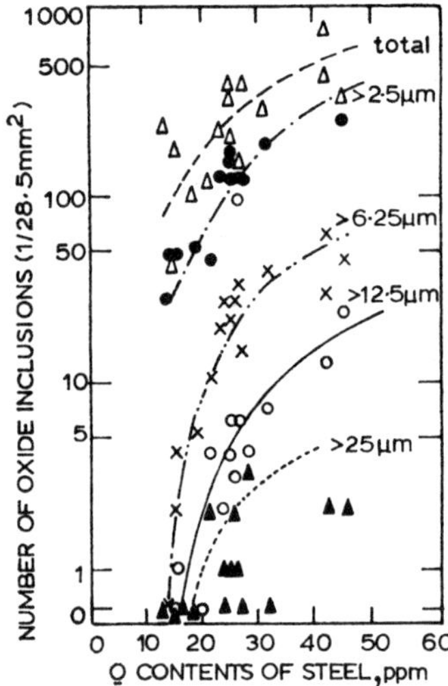

Fig. 28 - Relationships between oxygen contents of steel and size and number of oxide inclusions[4,7]

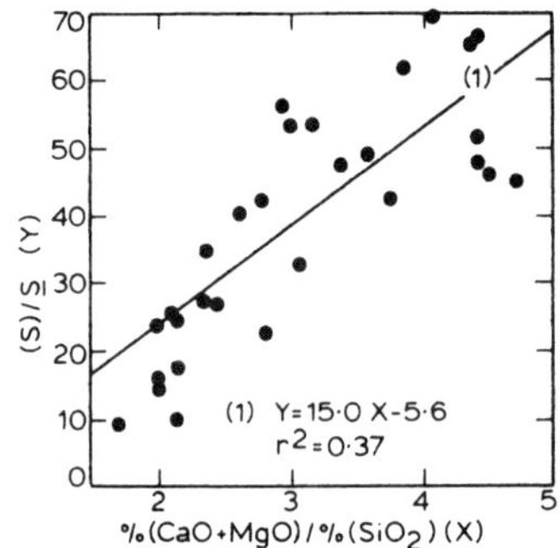
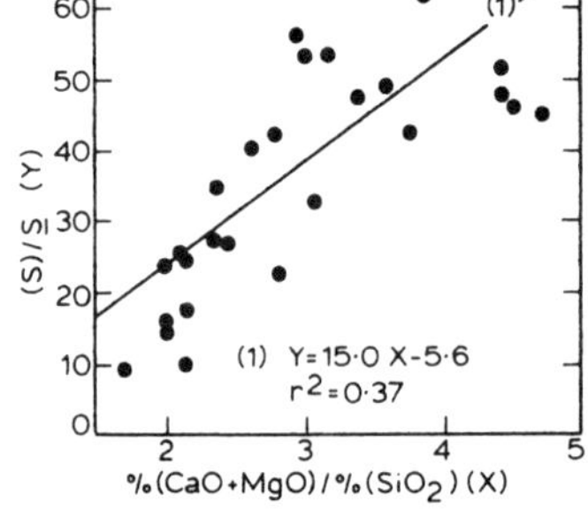

Fig. 29 - Sulphur distribution ratio (S)/S as function of slag basicity (Daido Ohmori)

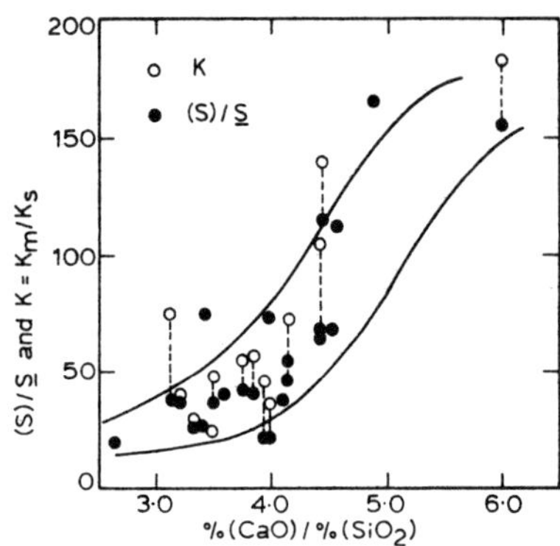

Fig. 30 - Sulphur distribution ratio (S)/S as function of slag basicity (NSC Yawata)[8]

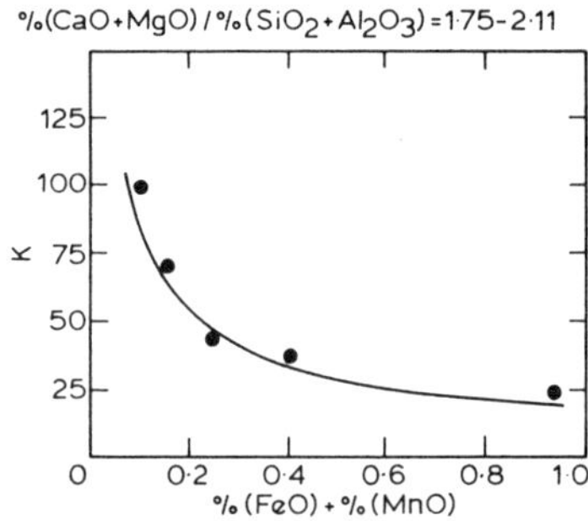

Fig. 31 - Effect of FeO + MnO on distribution ratio of sulphur (NSC Yawata)[3]

METALLURGICAL CONSIDERATIONS

The effects of several operating parameters on the behaviour of oxygen and sulphur in the LF process have been extensively investigated in NSC Yawata (2,4,7,8) using the 60t unit, and the oxygen and sulphur contents are related to the improvements in such mechanical properties as impact strength and reduction of area of heavy plates produced from an 18t ingot. Improvements in impact strength are also reported from Daido Ohmori, where inclusion shape is controlled by Ca addition at the end of LF processing, by means of Fe-clad Al-Ca wire. All the inclusions are converted into the globular type in the presence of 30-50ppm Ca and less than 30ppm S.

DEOXIDATION

The behaviour of oxygen in a 60t LF lined with either high-alumina or zircon bricks is shown in Fig. 25,[4] where the melt is deoxidized by either aluminium or silicon. In the alumina ladle, a lower oxygen content of 12ppm can be obtained and an apparent equilibrium can be attained within 30 min by Al, and within 50 min by Si deoxidation.

The deoxidation rate from around 100ppm to apparent equilibrium is faster by Al than by Si, 2·8ppm/min compares with 1·6ppm/min.

The factors affecting apparent equilibrium oxygen are shown in Fig. 26,[4,8] as a function of soluble Al content, in which stirring intensity varies from 50 to 150 l/min, i.e. from 127 to 382 W/m^3, in a zircon ladle. Apparent equilibrium can be attained at 0·03% Al or more in both the alumina and the zircon ladle. A lower oxygen content, however, is obtained from a higher stirring intensity. Not only brick quality and stirring intensity but also slag basicity have a marked effect on the apparent equilibrium; the higher the basicity the lower the oxygen, as is clearly shown in Fig. 27,[4] from which it can be seen that metallurgical relationships between metal and slag in the LF process are similar to those in the ESR process.

Figure 28[4,7] shows how the size and number of oxide inclusions decrease as the total oxygen decreases from 45 to 15ppm. The complete removal of inclusions larger than 12·5μm, at an oxygen level of 20ppm or less, should be noted.

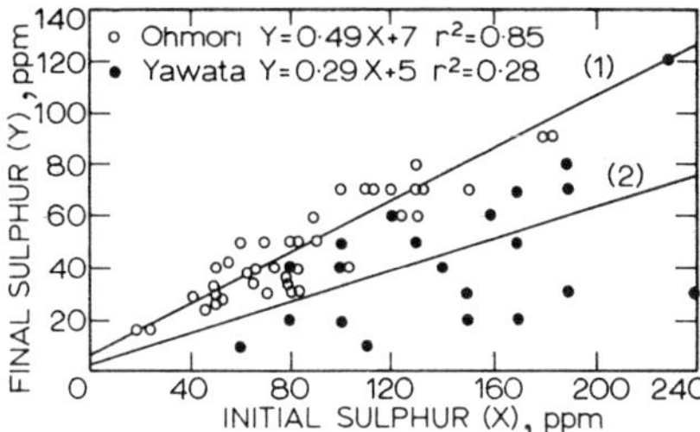

Fig. 32 - Empirical relationship between initial and final sulphur content (Daido Ohmori and NSC Yawata[4])

DESULPHURIZATION

In the LF process, desulphurization proceeds, in principle, by the mutual reaction between basic reducing slags and metal, and is therefore governed by the chemical and physical properties of the slag, especially its basicity, FeO and MnO contents, and the stirring energy. The effects of steel chemistry on the activity of sulphur in steel are not considered here.

Figure 29 shows an empirical relationship between sulphur distribution ratio (S)/$\underline{S}$ and slag basicity obtained from tool and constructional steels in the 20t unit in Daido. The amount of slag is about 4% and the bath is stirred by 30 l/min argon. In Fig. 30,[8] both the equilibrium distribution constant ($K=K_m/K_s$) and the ratio are plotted as a function of basicity (%CaO/%SiO_2) in the 60t unit in Yawata, from which it can be seen that desulphurization equilibrium cannot be attained in the ladle until a more intensified stirring is provided, as in the case of vacuum stirring (Fig. 20). Lower FeO and MnO contents, produced under a reducing atmosphere, give a higher K value as shown in Fig. 31.[8]

Figure 32[4] shows two empirical relationships between initial and final sulphur in Daido Ohmori and NSC Yawata, from which one can estimate the final sulphur content under fixed processing conditions in each steelworks. The better desulphurization in Yawata is probably due to stronger and more effective stirring intensity.

The sulphur content in steel is directly related to the total length of type A inclusions (JIS) in heavy plates in Yawata, as shown in Fig. 33,[4,7] in which results from three processes, namely, EAF, BOF, and BOF-LF, are summarized. The longer inclusion content from the BOF heats, shown as a broken line, is probably due to a higher oxygen content. The same strong correlation of $r^2=0.7$ between the sulphur and the type A inclusion content is obtained from the 20t heat in Daido.

IMPROVEMENTS IN MECHANICAL PROPERTIES

The decrease in type A inclusion length resulting from lower oxygen and sulphur increases the toughness of the steel, as shown in Figs. 34[4,7] and 35 from which the improvements in impact strength and reduction of area, respectively, in heavy plates can be clearly seen. In Fig. 34,[4,7] impact strength is measured on the samples taken from the surface of plates at a right-angle to the

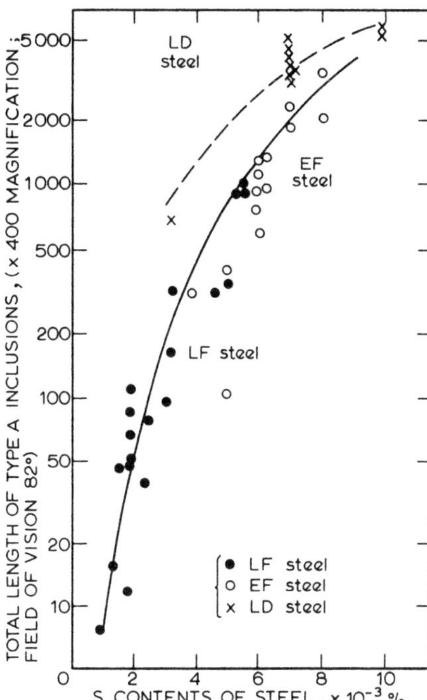

Fig. 33 - Relationship between sulphur contents in steel and total length of type A inclusions in products[4,7]

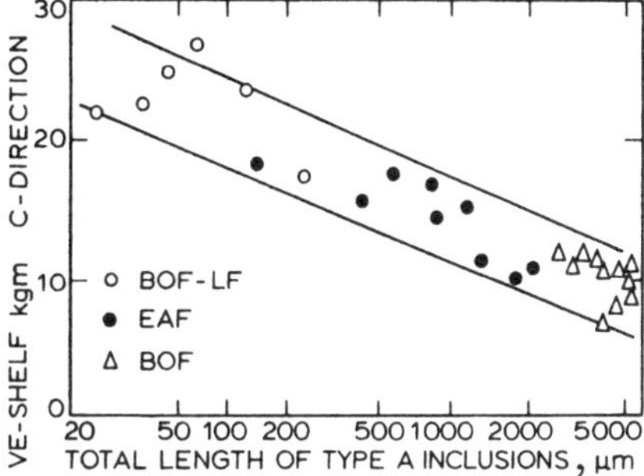

Fig. 34 - Relationship between impact strength and total length of type A inclusions: heavy plates (NSC Yawata)[4,7]

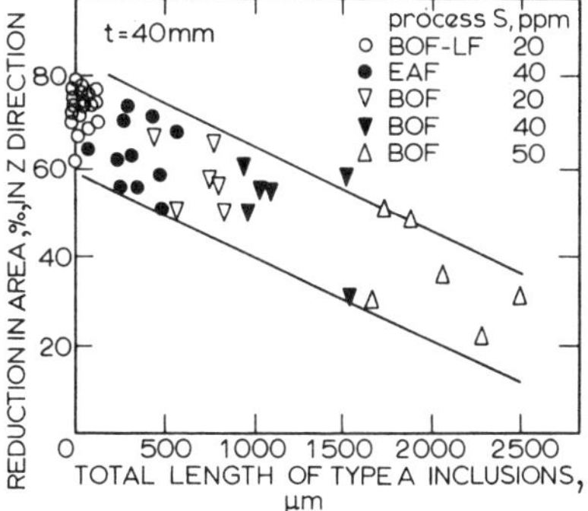

Fig. 35 - Relationship between reduction of area and total length of type A inclusions: heavy plates (NSC Yawata)[4]

Table 9 - Inclusion shape control by Ca addition in LF (Daido Ohmori)*

Heat no.	Contents, ppm Ca	S	Inclusion contents (JIS), % A	B	C	Total
50–7117	40	70	0·008	..	0·029	0·037
50–7132	30	60	0·017	..	0·013	0·030
51–790	40	30	..	..	0·046	0·046
51–795	60	20	..	..	0·033	0·033
			..	..	0·042	0·042
51–841	50	30	..	..	0·042	0·042

*Ca addition: 320 g/t; yield 15–25%.

Table 10 - Operating costs of 20t unit (Daido Ohmori, 1975)

Item	Quantity	Unit	Cost, £/t
Power	74·3	kWh/t	1·00
Electrodes	0·65	kg/t	0·42
Argon	113·0	l/t	0·7
Thermocouple	4·5	pieces/heat	0·15
Porous plug	3·8	heats/piece	0·19
Sliding-gate valve	2·6	heats/set	0·45
Refractories	12·54	kg/t	3·4
Refractories*	..	..	−1·38
Total			4·3

*Deduction of conventional ladle refractories cost including porous plug, pouring refractories, and personnel expenditure for bricklayers.

rolling direction and is almost double that produced by the EAF process. Elimination of anisotropy in heavy plates is clearly observed from the higher reduction of area in the Z-direction, perpendicular to the surface, in 40mm thick heavy plates. It should be noted that this value is nearly equal to those achieved by the ESR process.[9]

INCLUSION SHAPE CONTROL

To make large forgings for which impact strength at low temperature is specified, it is necessary to control the shape of inclusions by means of a Ca addition at the end of the LF processing. Ca is added in the form of Fe-clad Al-Ca wire through the slag layer in the last 2 min of processing. About 300 g/t Ca are added and 30-60ppm are detected in the samples after forging. Vaporization or oxidation losses of Ca during vacuum casting is not observed, and a mean yield of 15% can be expected.

Table 9 gives the Ca, S, and inclusion contents (JIS) in the forgings. In the range of sulphur contents $\leqslant$30ppm, all inclusions are converted into globular C type, containing mainly CaO and Al_2O_3. Al_2O_3 tends to precipitate in dendritic form in the absence of Ca, especially when sulphur decreases to less than 50ppm, and often causes the defects to be detected by ultrasonic inspection.

Through inclusion shape control by means of Ca addition, fracture appearance transition temperature (FATT) can be decreased by 10°-20°C and rejects by dendritic alumina can be eliminated.

OPERATING COSTS

The latest (1975) operating costs in the 20t unit in Daido Ohmori amount to £4·3/t, see Table 10; in that year, a total of 1131 heats were treated. The cost of refractories includes £0·43/t of personnel expenditure for bricklayers. The operating costs will become less if one takes the savings of power, electrodes, and refractories in the EAF by applying the LF process into account.

ACKNOWLEDGMENTS

The authors are indebted to Daido Steel Company Ltd. for permission to publish this paper and are also grateful for the consent of Nippon Steel Corporation, NSC Yawata Works, Japan Castings and Forgings Corporation, and Mitsubishi Steel Company Ltd. for the use of results of their investigations and practices. The authors would like to thank Professor S. Eketorp, Royal Institute of Technology, Stockholm, for his valuable advice in the early stages of developing the process.

REFERENCES

1. Paper presented by Japan Special Steel, No. 47 Special Steel Committee of ISIJ Joint Research Society, 17-18 May 1973.

2. Paper presented by NSC Yawata Works, No. 52 Special Steel Committee of ISIJ Joint Research Society, 16-17 Oct. 1975.

3. Y. Sundberg: personal communication, 1972.

4. H. Kajioka et al.: Nippon Steel Technical Report, to be published.

5. Paper presented by JCF Tobata Works, No. 54 Special Steel Committee of ISIJ Joint Research Society, 17-18 Oct. 1976.

6. Paper presented by JCF Tobata Works, No. 9 Electric Furnace Committee, 2nd Subcommittee, ISIJ Joint Research Society, 11-12 Nov. 1976.

7. N. Yamada et al.: Tetsu-to-Hagane (J. Iron Steel Inst. Japan), 1976, 62, (4), 241.

8. H. Kajioka et al.: ibid., 1976, 62, (11), 139.

9. K. Shimizu et al.: ibid., 334.

SECONDARY STEELMAKING BY THE ASEA/SKF- AND THE TN-PROCESS A COMPARISON

Bertil Tivelius Tomas Sohlgren

Dept of Process Metallurgy
SVENSKT STÅL
Oxelösund

This paper originally appeared in the Steelmaking Proceedings, Vol. 61, 1978, p. 154.

ABSTRACT

At the steelwork of Oxelösund secondary steelmaking is nowadays a very important tool for successful processing of a great deal of our steel program. The development of so called Z-steel, with good mechanical properties in the through thickness direction, started at the end of the sixties and required an extremely low sulphur content and good control of the amount and shape of the non-metallics. Ladle refining became necessary and the obvious choice of process was at that time the ASEA/SKF, which was taken into operation in 1971. The process was easy to install into an integrated steel production line because of the electric arc heating facilities. In 1974 we predicted that the ladle refining capacity had to be increased by the late seventies. At this time the injection technique had been satisfyingly developed and the positive effect of Ca on material properties and castability was well documented. Furthermore due to the significantly lower installation costs the choice of the TN system was made, this being commissioned in the beginning of 1977. We now operate both processes and have good possibilities of making relevant comparisons. Our experiences show that they have almost the same operating reliability and that the metallurgical results are equivalent and in some cases better for the TN-treated heats. In this paper we describe and discuss differences regarding oxygen- and sulphur refining, hydrogen removal during degassing in the ASEA/SKF and the nitrogen pick up during the TN-process. The possibility of obtaining the required analysis and temperature is still somewhat better for the ASEA/SKF but the difference is almost negligible. Refractory consumption is high for both processes but can be lowered by proper predeoxidation giving a low FeO-content in the slag before treatment. Rinsing with oxygen during casting has almost disappeared for the TN-treated heats and the material properties are probably better where toughness is concerned. In spite of significantly higher production costs for comparable grades we still use the ASEA/SKF due to our current limited diffusion heating capacity for hydrogen removal. Moreover the deep drawing grades requiring extremely low contents of interstitials, C and N, must be produced in the ASEA/SKF ladle furnace.

INTRODUCTION

Oxelösund Steelworks were erected in 1959 in order to supply the Swedish ship yards with heavy plate. The company is integrated and covers all production resources from mine to finished products. The Steel Mill is situated in Oxelösund on the Baltic and consists of coke plant, blast furnaces, steel melting shop and rolling mills.

In order to meet the increasing demand from the market for steel grades with tougher requirements we installed in April 1971 an ASEA/SKF ladle furnace in the Steel Mill.

Among steel grades processed in this unit can be mentioned Quenched and tempered (QT) steels, LPG-, LNG-grades, so called Z-grades (ref 1), pipe-line steels (X-70) and weather resistant steels.

To meet the severe requirements specified for these steel grades the conventional steel processes (BOF, electric arc furnace, Siemens Martin) are not well suited because of the following circumstances:

1. The accuracy and reproducability in the analysis obtained in the steelmaking are most essential. The conventional steelmaking processes do not fulfill these requirements.

2. To reach good cold bending, impact and through thickness properties it is necessary to control the content as well as the shape of non-metallic inclusions (oxides and sulphides). This is not possible using the conventional steelmaking processes.

To further increase the production capacity of such steel grades a Thyssen-Niederrhein, TN-process (ref 2), was installed in February 1977.

The purpose of this paper is to give a short comparison between these two ladle refining processes in respect of metallurgical results and technical economical considerations.

TECHNICAL LAY-OUT

The steel is nowadays supplied from a 150 ton LD-converter. The ASEA/SKF unit is of the two station type. One station is used for electric arc heating and alloying and the other for degassing. The ladle is transported between these two stations on a ladle car. To provide stirring of the steel bath the ladle car has two linear induction stirrers.

Vacuum is provided by a set of steam ejector pumps.

The TN unit is located on line with the LD-converter as opposed to the ASEA/SKF-unit, which stands half way between the LD and the continuous caster. The same ladles are used for both these secondary steel making processes. During TN-treatment the ladle is placed on the LD ladle car. At the TN-unit the steel is treated by calcium-injection with argon as carrier. Alloying and scrap cooling facilities are also provided at the TN station. Adjacent to the TN-unit we make the injection lances in a specially designed equipment.

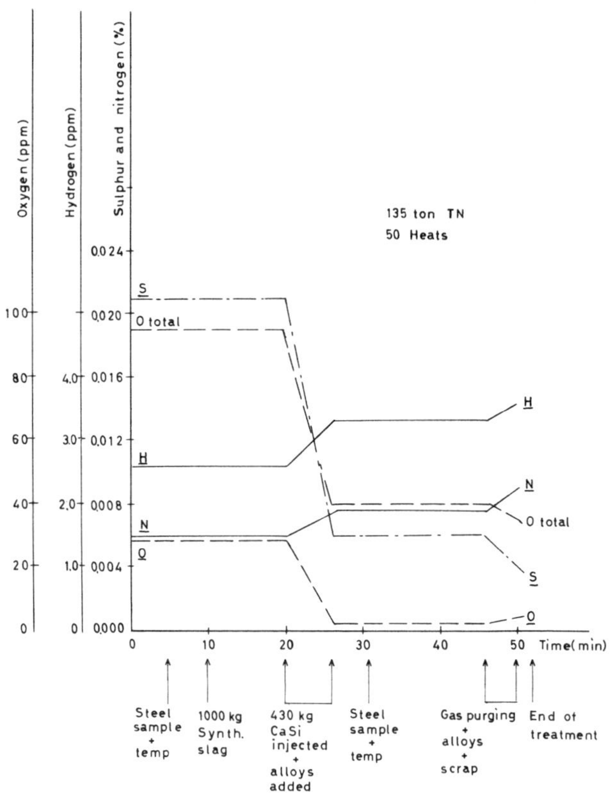

Fig. 2 - Hydrogen, oxygen, sulphur and nitrogen contents as a function of time

TYPICAL TREATMENT

The contents of hydrogen, oxygen and sulphur and their development related to time and the consecutive operating activities for a vacuum degassed ASEA/SKF heat are shown in figure 1.

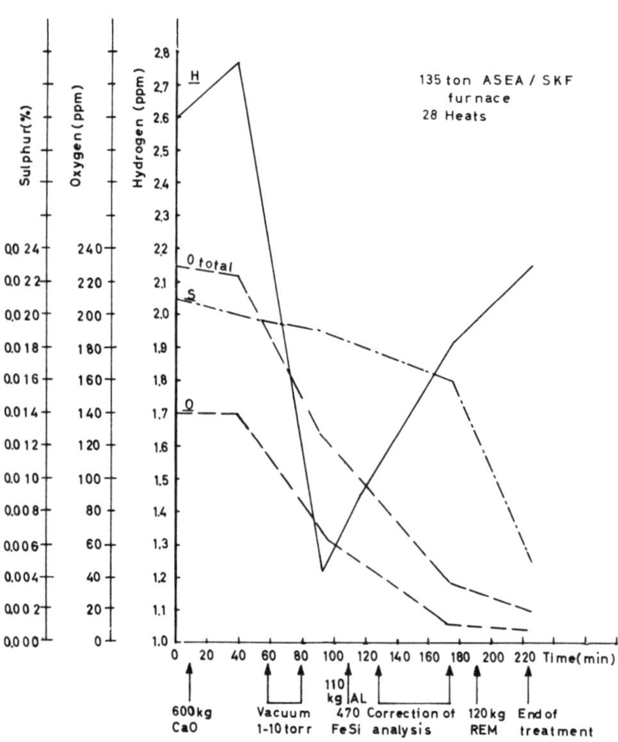

Fig. 1 - Hydrogen, oxygen and sulphur contents as a function of time

A corresponding diagram for a typical TN-heat is given in figure 2.

As can be seen from the diagrams there are large differences in processing time. This and the details in the process routes are discussed below.

OXYGEN - AND SULPHUR REFINING

In the ASEA/SKF-unit the steel is normally degassed in order to lower the hydrogen content. The prerequisite for hydrogen removal is the occurrance of a heavy carbon boil during the degassing. This is only achieved when the steel is completely unkilled on arrival at the ASEA/SKF-unit.

The carbon-oxygen reaction in molten steel under reduced pressure has been the subject of numerous investigations over many years. Most factors influencing this reaction are well known and will not be considered here. However it could be interesting to have a short look at the carbon-oxygen product (figure 3).

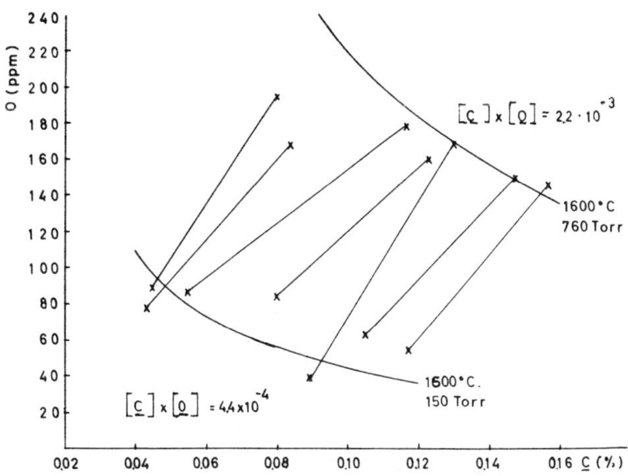

Fig. 3 - Soluble oxygen content before and after degassing as a function of carbon content ASEA - SKF - process

For some representative heats the carbon and oxygen contents are shown before and after degassing. Measurements pertaining to the same heat are connected with a straight line. Before degassing, the carbon and oxygen contents are in reasonable agreement with the equilibrium curve at 1600°C down to 0.12 % carbon. At lower levels manganese (>1 %) becomes a stronger deoxidizer than carbon. After degassing the obtained values correspond well to the carbon-oxygen equilibrium at 150 torr. This result does not differ significantly from other large-scale vacuum degassing methods. When the vacuum treatment is finished the steel is killed with aluminium.

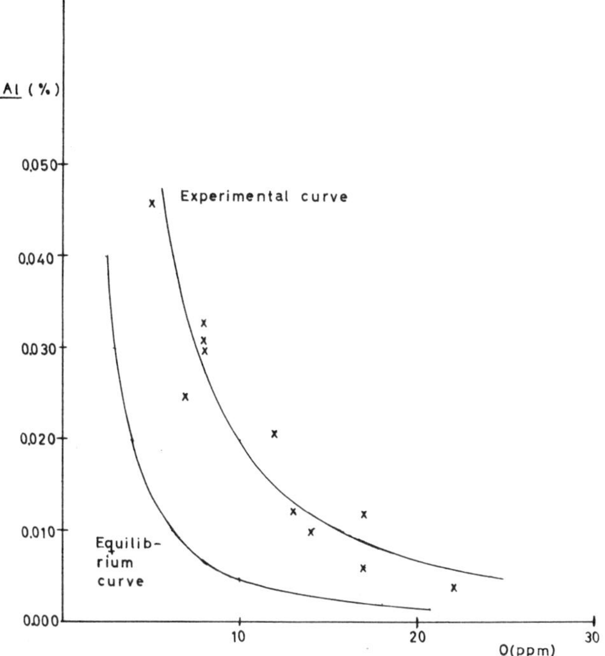

Fig. 4 - Soluble oxygen contents at different contents of soluble aluminium ASEA - SKF - process

The resulting oxygen content as a function of the aluminium content is shown in figure 4. The supersaturation of oxygen and aluminium is surprisingly low.

During sampling and analysis adjustment a certain desulphurization by slag refining occurrs according to the reaction:

$$\underline{S} + (O^{2-}) \rightleftarrows (S^{2-}) + \underline{O}$$

However this is a rather slow process in the ASEA/SKF ladle furnace.

The main removal of sulphur is therefore done by the addition of rare earth metals (REM).

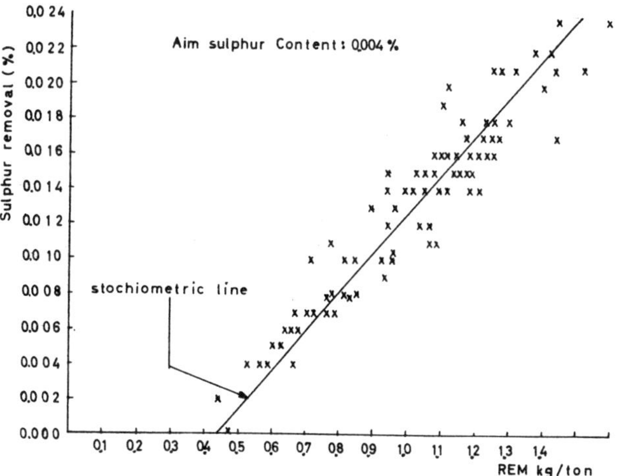

Fig. 5 - Sulphur removal as a function of REM addition ASEA - SKF - process

Figure 5 shows sulphur removal (ΔS %) against the specific amount of REM added. The stochiometric line represents the formation of (RE)S. According to the diagram the steel requires 0.44 kg REM/ton before reacting. We have calculated a need of 0.24 kg REM/ton in equilibrium with 0.004 % sulphur and the remaining 0.20 kg REM/ton is consumed by further reduction of the slag.

The removal of REM sulphides is a fairly slow process and takes about 30 minutes. This becomes clear from figure 6, which shows the change in composition of slag and steel during the stirring after REM addition. The consequence of further stirring is evident. Oxidation from air raises the FeO-content of the slag and resulphurizes the steel.

Heats to be treated at the TN-station are predeoxidized with Al during tapping from the LD-furnace (average 2 kg Al/ton).

When the steel arrives at the TN-station

after some minutes we measure the temperature and take a steel sample. The soluble oxygen content before injection versus the corresponding soluble Al-content is found in figure 7. Through improved deoxidation routines at the LD furnace we mostly get sol. Al-contents in the range of 0.025 % resulting in soluble oxygen contents around 30 ppm. After CaSi-injection the oxygen content falls below 5 ppm independently of the Al-content.

Contrary to the soluble oxygen the total oxygen content after CaSi-injection is strongly influenced by the oxygen content before the treatment (figure 8). One reason could be a limited condition for separation of oxides during the short time available.

On the other hand heats with high oxygen contents before treatment also have excessive contents of FeO in the ladle slag. This leads to a successive formation of oxide inclusions during a relatively long time of the treatment.

The latter explanation is probably the most plausible one.

With 30 ppm soluble oxygen before injection, which corresponds to 95 ppm total oxygen, we get an oxygen content of 35 ppm after CaSi-injection.

To sum up, the furnace is required to deliver a well deoxidized steel bath and a well reduced ladle slag. A successful desulphurization is then achieved when a proper amount of CaSi is injected and a calculated minimum amount of synthetic top slag is added.

A simple mass balance of sulphur gives the required amount of synthetic slag according to the four relationships:

$$M_{STEEL} \times \Delta S + M_{SYNTH} \times S + M_{FSLAG} \times S = M_{ESLAG} \times \left(\frac{(S)}{S}\right) \times S_{AIM} \quad (1)$$

$$\left(\frac{(S)}{S}\right) = K \times Bas \quad (2)$$

$$Bas = \frac{M_{CaO} + M_{SYNTH} \times CaO_{SYNTH}}{M_{SiO_2} + M_{SYNTH} \times SiO_{2\,SYNTH}} \quad (3)$$

$$M_{ESLAG} = M_{SYNTH} + M_{LINING} + M_{FSLAG} - M_{REACTION} \quad (4)$$

where

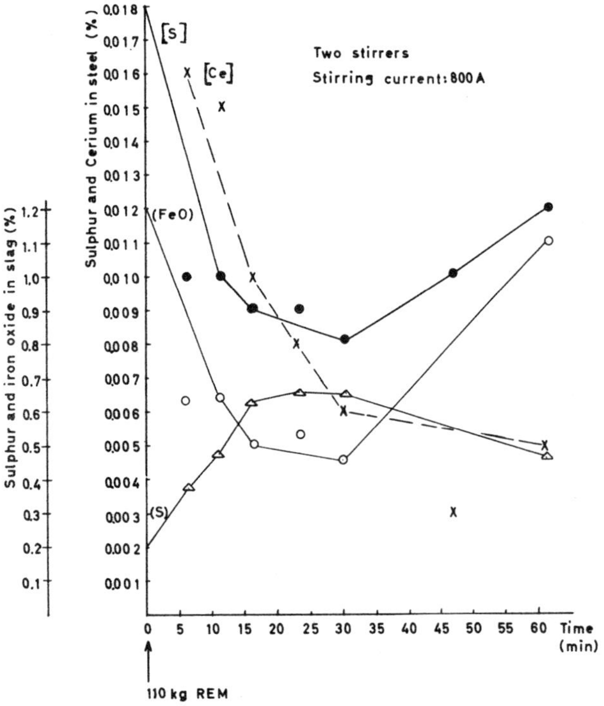

Fig. 6 - Changes in analyses of steel and slag after REM addition
ASEA - SKF - process

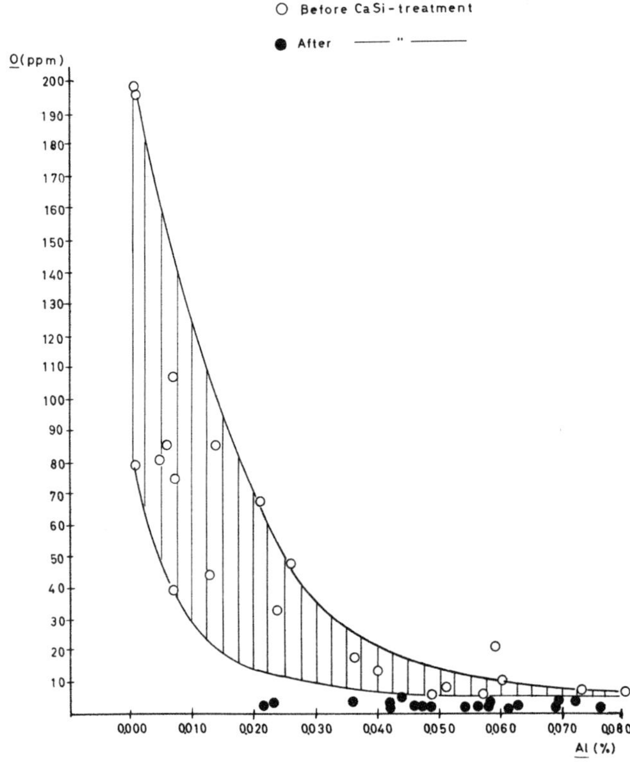

Fig. 7 - Oxygen (solute) as a function of Al (solute) before and after CaSi-treatment TN-process

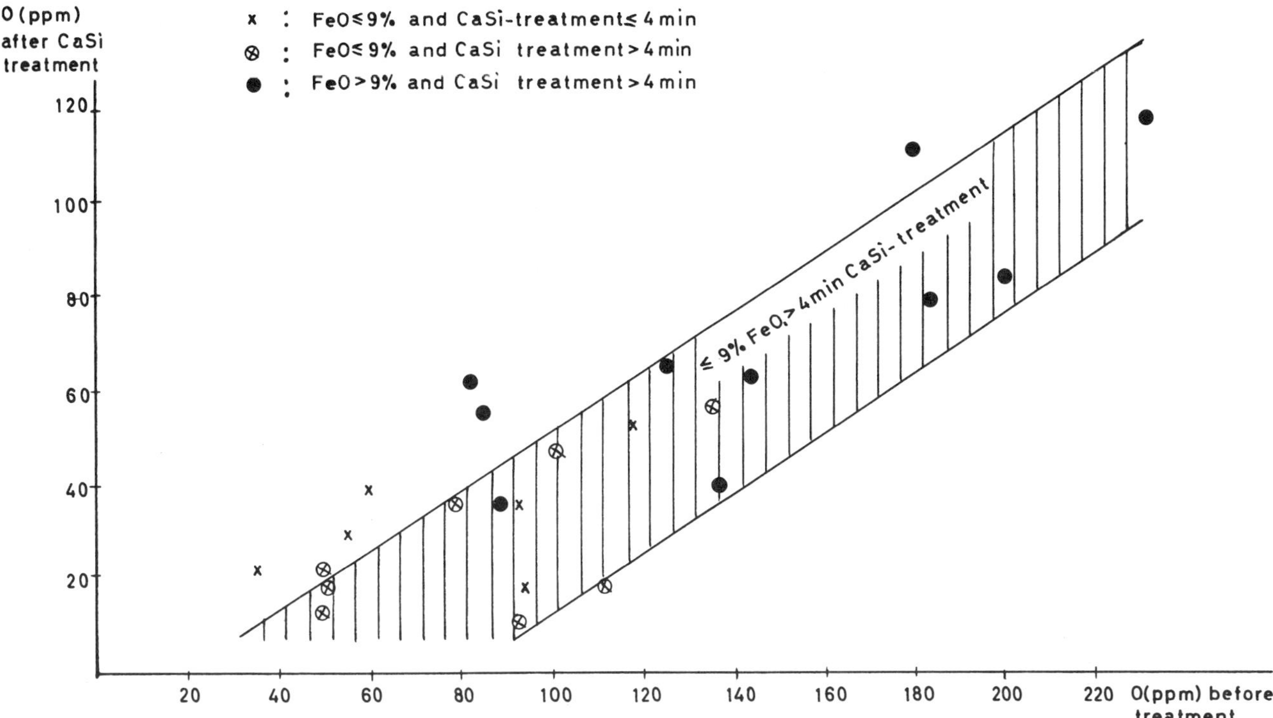

Fig. 8 - Relation between total oxygen content before and after treatment. TN process

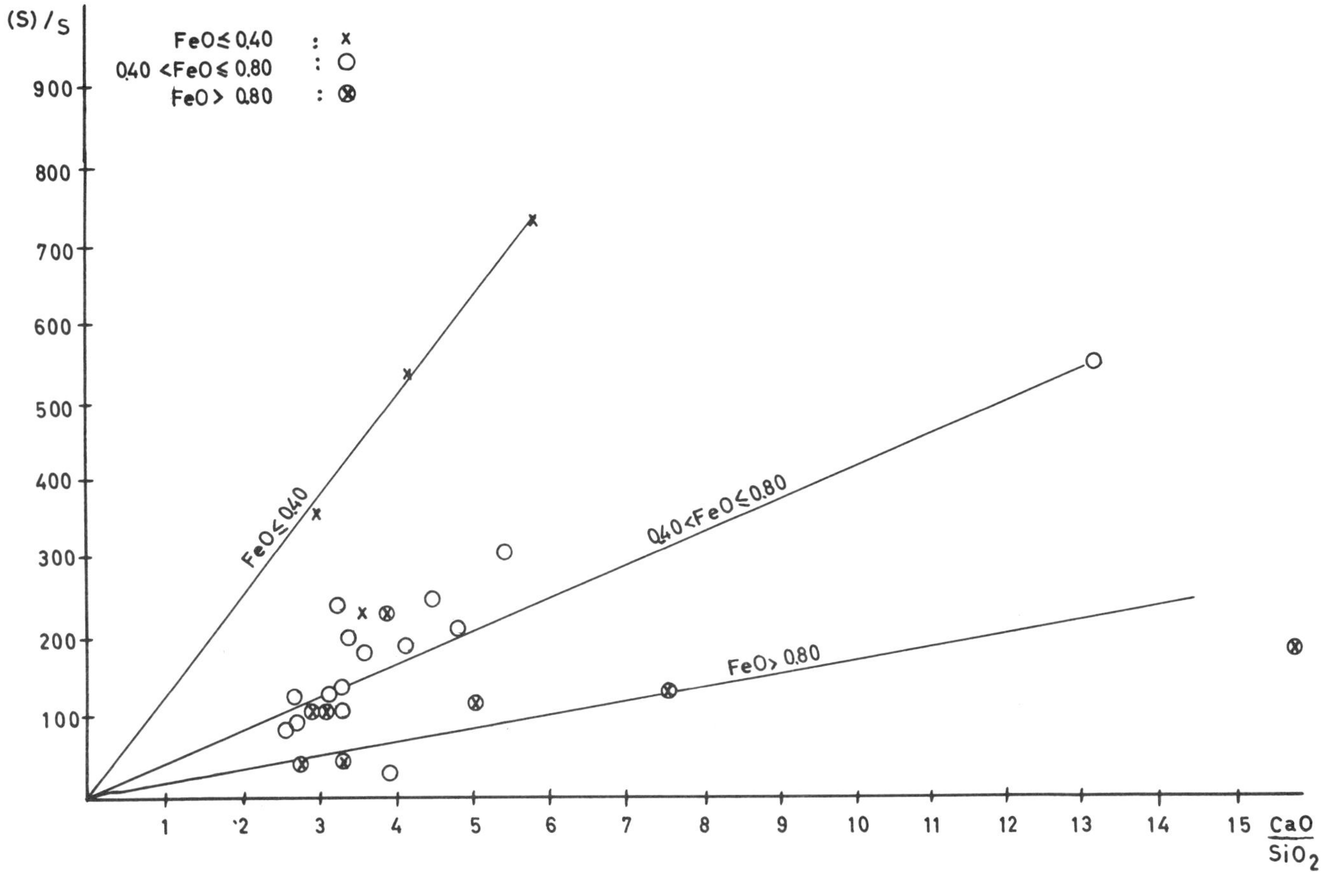

Fig. 9 - Sulphur ratio slag/steel as a function of basicity and FeO-content (sulphur saturated slag) TN-process

M_{STEEL} = Steel weight (kg)

M_{SYNTH} = Amount of synthetic slag (kg)

M_{FSLAG} = Amount of slag from LD-furnace (kg)

M_{ESLAG} = Amount of slag after treatment (kg)

M_{CaO} = Amount of lime from furnace slag and lining (kg)

M_{SiO_2} = Amount of SiO_2 from furnace slag and lining (kg)

M_{LINING} = Amount of lining wear (kg)

$M_{REACTION}$ = Decrease of slag amount on account of steel/slag reactions (kg)

(FeO and MnO reduction)

ΔS = Removed sulphur in steel (fraction)

S = Sulphur content in synthetic slag resp furnace slag (fraction)

$\frac{(S)}{S}$ = Sulphur distribution between slag and steel

S_{AIM} = Aimed sulphur content after treatment (fraction)

K = Inclination coefficient

Bas = Slag basicity

CaO_{SYNTH} = Lime content in synthetic slag (fraction)

SiO_{2SYNTH} = SiO_2 content in synthetic slag (fraction)

From the equations 1-4 the amount of synthetic slag M_{SYNTH} is solved. Hereby M_{ESLAG}, $\frac{(S)}{S}$ and Bas are regarded as dependent variables. The other parameters are to be handled as constants.

The inclination coefficient (K) is derived from the diagram in figure 9, which points out the strong influence from basicity and FeO-content on the final sulphur distribution between slag and steel.

The calculation of the needed amount of synthetic slag is computerized. Thereby the relationship in the figure 9 for $0.40\ \% < FeO \leq 0.80\ \%$ is normally used. In case of unsufficiently killed steel from the LD ($\leq 0.010\ \%$ Al) the curve corresponding to $FeO > 0.8\ \%$ is utilized. In table 1 is shown the effect of varying degree of predeoxidation and amount of LD-slag when desulphurizing 135 ton of steel from 0.026 % down to 0.004 % S.

The need of Ca injected at various degrees of desulphurization is shown in figure 10. The effect of insufficient amounts of synthetic slag added according to the above mentioned formula is evident.

Table 1

Al (%)	Slag from LD (kg)	Calculated synth slag (kg)
> 0.010	500	1125
> 0.010	2000	1300
≤ 0.010	500	1900
≤ 0.010	2000	2475

Table 2

		ASEA/SKF	TN
Oxygen total after treatment	ppm	25(10-55)	35(10-65)
Sulphur total after treatment	%	0.004 ± 0.002	0.004 ± 0.002
Lowest achieved sulphur content	%	0.001	0.001
REM (100 %)	kg/ton	1.4	-
CaSi (29 % Ca)	kg/ton	-	3.7
Desulphurization time	min	30	10*/
Costs for desulphurizer	$/ton	9.20	4.70

*/ Including 4 min argonpurging after finished CaSi-injection

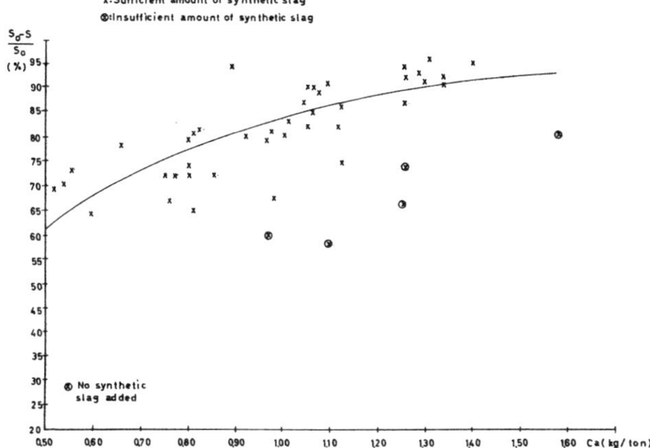

Fig. 10 - Desulphurization as a function of amount of Ca blown. TN process

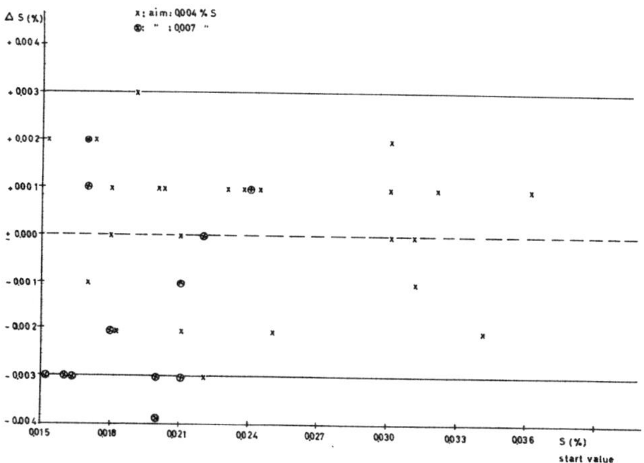

Fig. 11 - Deviation from aim sulphur content as a function of start sulphur content. TN-process

With the described procedure we achieve good precision in obtaining the aimed sulphur content, ± 0.002 % (aim 0.004 %), independently of the start values (figure 11).

A comparison between the results obtained when using Rare Earth Metals in the ASEA/SKF and CaSi in the TN regarding deoxidation and desulphurization from 0.025 to 0.004 % S is given in table 2.

STIRRING

Stirring is without any doubt the most important tool of ladle metallurgy. The forced stirring of molten metal provides rapid and efficient slag-bath reactions, homogenization and removal of non-metallic inclusions. A short comparison between our two processes in this respect can therefore be interesting as the stirring is carried out in quite different ways.

In the ASEA/SKF ladle furnace the stirring is provided by two straight induction stirrers, both operating upwards. The stirring in the TN-injection ladle is provided by the carrier gas, argon, and the evaporation of Ca-metal into bubbles.

From a metallurgical point of view it is also desirable to define a quantity describing the stirring intensity irrespective of the stirring method used. One way is to use the volume power density i.e. the specific stirring power of the melt. When calculating the specific stirring power according to Sundberg (ref 3) and Lindskog (ref 4) we found the figures in table 3 below.

The effect of stirring through evaporation of Ca is difficult to evaluate because of gas/steel reaction phenomena.

The total specific stirring power in TN is probably in the order of 1000-2000 W/m^3.

Our metallurgical results indicate quite clearly that power density does not generally describe the quality of the stirring.

The large differences in specific stirring power and type of stirring between ASEA/SKF and the TN-processes explain to a certain extent the obvious difference in desulphurization rate in table 2. However the sulphur is removed by means of different desulphurization agents, the reaction products of which probably have different separation rates.

The heavy agitation when using injection technique thus favours efficient slag/steel reactions, but has the disadvantage of higher pick up rate of hydrogen and nitrogen from air.

HYDROGEN

Hydrogen is efficiently removed during the vacuum degassing period in the ASEA/SKF-ladle furnace. Carbon monoxide bubbles form at lining cavities in the upper part of the ladle wall and control to a great extent the hydrogen removal (figure 12). In average, carbon drops by 0.04 % and hydrogen from 2.8 to 1.2 ppm (figure 1).

Table 3

Process	Specific stirring power W/m^3
ASEA/SKF	100
TN	Carrier gas 800 + Ca-bubbles max 1200

Table 4

Process	Hydrogen (ppm) in liquid steel
ASEA/SKF and vacuum	2.2
ASEA/SKF without vacuum	3.6
TN	3.6

Lowest achieved carbon and hydrogen contents in our ASEA/SKF-ladle furnace are 0.010 % and 0.5 ppm respectively.

However an increase of the hydrogen content after degassing is hard to avoid due to a hydrogen pick up from the atmosphere and the alloys.

Anyhow, a low content after degassing corresponds to a low final content (figure 13). The pick-up of hydrogen from the atmosphere is significantly larger, when only one stirrer is used. The reason is that the slag is driven aside exposing the steel surface to the air, which does not happen when both stirrers are in operation. At the end of the treatment we reach an average of 2.2 ppm due to the above mentioned pick up.

At the TN-station the hydrogen pick up averages 1 ppm. The comparable pick up at the ASEA/SKF after degassing is also 1 ppm (figure 1). A comparison regarding the final contents of hydrogen for the two processes is found in table 4 below.

The amounts of alloys and slag added are the same at the two equipments. Thus the hydrogen pick up from air is the same for both processes (0.6 ppm) despite 120 minutes of stirring at the ASEA compared with only 10 minutes stirring at the TN. The most plausible reason for the higher rate of hydrogen pick up from air for TN is the large difference in specific stirring power and type of stirring discussed earlier.

In Oxelösund it has been shown that hydrogen contents exceeding 1.5 ppm in solid material cause a detrimental increase of ultrasonic rejections for heavy plates. Small hydrogen cracks then appear in the centre and they are easily detected at ultrasonic testing.

According to table 4 it is necessary to perform a hydrogen diffusion treatment on solid material in order to avoid this cracking. Depending on the higher hydrogen content in TN-treated material the diffusion time must be 60 % longer compared to vacuum degassed steel.

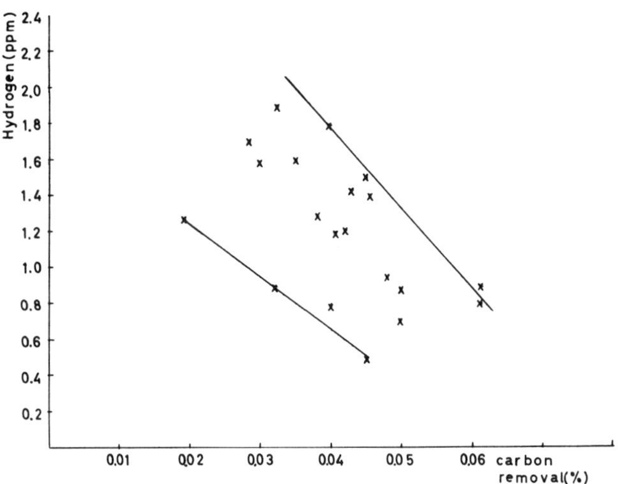

Fig. 12 - Hydrogen contents immediately after degassing as a function of carbon removal
ASEA - SKF - process

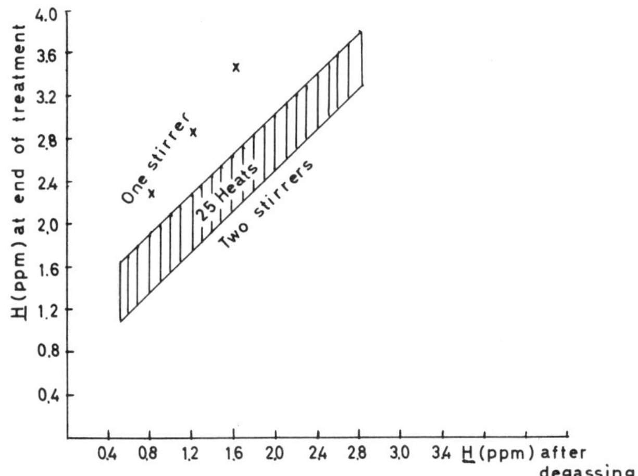

Fig. 13 - Hydrogen content at end of treatment as a function of hydrogen content immediately after degassing
ASEA - SKF - process

NITROGEN

The heavy stirring effect from the injection of argon and CaSi at the TN-treatment leads to a certain increase in nitrogen content. The amount of slag in the ladle strongly influences this pick up of nitrogen from air on account of its protective effect. This can be seen in figure 14 from which it's also evident that low sulphur content strongly promotes nitrogen pick up. This can be brought back to the well-known property of sulphur and oxygen as surface active elements.

For ASEA/SKF-treatment the pick up of nitrogen according to table 5 is negligible, which is not the case for TN-heats.

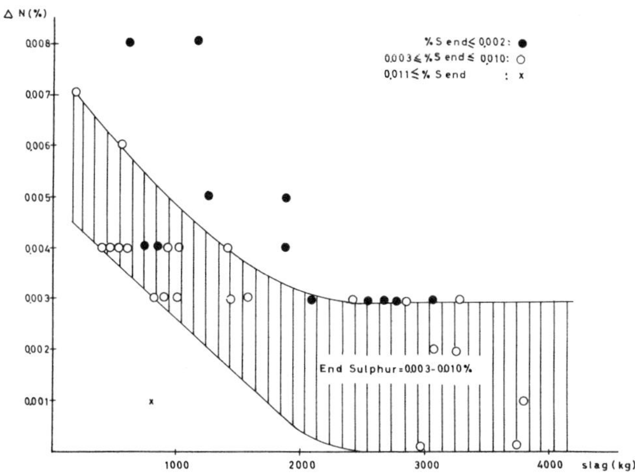

Fig. 14 - Nitrogen pick up during CaSi-injection as a function of the amount of furnace slag in the ladle. TN-process

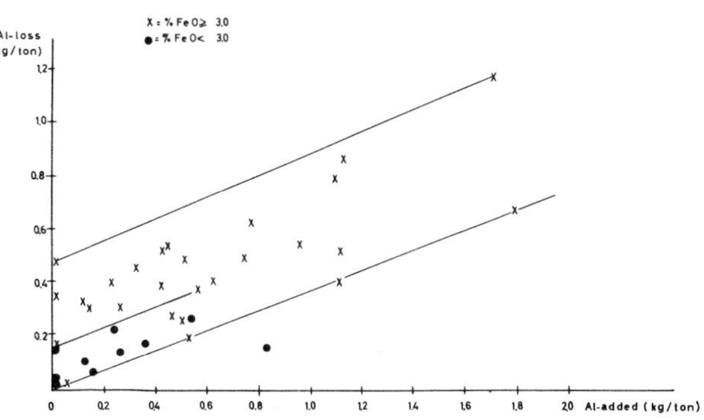

Fig. 15 - Al-loss as a function of added amount of Al/during CaSi injection. TN-process

Table 5

Process	Nitrogen (%) in liquid steel
Normal gas purged LD-steel	0.006
ASEA/SKF treated steel	0.007
TN treated steel	0.009

Table 6

Process	Element % standard deviation 2σ									
	C	Si	Mn	S	Cr	Mo	Cb	B	Al	Ti
ASEA/SKF	0.02	0.08	0.09	0.002	0.05	0.04	0.006	0.001	0.016	0.007
TN	0.01	0.05	0.10	0.002	0.08	0.04	0.006	0.001	0.014	0.008
Aim analysis	0.14	0.25	1.35	0.004	0.40	0.25	0.025	0.002	0.030	0.040

Table 7

Process	Temperature standard deviation 2σ °C
ASEA/SKF	4
TN	8
Aim temperature °C	1580

CORRECTION OF ANALYSIS AND TEMPERATURE

The precison in chemistry and teeming temperature is compared in table 6 for our ladle refining processes.

There is no obvious divergence in analysis precision between the two processes, which is somewhat surprising. One reason is the computerized alloy calculations at the TN compared to manual calculations at the ASEA/SKF. The greatest analysis problem for both processes is to hit the aimed Al-content. Hereby the FeO-content and the amount of slag control the burn off of aluminium. This is evident from figure 15, which for TN-heats shows that a small and even Al-loss is achieved when FeO-content is low in the ladle slag before treatment.

As far as temperature is concerned the advantage of energy supply in the ASEA/SKF ladle furnace is clearly demonstrated in table 7.

The temperature precision at TN is for the moment considered satisfactory and is achieved by preheating the ladles to 1000-1200°C and superheating the steel in the LD-furnace by 35°C relative to normal gas-purged heats. Theoretically we need only 20°C superheating but to get a satisfying accuracy in teeming temperature we add a further 15°C and scrap cool at TN when necessary.

REFRACTORY CONSUMPTION

We use the same ladles at TN- and ASEA/SKF-treatments for practical and metallurgical reasons. Bottom, underwall and slag line are bricked with tar or ceramic bonded dolomite. The best result is obtained with the tar bonded, tempered, variant. Above the slag line we use high alumina bricks. Between these two materials we have a layer of magnesite-chrome in order to avoid the dolomite to react with the high alumina. We have found this lining composition to be the best one regarding process requirements and refractory costs. However these dolomite bricked ladles must be well preheated and may not be cooled down between the treatments below 800°C in order to avoid steel penetrations into the joints, which detrimentally reduces the lining life.

The average costs are compared in table 8.

The relationship between lining wear and FeO-content in the slag before treatment is plotted in figure 16 for TN-heats. Basically the same relation should be valid also for ASEA/SKF heats.

The steel is unkilled at the vacuum degassing why the FeO-content is very high (about 20 %) and thus explains a great deal of the difference in lining wear compared to the TN-treatment where FeO-content is in the order of 7 % before the injection.

The longer oparating time and the influence from the electric arc heating can also have some effect on the refractory consumption.

The lance for injection of CaSi at the TN is lined in order to resist one treatment. The lining consists of fire clay stopper sleeves and is in the slag line upgraded with high alumina. The total cost for lances are given in table 8 below.

CASTING

Aluminium fine grained steels have a strong tendency to give casting problems owing to the existence of aluminium oxides.

By the injection of Ca these oxides are transformed into calcium aluminates and many of the previous problems with clogging in the nozzles are eliminated.

This has also been confirmed at our plant and is illustrated for up-hill ingot casting in table 9.

The difference in castability in favour of TN is to a great extent explained by the absence of oxygen rinsing due to clogging in the nozzle.

MATERIAL PROPERTIES

The most frequent grades processed in TN have so far been the LPG (Liquid Petroleum Gas) materials. A very important property in these grades is the notch toughness as these steel grades are used for low temperature applications. A comparison

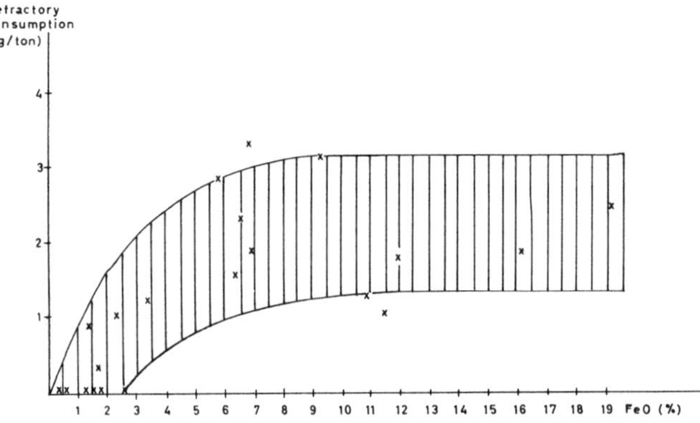

Fig. 16 - Ladle refractory consumption as a function of FeO-content in slag before treatment. Blowing time not exceeding 9 min. TN-process

Table 8

Process	Average number of heats	Lining costs $/ton	Lance costs $/ton
ASEA/SKF with vacuum	14	2.77	-
ASEA/SKF without vacuum	18	2.13	-
TN	18	2.13	0.69

Table 9

Process	Tonnage teemed without problem %	Oxygen rinsing of ladle nozzle %
ASEA/SKF (REM)	88.0	4.4
TN (CaSi)	93.1	0.0

Table 10

Process	Number of heats/tests	Toughness average (Joule)	Standard dev. 1 σ (Joule)
ASEA/SKF (REM)	177/3495	96	44
TN (CaSi)	41/756	121	59

Table 11

Type of cost	Costs $/ton liquid steel	
	ASEA/SKF + vacuum	TN
Refractories ladle	2.77	2.13
lance	-	0.69
Energy el power	1.66	-
electrodes	0.89	-
Superheating in LD-converter	-	1.23
Slag lime	0.10	-
synthetic	-	0.96
Desulphurizer REM	9.20	-
CaSi	-	4.70
Media water, steam	0.43	-
argon	-	0.41
Maintenance	0.96	0.64
Labour	1.17	0.30
Diffusion heating	0.16	0.26
Production costs	17.34	11.32
Capital costs at 200.000 ton/year	6.65	1.33
Total costs	23.99	12.65

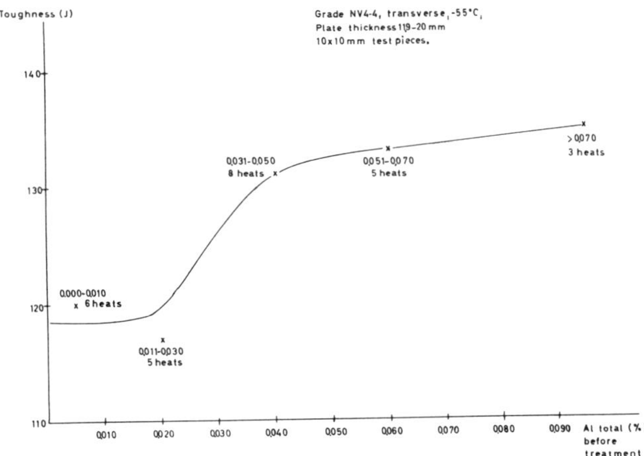

Fig. 17 - Toughness as a function of aluminium content before treatment. TN-process

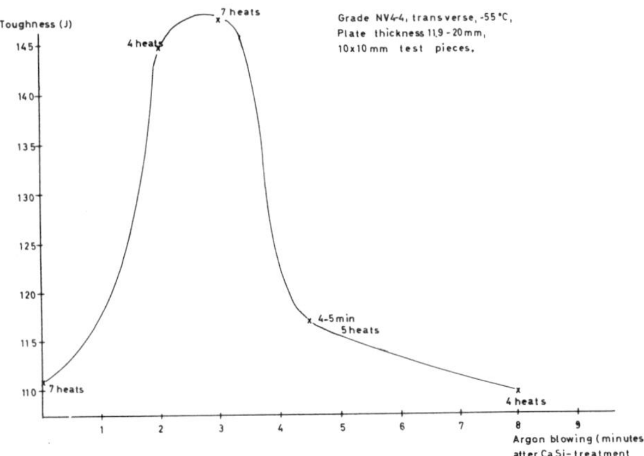

Fig. 18 - Toughness as a function of the time argon is blown after CaSi-treatment. TN-process

between our two ladle refining processes regarding toughness properties is found in table 10 above. The results refer to grade NV 4-4, transverse, -55°C, plate thickness 11.9-20 mm and 10 x 10 mm ISO-V test pieces.

As far as TN-treatment is concerned we have found some interesting relationships regarding toughness and production parameters. Insufficient Al-deoxidation before the CaSi-injection gives somewhat lower toughness values according to figure 17. Evidently 0.040 % Al is needed before injection to get optimum toughness properties.
Due to the scatter in Al-content from LD we today average 0.030 % total Al (equal to 0.025 % soluble aluminium). At this deoxidation degree we have obtained satisfying toughness properties down to 0.54 kg Ca/ton, which is the lowest amount of desulphurizer blown so far. According to Motte and Cordier (ref 5) a complete modification of the inclusions is achieved when injecting a minimum amount of 0.45 kg Ca/ton of steel.

After CaSi-injection the analysis and temperature are finally adjusted. The necessary gas purging at this correction influences the toughness properties according to figure 18. Obviously there is an optimum purging time regarding the toughness value. At short purging times inclusions left from the CaSi-injection are not satisfyingly separated from the steel. At longer purging times the steel reoxidation becomes predominant, which happens already after 3 minutes. For the ASEA/SKF this happens as late as 30 minutes after REM addition.

TOTAL COST COMPARISON

It is hard to do full justice to a comparison of production costs for two processes working under not quite equal conditions. However we have made an attempt and the estimates are found in table 11 below. The comparison is valid for grades requiring low hydrogen content.

Differences in alloying costs and yields from liquid steel to heavy plate have not been taken into account in the calculation below. Probably these differences are in the favour of the TN-process. In order to further decrease the costs of desulphurization by injection we are now making trials to replace CaSi by lime/fluorspar.

In spite of the difference in costs we operate both processes. The reason our still using the ASEA/SKF is that we have a limited diffusion heating capacity for hydrogen removal and that the deep drawing grades require extremely low contents of the interstitials (C, N).

REFERENCES

1. Tivelius, Sohlgren, Wretlind, "The processing of Z-steel in an ASEA/SKF ladle furnace". Metals Society Conference, London, England, May 1977
2. Förster et al., "Deoxidation and desulphurization by blowing of calcium compounds into molten steel and its effects on the mechanical properties of heavy plates". Stahl und Eisen 94 (1974) Nr 11
3. Sundberg Y., "Mechanical stirring power in molten metal in ladles obtained by induction stirring and gas blowing". ASEA Information FAU Nov, 1976
4. Lindskog N., "A study regarding separation of inclusions in ladle" (In Swedish). Jernkontorets Annaler 159:13-18, 1975
5. Motte J P. and Cordier J., "In-depth powder injection in liquid iron and steel". International Conference on Injection Metallurgy, Luleå, Sweden, June, 1977. Jernkontoret Preprint.

THE INFLUENCE OF DIFFERENT DESULPHURIZERS ON THE PROCESS PARAMETERS AND THE PROPERTIES OF SOLID MATERIAL DURING TN LADLE TREATMENT

Bertil Tivelius and Tomas Sohlgren

B. Tivelius
Manager of Process Research
Svenskt Stal AB Oxelosund

T. Sohlgren
Process Metallurgist
Svenskt Stal AB Oxelosund

This paper originally appeared in the Symposium Proceedings, McMaster University, Hamilton, Ontario, May 1979.

Ladle treatment of carbon steel is an established technology in the steel mill at Oxelosund since April 1971 when an ASEA/SKF ladle furnace was taken into commission. To further increase the capacity of secondary steelmaking a Thyssen-Niederrhein (TN) equipment for ladle injection was installed in February, 1977.

The TN-plant was put into service entirely according to the know-how and recommendation from the supplier. In general, the results turned out well and were in good agreement with earlier published results.(1)

The TN-process is based upon the injection of CaSi. However, several steel mills have reported good results from injections of other types of desulphurizers.(3,4,5,) Some mills also report similar results achieved just by gas purging of a well-deoxidized bath covered with a top slag of high basicity.(3,6)

As a change from CaSi to a lime powder mix would mean a considerable cost reduction we have run a series of tests with different desulphurizers.

In this paper we have compiled the test results regarding influence on the process parameters and the properties of solid material.

TESTING PROGRAMME

All heats involved in the testing programme had a chemistry corresponding to ASTM A 633, grade C intended for heavy gauge plate. The analysis is given in Table 1.

The tests were carried out in our TN-plant according to normal shop procedure for CaSi treated heats (Figure 1).

The trials have all been performed in dolomite-lined ladles. Before treatment the steel was deoxidized. The total aluminium content and the FeO content of the slag before treatment have roughly scattered between 0.010-0.100% and 1-10% respectively. Synthetic top slag, containing 85% CaO + 15% CaF_2, was added in such an amount that the sum of injected powder and the top slag corresponded to the normal quantity of synthetic slag used at a normal CaSi injection.(2)

The different desulphurizers had compositions according to Table 2 and were mechanically mixed from lime-, flourspar-, aluminium and CaSi-powders respectively.

The grain size distributions for the two most frequently injected desulphurizers are compared in Table 3.

TABLE 1

Element	C	Si	Mn	Pmax	Smax	Al	Cb
Aim value	0.16	0.45	1.35	0.020	0.025	0.025	0.020

TABLE 2

Powder composition	Number of heats
100 % CaO	2
80 % CaO + 20 Al	2
85 % CaO + 15 % CaF_2	14
65 % CaO + 15 % CaF_2 + 20 Al	2
65 % CaO + 15 % CaF_2 + 20 CaSi	1
100 % CaSi	Heats selected at random from normal production.
Top slag desulphurization*	3

*No injection, just argon gas purging of a steel bath covered with a top slag containing 85% CaO and 15% CaF_2.

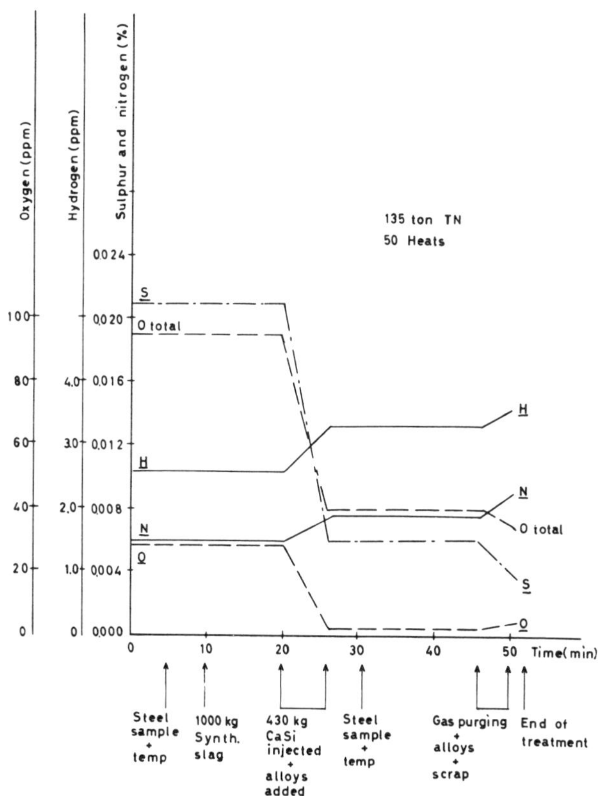

Fig. 1 - Hydrogen, oxygen, sulphur and nitrogen contents as a function of time

TABLE 3

Grain size mm	Weight % 85% CaO / 15 % CaF_2	CaSi
2.0 - 1.0	0.1	—
1.0 - 0.50	0.3	0.3
0.50 - 0.25	47.5	36.3
0.250 - 0.125	33.0	34.0
0.125 - 0.063	9.9	18.0
< 0.063	9.2	11.4

BLOWING TECHNIQUE

At a normal TN treatment 60-70 kg CaSi per minute is injected together with up to 1.5 Nm^3 Ar per minute through a 1/2" ID pipe. The carrier gas is supplied at a pressure of 14-15 atmospheres. The tested lime based mixtures were injected in roughly the same manner. At the trials with top slag desulphurization we have blown the same amount of argon as for the injected heats.

In general it has been considerably harder to inject lime based powder mixes than CaSi in view of the tendency of stopping up of powder feed (Table 4).

TABLE 4

Type of treatment	Number of heats	Heats with clogging %
100 CaO	2	100
80 CaO + 20 Al	2	0
85 CaO + 15 CaF_2	14	50
65 CaO+ 15 CaF_2 + 20 Al	2	100
65 CaO+ 15 CaF_2 + 20 CaSi	1	0
100 CaSi (1978)[2]	515	12
Top slag desulph	3	0

The significant difference in tendency to clogging is due rather to differences in physical properties between CaSi and lime materials than to the documented little difference in grain size distribution according to Table 3.

However, there are ways to blow lime based mixtures without clogging. Higher gas rates must be used compared with CaSi. This is illustrated for the 14 lime/fluorspar heats in Table 5.

TABLE 5

Type of treatment	Number of heats	Gas rate Nm^3/min	Clogging %
85 CaO + 15 CaF_2	8	1.00-1.50	88
85 CaO + 15 CaF_2	6	1.67-2.00	0

These higher gas rates were in our trials achieved by a careful rinsing of the gas feeding system ahead of the powder dispenser.

OXYGEN AND SULPHUR REFINING

The total oxygen content in liquid steel after treatment for CaSi-injection is strongly influenced by the total oxygen before the treatment according to an earlier publication.(2)

With the use of lime powder mixtures or top slag desulphurization this relationship no longer seems valid, as shown in Figure 2, where the above mentioned CaSi- results have also been included.

Contrary to CaSi, the lime powders or top slag desulphurization give low and uniform total oxygen contents after treatment (independent) of the initial content. The reason for this difference is that before the treatment a high total oxygen content corresponds to high FeO content in the ladle slag. Due to the stronger stirring effect emanating from Ca-evaporation at CaSi injection, the ladle slag is more efficiently reduced.

Unfortunately this leads to a continuous formation of oxide inclusions, which are not completely separated during the short time available.

The total oxygen contents before and after the treatment are compiled in Table 6 for the different desulphurizers. Lime without fluorspar seems to give the highest total oxygen contents after treatment probably on account of a stiffening effect on the top slag. The use of CaSi produces higher total oxygen contents than lime/fluorspar depending on the above mentioned phenomenon.

TABLE 6

Type of treatment	Number of heats	Total oxygen content ppm before/after treatment			
		CaO	CaF_2	CaSi	Top slag
100 CaO	2	72/43			
80 CaO + 20 Al	2	168/37			
85 CaO + 15 CaF_2	14		85/25		
65 CaO + 15 CaF_2+20Al	2		100/25		
65 CaO + 15CaF_2+20CaSi	1			79/31	
100 CaSi (Ref 2)	24			89/35	
Top slag desulph.	3				134/26

However, the behaviour of oxygen activity does not follow the same pattern (Table 7). CaSi or CaSi-containing powders thus showed the lowest activities after treatment. Injection of lime based powders gave a definitely higher oxygen activity than CaSi and the poorest result in this respect was attained with the top slag desulphurization.

The sharp difference in oxygen activity between the injection of lime based powders and

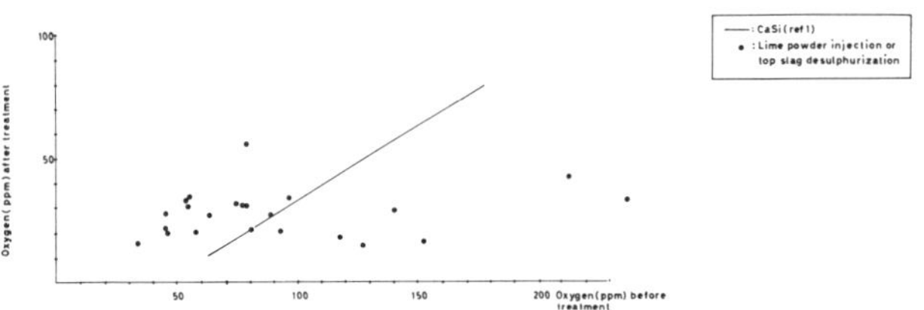

Fig. 2 - Total oxygen content after treatment as a function of total oxygen content before treatment

TABLE 7

Type of treatment	Number of heats	Oxygen activity ppm before/after treatment			
		CaO	CaF_2	CaSi	Top slag
100 CaO	2	14/5			
80 CaO + 20 Al	2	43/9			
85 CaO+15 CaF_2	14		18/5		
65 CaO + 15 CaF_2 + 20Al	2		31/12		
65 CaO + 15 CaF_2 + 20CaSi	1			9/3	
100 CaSi (Ref. 2)	24			30/3	
Top slag desulph.	3				23/16

pure top slag desulphurization can be a matter of favoured alumina nucleation, when blowing powders through the steel bath.

The desulphurization is to a great extent controlled by the degree of deoxidation before injection.

Unfortunately the scatter in aluminium content was relatively high during these trials. To ensure a proper evaluation of the effect of the different desulphurizing techniques, we constructed Figure 3, in which the degree of desulphurization is plotted against the aluminium content before injection. At a given Al-content powders containing CaSi are the most efficient. This is most certainly due to the fact that calcium itself is a very strong deoxidizer and moreover contributes to the stirring efficiency in a way discussed earlier. The use of CaF_2 in the injected powder also seems to have a positive effect.

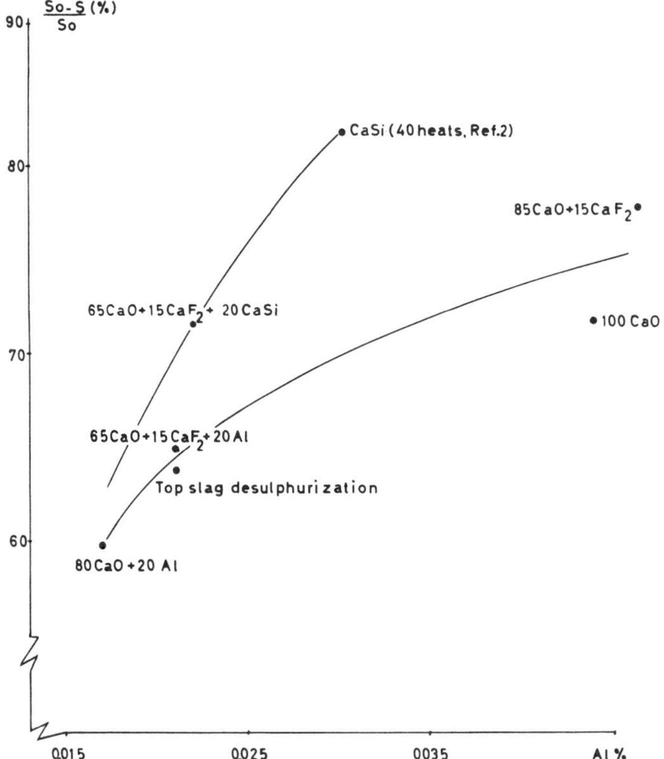

Fig. 3 - Desulphurization as a function of Al content before treatment

HYDROGEN, NITROGEN AND ALUMINIUM

Hydrogen pick-up during the treatment has been studied for heats injected with lime/fluorspar. A comparison with CaSi-treated heats has been made (Table 8).

Apparently the hydrogen pickup is 0.5 ppm higher for lime/fluorspar than for CaSi. This difference is partly a consequence of the fact that the lime/fluorspar mix holds 0.3% crystalline bound water. However, this can increase the hydrogen content in the steel bath with up to 0.9 ppm. As CaSi holds no water at all it is evident that the hydrogen pick up from the atmosphere is substantially larger at CaSi injection. Thus it seems possible to limit the hydrogen pick up when using lime/fluorspar to at least the same level as for CaSi if the water content of the lime powder mix could be lowered by the supplier.

As far as nitrogen is concerned the use of lime/fluorspar leads to a smaller pick-up from air compared with CaSi (Table 9). This is in good agreement with the hydrogen behaviour and must be due to the lower oxygen activity attained with CaSi and more vigorous stirring caused by the evaporation of calcium.

The burn off of aluminium during the injection has been investigated for heats treated with lime/fluorspar. A comparison with ordinary CaSi injected heats reveals a considerably larger burn off of aluminium for the lime/fluorspar heats. Moreover the scatter was unacceptably high for this variant (Table 10). The reason for this very significant difference is certainly the strong deoxidizing effect of calcium in CaSi. When using CaSi at the injection the steel and slag are supplied with an excess of deoxidizer,

TABLE 8

Type of treatment	Number of heats	Hydrogen ppm Before injection	After injection	Δ H ppm
85 CaO + 15 CaF_2	6	2.5	3.7	+1.2
100 CaSi	5	3.2	3.9	+0.7

TABLE 9

Type of treatment	Number of heats	Nitrogen % Before injection	After injection	Δ N %
85 CaO + 15 CaF_2	9	0.005	0.006	+0.001
100 CaSi	50	0.005	0.007	+0.002

TABLE 10

Type of treatment	Number of heats	Burn off of aluminum % Average	Scatter 2σ
85CaO + 15CaF_2	14	0.032	0.034
100 CaSi	50	0.014	0.014

TABLE 11

Type of treatment	Number of heats	Tonnage teemed without problem %
85CaO + 15CaF_2	8	92.9
CaSi normal production	29	88.8

which effectively levels out the effects of even large differences in initial oxygen potential. In this way calcium protects the aluminium from heavy burn-offs.

If CaSi is to be exchanged for lime/flourspar a prerequisite is a much better control and knowledge of the amount of oxygen in ladle slag and steel before treatment.

CASTING AND SURFACE CONDITIONING

The main parts of the heats have been continuously cast to slab size 220 or 270 x 1570 mm. In no case has there been any problem with oxygen rinsing due to clogging in spite of a sometimes too low teeming temperature. According to this experience heats treated with lime or lime mixtures have just as good or better teemability as CaSi treated heats (Table 11).

Surface conditioning of slabs is in our mill performed by flame scarfing. The difference measured as material loss is very small and favours the lime fluorspar heats compared with normal CaSi treated heats of the same grade. On the other hand surface defects on correspondingly rolled heavy gauge plates seem to appear somewhat more frequently in lime/fluorspar heats according to Table 12. Probably there is no difference in surface standard between the two groups of material.

TABLE 12

Type of treatment	Number of heats	Flame scarfing % scatter	average	Plates with surface defects %
$85CaO+15CaF_2$	8	0.00-1.44	0.54	17.9
100 CaSi	29	0.00-2.01	0.63	14.2

The numbers in the staples refer to following types of inclusions:

1 : Sulphides	without Ca	5 : Oxides	without Ca
2 : "	with Ca	6 : "	with Ca
3 : Duplex	without Ca	Re : Remainder	
4 : "	with Ca	L/W : Sulphide shape Factor	

EFFECT OF DIFFERENT DESULPHURIZERS ON MECHANICAL PROPERTIES

Slabs from all heats have been rolled to heavy gauge plate in a four-high rolling mill. Samples have been taken in the as-rolled condition and have then been subjected to a normalizing treatment before testing. To map out the extent of relationships between desulphurizing method and properties in solid material, we have tested the plates regarding nonmetallic inclusions, reduction of area in the through thickness direction and toughness in the rolling and transverse directions. This special part of the investigation comprises heats from the different desulphurizing methods together with one CaSi treated heat chosen at random from the normal production. It is well known that the sulphur content plays a predominating role for the mechanical properties. Consequently, to isolate the true effect of the different desulphurizers these investigations are limited to heats with 0.004-0.006% of sulphur. Furthermore, the plate thicknesses are in the range of 20-30 mm.

Non-metallic inclusions

During hot working inclusions are more or less elongated to flat plates. These stringers seriously reduce the ductility in the transverse and particularly in the short-transverse direction. Thus we found it necessary to scrutinize how the desulphurizers influenced the non-metallics. The area fraction of sulphides and oxides has been determined by the optical image analysis system QTM 360 and the length/width ratio by QTM 720. A large number of samples from different positions in the plates have been investigated. The inclusions have been studied in the rolling direction.

According to this analysis all heats examined have a good micro cleanliness and an area fraction of inclusions which does not diverge conspicuously between the heats (Table 13).

TABLE 13

Type of treatment	Number of heats	test-samples	Area % Sulph.	Oxides	Tot	Sulphide shape fac.*
100 CaO	1	18	0.006	0.014	0.020	2.55
80 CaO + 20 Al	1	18	0.008	0.019	0.027	2.90
85 CaO + 15 CaF_2	2	24	0.005	0.014	0.019	2.32
65 CaO + 15 CaF_2 + 20 Al	2	29	0.010	0.017	0.027	3.20
65 CaO + 15 CaF_2 + 20 CaSi	1	11	0.010	0.016	0.026	2.22
100 CaSi	1	6	0.007	0.017	0.024	1.79
Top slag desulph.	1	18	0.006	0.017	0.023	2.43

*The sulphide shape factor (L/W) is defined according to: $L/W = \frac{\text{Length x Width}}{\frac{\pi \cdot \text{Width}^2}{4}}$

For wholly globular inclusions thus L/W is 1.27

If the area fraction of inclusions cannot reveal anything the sulphide shape factor, defined as length/width ratio of sulphides, in any case indicates that there is a difference. Calcium treated heats thus show the best sulphide shape control.

From each heat one plate sample was particle analysed with PASEM in order to determine the composition of present inclusions and their relative frequency. This micro-analysis confirms that the injection of CaSi results in a complete modification of MnS inclusions to calcium containing sulphides and duplex oxides (Figure 4). The purer the MnS, with respect to calcium the more plastic it becomes. This is also confirmed here as the sulphide shape factor (L/W) increases with higher content of pure MnS (inclusion type 1 according to Figure 4). In this connection it should be mentioned that the 100% CaSi and the 20% CaSi heats hold the highest percentage of calcium in solid material (0.004 resp 0.002% Ca).

Reduction of area

The reduction of area in the short transverse direction (RAZ) is a very important property in our so-called Z-grades. These steels are recommended for applications where plates are welded perpendicular to each other. Normally fissures and cracks are likely to form in the vicinity of the welds and reduce the ductility of the steel through its thickness. This is known as lamellar tearing and is mainly a result of the presence in the steel of varying amounts of non-metallics.(7) The reduction of area at mechanical testing in the short transverse direction of the plates significantly reflects the susceptibility to lamellar tearing. The RAZ has to be at least 25% to eliminate this risk. In Table 14, the RAZ-results are compiled from the actual investigation.

It is obvious that all variants of desulphurizers have given high and relatively even Z-values with reasonably good margin to the critical level of 25%. The influence of the inclusion morphology for the heats in Table 14 is illustrated in Figure 5, which shows that the RAZ drops with increasing projected length of inclusions. The CaSi treated heat and the top slag desulphurized heat fit well into this pattern showing lowest and highest projected lengths respectively. The projected length of inclusions were measured by PASEM.

TABLE 14

Type of treatment	Number of heats	test-samples	Average	RAZ % Range of variat.
100 CaO	1	18	67	54-75
80 CaO + 20 Al	1	17	61	53-75
85 CaO + 15 CaF_2	2	22	63	46-74
65 CaO + 15 CaF_2 + 20 Al	2	29	66	31-78
65 CaO + 15 CaF_2+20 CaSi	1	11	63	50-70
100 CaSi	1	6	68	67-70
Top slag desulph.	1	18	59	42-72

Toughness

The effect of the desulphurizers on transverse impact properties is demonstrated in Figure 6. With the exception of the CaSi-treated heat there are very small differences at low temperatures.

From Figure 6 the toughness at -50°C is evaluated (Table 15). The CaSi treated heat shows a 100% better value than the poorest one.

Of great significance is the effect of calcium on the impact shelf energy here defined as the toughness at +20°C. The influence of the different methods of desulphurization with regard to longitudinal and transverse shelf energies are clearly demonstrated in Figure 7. This is very convincing evidence of the ability of calcium to improve the transverse impact properties of steel to levels almost equalling the longitudinal properties. Moreover, the figure reveals that the degree of anisotropy increases with higher sulphide shape factor.

MICROGRAPHS OF INCLUSIONS

To further visualize the difference in sulphide shape control between CaSi and lime/fluorspar treated heats and the influence of sulphur content on the inclusion morphology the Figures 8 and 9 were constructed. Here 100 fields of view have been accumulated for three heats of each

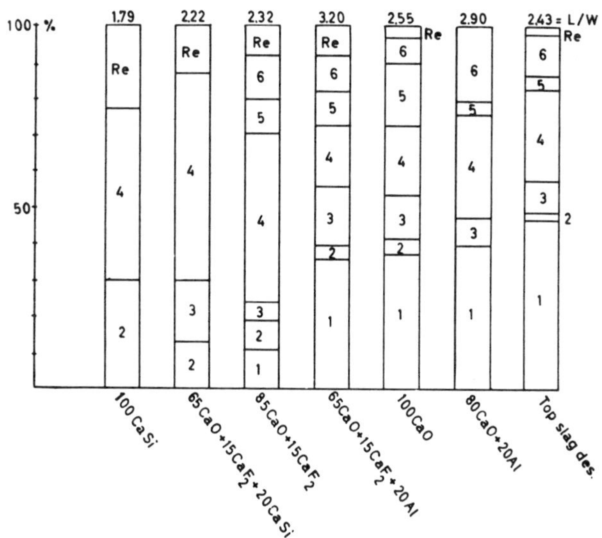

Fig. 4 - Distribution of non-metallic inclusions regarding morphology for the different desulphurizing methods

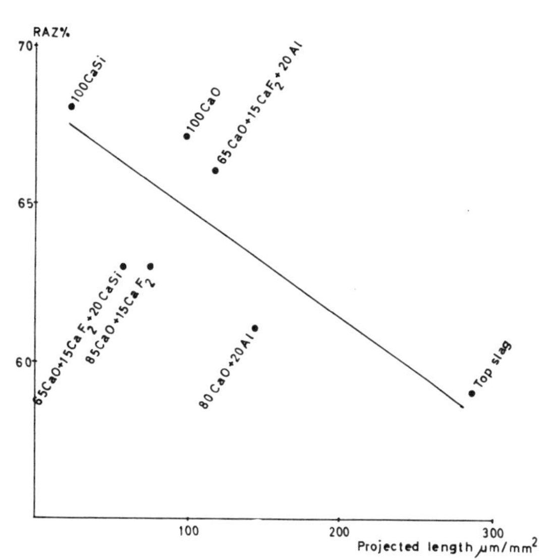

Fig. 5 - Reduction of area as a function of projected length of inclusions in the rolling direction

TABLE 15

Type of treatment	Number of heats	Toughness Joule at -50°C
100 CaO	1	74
80 CaO + 20 Al	1	50
85 CaO + 15 CaF_2	6	55
65 CaO + 15 CaF_2 + 20 Al		76
65 CaO + 15 CaF_2 + 20 CaSi	1	69
100 CaSi	1	100
Top slag desulph.	1	63

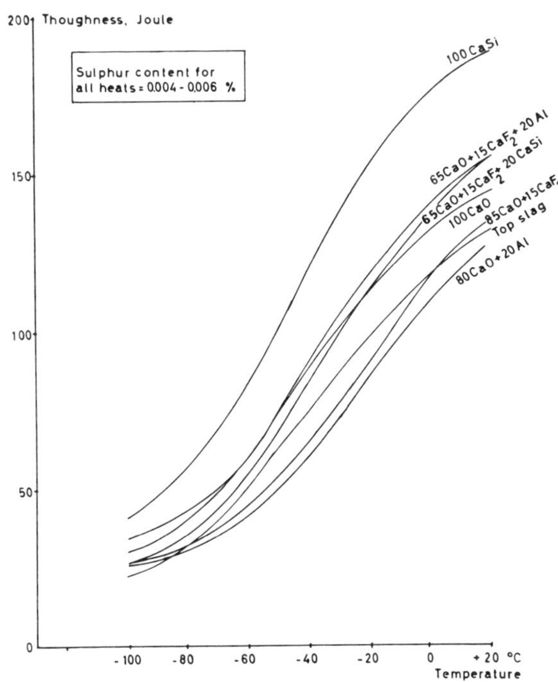

Fig. 6 - Toughness, transverse, 20-30mm plate thickness and 10 x 10 mm ISO-V test pieces

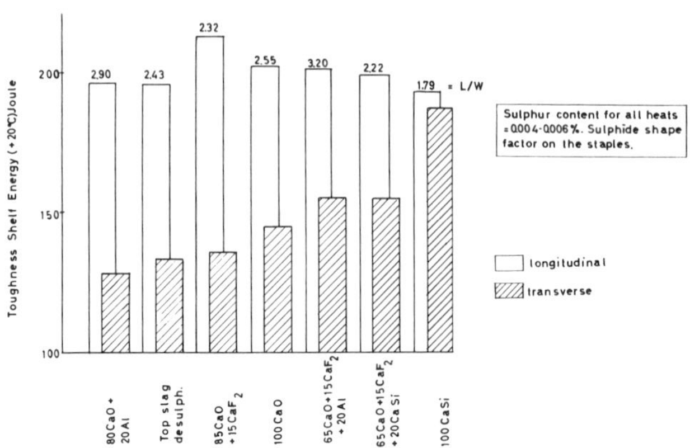

Fig. 7 - A comparison of toughness in longitudinal and transverse direction for the different desulphurization variants. Plate thickness 20-30 mm. 10 x 10 mm JSO-V test pieces

TABLE 16

Type of cost	CaSi	Lime/Fluorospar	Top slag desulph.
	Costs $/ton liquid steel		
Desulphurizer	3:20	0:94	—
Top slag	1:05	0:69	1:05
Aluminium burn-off	0:18	0:44	0:34
Silicon alloying	—	1:24	1:24
Lance	0:76	1:12	0:69
Total	5:19	4:43	3:32

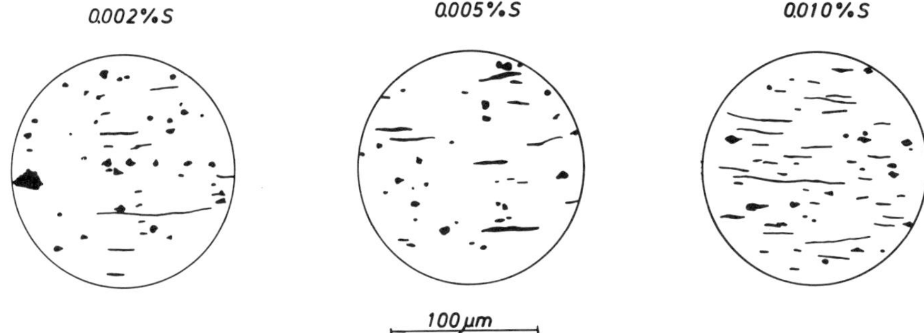

Fig. 8 - Slag inclusion morphology for heats treated with 85% CaO + 15% CaF_2, 100 accumulated fields of view equal to a total investigated area of 1.89 mm^2

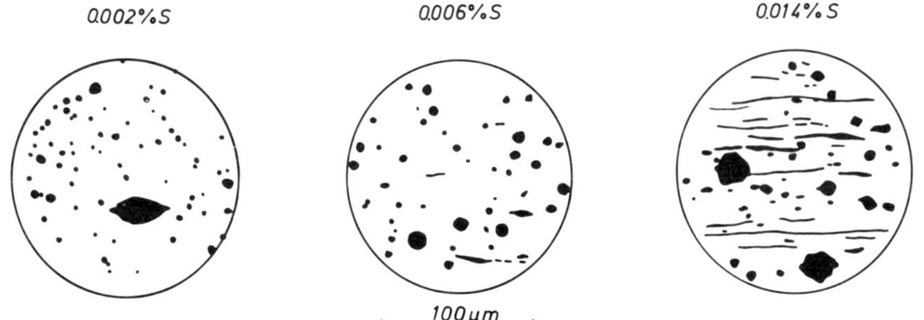

Fig. 9 - Slag inclusion morphology for heats treated with 100% CaSi, 100 accumulated fields of view equal to a total investigated area of 1.89mm^2

type. It is evident that when using CaSi the sulphides are almost globular for sulphur contents not exceeding 0.006%, which is not the case when using lime/fluorspar.

A COST COMPARISON

With a limited number of heats it is difficult to make a fair comparison of production costs. However, with the experience gained in this investigation, an attempt has been made and the estimates are found in Table 16.

The cost analysis includes only operating prime costs, which differ between the variants, and is valid for heats desulphurized from 0.019 to 0.005% S in average. It is clear that the cost for desulphurizer is substantially reduced, when CaSi is replaced by lime powders or top slag desulphurizing, but this difference is for our application reduced by higher alloying costs for the other two methods. Still there is a net balance in favour of the lime and top slag variants.

SUMMARY AND CONCLUSIONS

1. From the process evaluation we have experienced that:
   a) lime powder based mixes are harder to inject compared with CaSi in view of the clogging risk
   b) contrary to the use of CaSi the other desulphurizers give a low and uniform total oxygen content after the treatment and mostly independent of the initial level
   c) the lowest oxygen activity is achieved with the CaSi or CaSi containing powder
   d) the desulphurization is strongly controlled by the degree of deoxidation before the treatment. However, at a given Al-content CaSi or CaSi containing powder are the most efficient
   e) the pick up of hydrogen is somewhat higher for lime powders than for CaSi and vice versa for nitrogen
   f) burn off of aluminium is considerably higher when using lime powders
   g) castability of lime powder treated heats is as good or better than for CaSi heats
   h) there are no differences in surface standard of slabs and plates between the variants of desulphurizers

2. From the micro analysis of heavy plates with roughly the same sulphur content it has been demonstrated that:
   a) the difference in total area fraction of oxides and sulphides is negligible
   b) the sulphide shape factor is very low for the CaSi treated heat, i.e. the sulphides are almost globular
   c) CaSi or CaSi containing powder results in a complete modification of MnS to calcium containing sulphides and duplex oxides
   d) MnS inclusions remain in the steel for all other variants

3. The testing of mechanical properties has shown that:
   a) the reduction of area in the short transverse direction drops with increasing projected length of inclusions and is highest for CaSi and lowest for top slag desulphurized heats
   b) CaSi improves the transverse impact properties to levels almost equal to the longitudinal properties

To sum up, reasonably good results have been attained regarding sulphur refining and material properties with all variants of desulphurizing methods. However, there are always differences in favour of CaSi. On the other hand a certain cost reduction can be obtained using the other desulphurizing methods. Consequently, the final choice of desulphurizer must be governed by the requirements for the properties of the finished product.

REFERENCES

1. Forster et al., "Deoxidation and deoxidation and desulphurization by blowing of calcium compounds into molten steel and its effects on the mechanical properties of heavy plates," Stahl and Eisen 94 (1974) No. 11.

2. Tivelius-Sohlgren. "Secondary Steelmaking by the ASEA/SKF- and the TN-process: a comparison." 37th Ironmaking Conference Chicago April, 1978, I & SM November 1978.

3. Gruner, H et al., "Steel desulphurization subsequent to the melting process," SCANINJECT - Proceedings, International Conference on Injection Metallurgy, Lulea, Sweden, June 1977.

4. Langhammer, H.J. et al., "Influence of injection of powdered materials on oxide and sulphide cleanliness of steel grades," SCANINJECT Proceedings, International Conference on Injection Metallurgy, Lulea, Sweden, June 1977.

5. Saxena S.K. et al., "Effect of injection of CaO-Containing Slab Powder on Deoxidation Characteristics of Aluminium-Killed Steel." Scandinavian Journal of Metallurgy 4 (1975) 42-48.

6. Suzuky Y. et al., "Secondary Steelmaking: Review of present situation in Japan," Secondary Steelmaking Conference, London, The Metals Society, May 1977.

7. Tivelius-Sohlgren, "The processing of Z-steel in an ASEA/SKF ladle furnace," Secondary Steelmaking Conference, London, The Metals Society, May 1977.

THE USE OF THERMODYNAMICS AND PHASE EQUILIBRIA TO PREDICT THE BEHAVIOR OF THE RARE EARTH ELEMENTS IN STEEL

W. G. Wilson, D.A.R. Kay and A. Vahed

W.G. Wilson
Molycorp of America

D.A.R. Kay
Professor of Metallurgy & Materials Science
McMaster University, Hamilton, Ontario

A. Vahed
Research Asst.
McMaster University, Hamilton, Ontario

This paper was presented at the 103rd TMS-AIME Annual Meeting, Dallas, Texas, February 1974.

What are the conditions necessary for the formation of rare earth oxides?

Why is there an enrichment of lanthanum in oxysulfide type inclusions?

Why is there an apparent sequence of formation of the various types of rare earth inclusions in steel?

The examination of the available standard free energy of formation data[1,2] for the various rare earth (RE) compounds makes it possible to make some predictions about the reactions of rare earths with oxygen, sulfur, and white metals. Most results are compatible with the theoretical predictions, whereas other predictions appear to be erroneous, and finally some of the reactions that are observed can not be explained by examination of the standard free energies of formation alone.

From the phase diagrams[3] of some of the rare earths and the "white metals" which include lead, arsenic, etc. it is possible to predict that rare earths should be able to obviate the deleterious effect of some of these elements on the mechanical properties of steel.

STANDARD FREE ENERGY DATA

Comparison of the standard free energies of formation of the oxides and sulfides of the RE's and other elements and RE oxysulfides are given in the accompanying boxes.

The standard free energies of formation of the rare earth compounds of interest are compared in Fig. 1. Based on this comparison, one would expect that with the addition of rare earths in steelmaking, the first compounds that would form would be oxides, then oxysulfides, then $(RE)_xS_y$ compounds, then (RE) S compounds, then the compounds of rare earths with the white metals (arsenic, lead, antimony, etc.) then rare earth nitrogen compounds, and finally rare earth carbide compounds. Stated another way, increasing amounts of rare earths are necessary to form compounds with more positive free energies of formation when rare earths are added to steel.

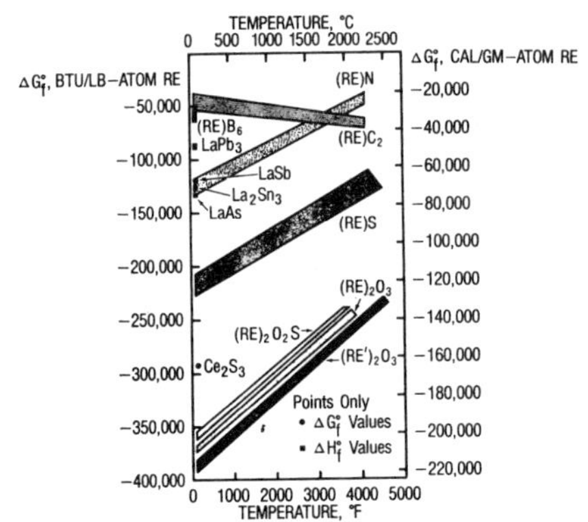

Fig. 1 - Comparison of the standard free energies of formation of a number of RE compounds as a function of temperature

PREDICTIONS FROM PHASE DIAGRAMS

The cerium-lead diagram given in Fig. 2 indicates that some of the cerium-lead compounds have melting points which are several times higher than the melting point of either cerium or lead. In addition, the melting points of some of the cerium-lead compounds are higher than the ordinary steel rolling temperatures. Similar phase diagrams for rare earths with arsenic, and antimony show similar high melting points for these compounds. Therefore, the detrimental effects of some of these white metals on steels should be minimized by the addition of rare earths.

PREDICTIONS FROM MORE RIGOROUS EXAMINATION OF RE THERMODYNAMIC DATA

In order to calculate equilibrium constants of use in iron and steelmaking, it is necessary to know the standard free energy of formation of the RE compound as well as the free energies of solution of oxygen, sulfur, and the rare earths in iron. The free energies of solution for oxygen and sulfur are well known.

The free energies of solution of the rare earths in iron can be calculated from the relevant phase diagrams and it is therefore possible to calculate the standard free energies of formation of the RE compounds using rare earths, oxygen, and sulfur dissolved in steel as reactants instead of

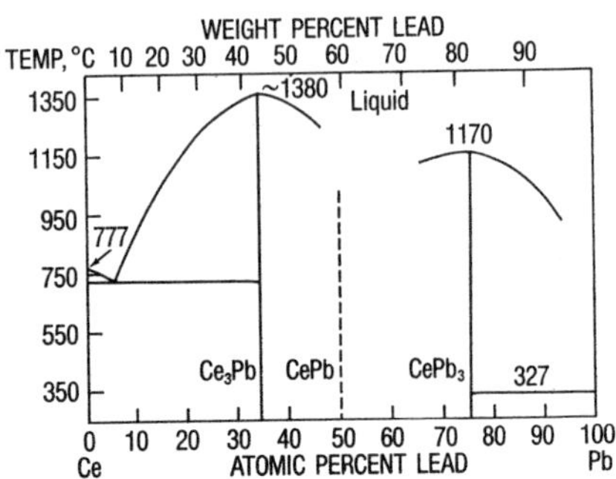

Fig. 2 - The cerium-lead phase diagram

pure liquids, solids and gases at one atmosphere pressure.

The standard free energies of formation of some RE compounds in steelmaking are given in Table I. Equilibrium constants may be calculated for the reactions given in Table I. The reciprocal of these constants, K', gives a "deoxidation constant" in terms of Henrian activities. For example, in the case of Ce_2O_3:

$$K' = h^2_{Ce} \times h^3_O = 2.5 \times 10^{-21}$$

Developing an inclusion precipitation diagram

The Henrian activity h_i, of a component i, is related to the w/o (weight percent) i through the activity coefficient f_i:

$$h_i = f_i(\text{w/o } i)$$

The above calculations, however, are limited as to their practicality because most of the activity coefficients are unknown.

Table II gives the calculated K' values for some of the rare earth compounds at 1900°K.

Using these K' values calculated for cerium oxides, sulfides, and oxysulfides, a diagram may be constructed which will show the precipitation of the compounds that will be formed when rare earths are added to steel.

The first step is to plot the Henrian activity of oxygen vs. the Henrian activity of cerium (h_O vs. h_{Ce}) assuming the precipitation of Ce_2O_3 and this is shown schematically in Fig. 3.

The next step would be to show the Henrian activity of cerium vs. the Henrian activity of sulfur (h_{Ce} vs. h_S) for the three types of cerium sulfide that might be formed. That is -CeS, Ce_3S_4, and Ce_2S_3. This is shown schematically in Fig. 4.

Table I—Standard free energies of formation of some RE-compounds in steelmaking: $\Delta G° = -X + YT$ cal/g.f.w.

Reaction	*X*	*Y*
$[Ce]_{1w/o} + 2[O]_{1w/o} = CeO_{2(s)}$	204000	59.8
$2[Ce]_{1w/o} + 3[O]_{1w/o} = Ce_2O_{3(s)}$	341810	86.0
$2[La]_{1w/o} + 3[O]_{1w/o} = La_2O_{3(s)}$	344865	80.5
$[Ce]_{1w/o} + [S]_{1w/o} = CeS_{(s)}$	100980	27.9
$3[Ce]_{1w/o} + 4[S]_{1w/o} = Ce_3S_{4(s)}$	357180	105.1(°)
$2[Ce]_{1w/o} + 3[S]_{1w/o} = Ce_2S_{3(s)}$	256660(°)	78.0(°)
$[La]_{1w/o} + [S]_{1w/o} = LaS_{(s)}$	91750	25.5
$2[Ce]_{1w/o} + 2[O]_{1w/o} + [S]_{1w/o} = Ce_2O_2S_{(s)}$	323300	79.2
$2[La]_{1w/o} + 2[O]_{1w/o} + [S]_{1w/o} = La_2O_2S_{(s)}$	320340(°)	71.9(°)

[ ] Represents a component in solution in steel and the subscript 1w/o the standard state.
(°) Estimated.

Table II—Deoxidation-desulfurization constants for some RE-compounds

Compound	*Activity Product*	*$K'_{(1900°K)}$*
CeO_2	$h_{Ce} * h_O^2$	$4.0 * 10^{-11}$
Ce_2O_3	$h^2_{Ce} * h_O^3$	$3.0 * 10^{-21}$
CeS	$h_{Ce} * h_S$	$3.0 * 10^{-6}$
Ce_3S_4	$h^3_{Ce} * h_S^4$	$7.6 * 10^{-19}$ (°)
Ce_2S_3	$h^2_{Ce} * h_S^3$	$3.3 * 10^{-13}$ (°)
LaS	$h_{La} * h_S$	$1.0 * 10^{-5}$
La_2O_3	$h^2_{La} * h_O^3$	$8.4 * 10^{-23}$
Ce_2O_2S	$h^2_{Ce} * h_O^2 * h_S$	$1.3 * 10^{-20}$
La_2O_2S	$h^2_{La} * h_O^2 * h_S$	$7.3 * 10^{-22}$ (°)

(°) Estimated

OXIDES

The latest compilation of free energy data for the rare earth oxides is shown below. Standard free energies of formation of the RE elements that represent >98% of those contained in mischmetal are represented by the dotted band. The free standard energies of formation of the other rare earths, shown in the lower hashed band, have little or no effect on the result expected in steelmaking because of the small volume fraction of these elements in normal steel additives.

Based on the standard free energies of formation of the common rare earth elements (Ce, La, Pr, Nd) it can be inferred that their behavior should be quite similar when reacting with oxygen, and that they should form a $(RE)_2O_3$ oxide. If $(RE)_2O_3$ oxides are formed, the rare earths should be the strongest deoxidizers available in steelmaking because only the rare earths have boiling points which are high enough to allow retention of effective residuals in solution. On the other hand, calcium and magnesium have very high negative standard free energies of formation but their usefulness, compared to the rare earths, is limited by their low boiling points which do not allow retention of residuals greater than 0.010%.

The standard free energy of formation of CeO_2 is appreciably more positive than the standard free energy of formation of Ce_2O_3. From inspection of the standard free energy values alone, it is impossible to predict the conditions under which cerium will react to form either CeO_2 or Ce_2O_3.

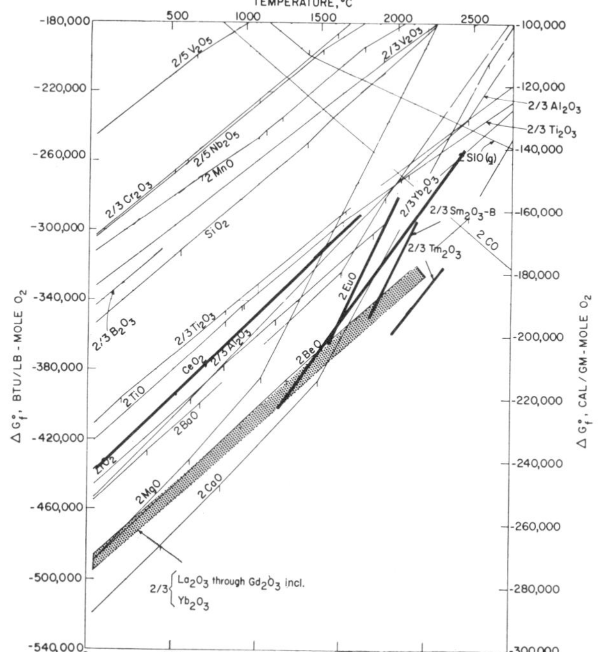

OXYSULFIDES

Data on the standard free energies of formation of the rare earth oxysulfides are presented in the illustration. These have the general formula of $(RE)_2O_2S$. No compounds of this form with other elements have come to the writer's attention. Therefore, their existence seems to be unique to the RE system. The negative standard free energies of formation of these oxysulfides are less than 10 kilocalories less negative than the free energies of formation of the common RE oxides at 3000°F, and therefore if the oxygen with which the rare earths react in steelmaking is not in the form of an $(RE)_2O_3$, then it is quite likely that it is in the form of an oxysulfide.

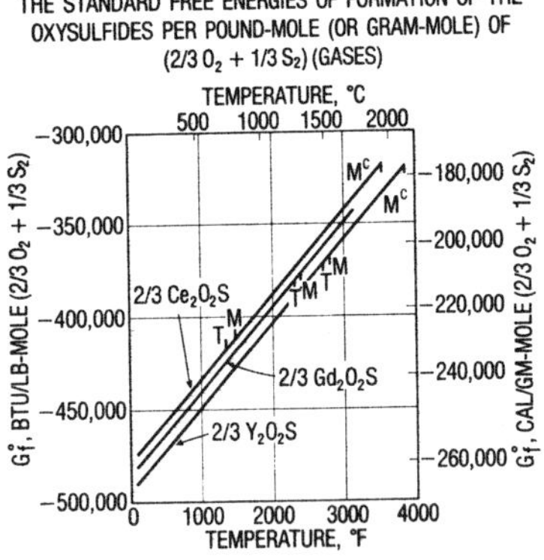

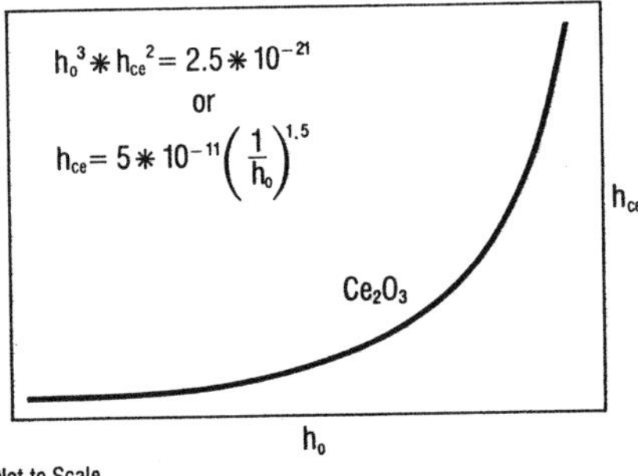

Fig. 3 - The Henrian activity of cerium as a function of the Henrian activity of oxygen

SULFIDES

Cerium sulfide RE S type has a more negative standard free energy of formation than any other sulfide, Fig. A, below. The other sulfides have more positive standard free energies of formation that overlap calcium sulfide of lower temperatures. A portion of the cerium-sulfur diagram, Fig. B, bottom right, shows that there are other known cerium sulfides which might form that have standard free energies which are more negative than that for cerium sulfide, Table III.

The standard free energies of formation of either the $(RE)_xS_y$ or (RE) S compounds are more negative than the free energies of formation of manganese sulfide, zirconium sulfide, and titanium sulfide. This is readily apparent by examination of Fig. A. Therefore, it would be anticipated that at some rare earth level there should be no manganese sulfide found in steel. Because the standard free energies of formation of $(RE)_xS_y$ sulfides are the more negative of the sulfides, it would seem likely that these would be the first of the sulfides to form after the rare earth addition, and then at the higher levels of rare earth addition, the (RE) S type should appear.

Magnesium, sodium, and calcium have been used in the manufacture of iron and steel for the removal of sulfur. Therefore, it seems likely that the elements with more negative standard free energies of formation of their sulfides, such as the rare earths, would desulfurize both iron and steel under the proper conditions.

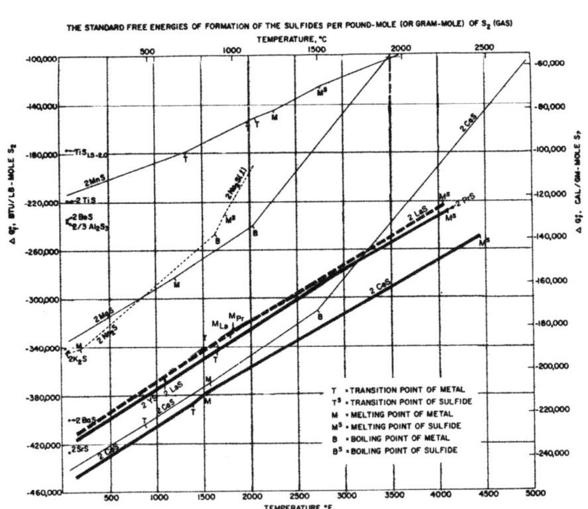

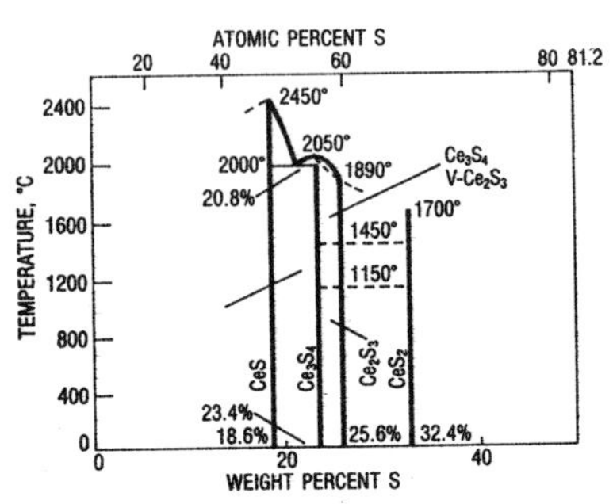

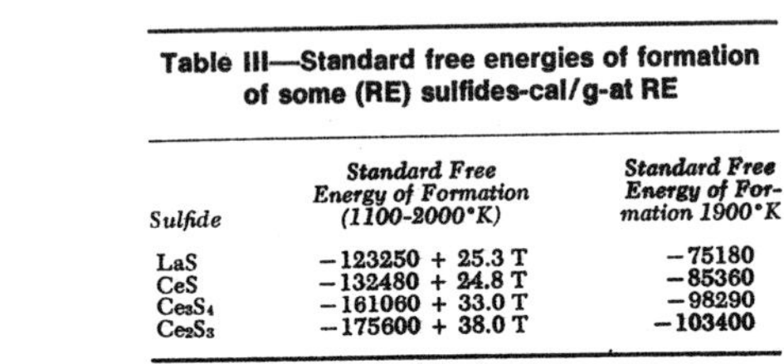

Table III—Standard free energies of formation of some (RE) sulfides-cal/g-at RE

Sulfide	*Standard Free Energy of Formation (1100-2000°K)*	*Standard Free Energy of Formation 1900°K*
LaS	−123250 + 25.3 T	−75180
CeS	−132480 + 24.8 T	−85360
Ce_3S_4	−161060 + 33.0 T	−98290
Ce_2S_3	−175600 + 38.0 T	−103400

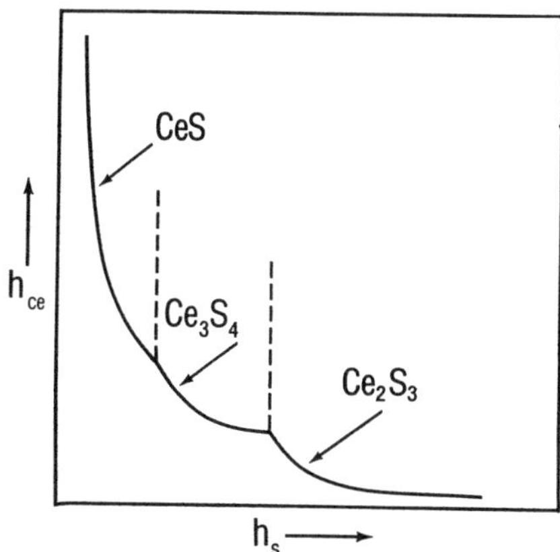

Fig. 4 - The Henrian activity of sulfur as a function of the Henrian activity of cerium

Fig. 4 is a combination of the CeS, Ce_3S_4, and Ce_2S_3 diagrams which result in a single curve that is not continuous. The intersection points signify the equilibrium between the corresponding phases.

Fig. 5 is produced by combining Fig. 3 and Fig. 4 using the Henrian activity of cerium as the common axis. This now is the start of a three dimensional model which will enable us to understand the precipitation of the inclusions.

The next step is to construct the Ce_2O_2S diagram (Fig. 6 & 7) which will be shown to be a function whose surface is in the h_{Ce}, h_O, h_S space. To plot it, one may imagine iso-cerium, iso-oxygen, and iso-sulfur planes thus:

h_O = constant;

$$h_{Ce} = \frac{1.14 \times 10^{-10}}{(h_O)} \left(\frac{1}{(h_S)}\right)^{1/2} \qquad A$$

h_S = constant;

$$h_{Ce} = \frac{1.14 \times 10^{-10}}{(h_S^{1/2})} \frac{1}{(h_O)} \qquad B$$

h_{Ce} = constant;

$$h_S = \frac{1.3 \times 10^{-20}}{(h^2_{Ce})} \frac{1}{(h^2_O)} \qquad C$$

In order to get an idea of the types of inclusions involved in the Fe-Ce-O-S system, one may combine Fig. 5 and 7 to produce Fig. 8. This composite precipitation diagram gives a clear, qualitative picture of the conditions of phase formations and transformations.

As suggested by the form of the deoxidation and desulfurization relations the Ce_2O_3, CeS, Ce_3S_4, and Ce_2S_3 surfaces extend inward parallel to the axis of which they are independent. That is, Ce_2O_3 extends inward parallel to the h_S axis. These surfaces can extend until they meet the oxysulfide surface. Beyond this point, they enter an unstable region and the oxysulfides form. The intercept between each two phases is a line or a curve and the intercept between three phases is a point.

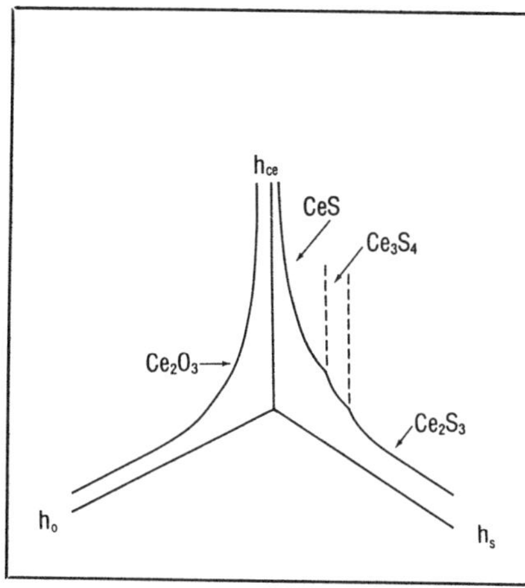

Fig. 5 - The Henrian activity of sulfur and oxygen as a function of the Henrian activity of cerium

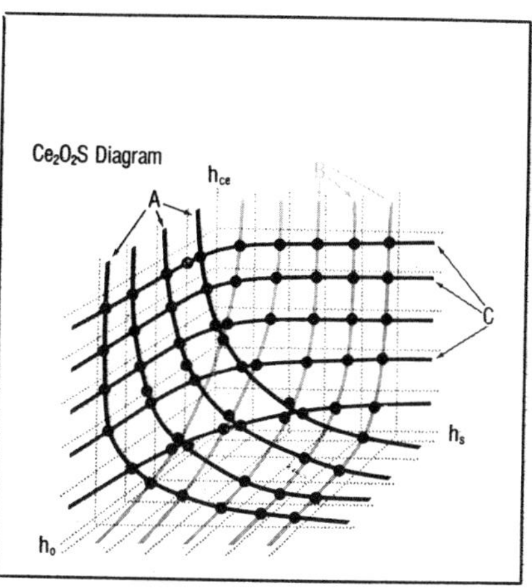

Fig. 6 - Schematic presentation of the method of construction of the Ce_2O_2S surface of the h_{Ce}-h_O-h_S diagram

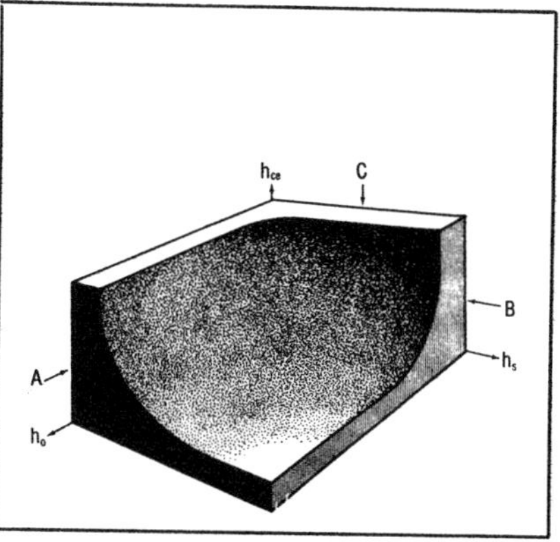

Fig. 7 - Schematic presentation of the oxysulfide surface of the h_{Ce}-h_O-h_S diagram

As an illustration of the usefulness of this diagram, assume a situation in a heat of steel where the cerium content and the sulfur content are high but the oxygen content is low. This composition is above the Ce_2O_2S surface and therefore there would be precipitation of this compound. The precipitation of Ce_2O_2S would proceed with the concentrations of cerium, sulfur, and oxygen diminishing until the amount of cerium, sulfur, and oxygen would be represented by a point on the Ce_2O_2S surface. If this point on the Ce_2O_2S curve is below the level of the curve on the h_{Ce} - h_S surface then that amount of cerium and sulfur will remain in solution in the steel. However, if after the precipitation of Ce_2O_2S is completed, the cerium and sulfur contents are above either the Ce_2S_3, the Ce_3S_4, or the CeS portion of the h_{Ce} - h_S diagram then a Ce_xS_y of some form would precipitate until the composition had reached a point on the line DEF of the precipitation diagrams.

Computations based on the transformation of Ce_2O_3 to Ce_2O_2S indicate that no Ce_2O_3 will be formed until the ratio $h_S/h_O = 4.3$. It is impossible to calculate the exact value for this ratio because of the cumulative error bars resulting from the experimental errors inherent in the determination of the free energies, heats of solution etc. However, this ratio indicates that the sulfur content of the steel, to which rare earths are added, may have to be as much as five times greater than the oxygen content to prevent the precipitation of Ce_2O_3. As an example, if we assume a sulfur content of .020%, it may be necessary to have more than .0047% oxygen or 47 ppm in order to precipitate Ce_2O_3. Contents of this magnitude may be encountered in the melting of steel. Therefore, present thermodynamic calculations indicate the possibility of the precipitation of Ce_2O_3.

Possibility of formation of CeO_2

In discussing the boxes, the possibility of the formation of CeO_2 was mentioned. Calculations based on the free energy change for the transformation $CeO_2 \rightleftharpoons Ce_2O_3$ indicate that h_O would have to be equal to .50 in order for CeO_2 to form at 1873°K. Since the above oxygen activity is not realistic in steelmaking conditions, the possibility of CeO_2 formation may be overlooked.

Possibility of formation of Ce_2S_3

Calculations based on the free energy change for the transformation $Ce_3S_4 \rightleftharpoons Ce_2S_3$ indicate that the probability of formation of Ce_2S_3 with the amount of sulfur ordinarily found in conventional steelmaking is remote.

Lanthanum content of $(RE)_2O_2S$ inclusions

Calculations based on the transformation of Ce_2O_2S to La_2O_2S reveal some interesting information on the chemical composition of the inclusions that should be formed with the addition of mixed rare earth metals in steelmaking. The results of this calculation combined with some analyses of rare earth oxysulfide inclusions indicate that the concentration of cerium should be 4.2 times

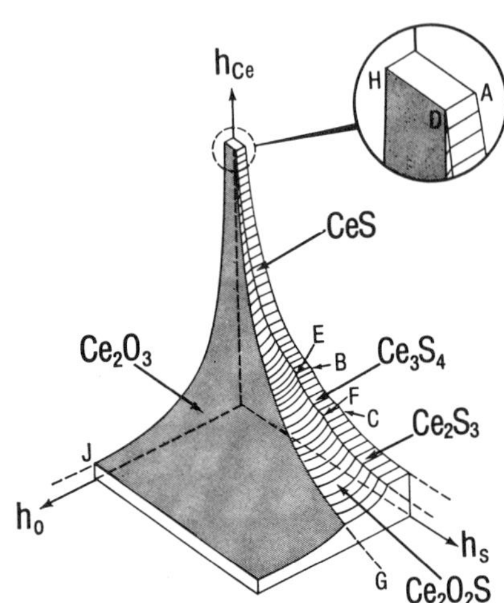

Fig. 8 - Inclusion precipitation diagram for Fe-Ce-O-S system

higher that the lanthanum concentration to give the same power of oxysulfide formation. Therefore, we would anticipate a concentration of lanthanum in the oxysulfide inclusion.

Lanthanum content of (RE)S inclusions

Further calculations based on the transformation of CeS to LaS, coupled with some microprobe inclusion analyses previously obtained, indicate that lanthanum is weaker than cerium as far as rare earth sulfide formation is concerned.

CONDITIONS FOUND IN RE TREATED STEEL

Deoxidation constants

Steelmaking investigations have confirmed the superiority of rare earths as deoxidizers. In ordinary refractories, a deoxidation constant was determined by Leary[4] (based on the formation of $(RE)_2O_3$) to be: $\% RE^2 \times \%O^3 = 3.5 \times 10^{-11}$. This relationship is plotted in Fig. 9 along with the deoxidation relationships for carbon, aluminum, and silicon. The latter two were determined in the presence of .50% manganese.

Leary's results show that rare earths can reduce oxygen levels in steel to values below those obtained with other deoxidizers or with vacuum carbon treatment (< .30% carbon). His findings have been verified by a number of laboratory and plant trials as indicated by the points on the graph. Leary's constant was developed in steel melts where the refractories contained > 30% SiO_2. The standard free energy of formation of silica is small as compared to the standard free energy of formation of Ce_2O_3. This was shown in the accompanying boxes. Therefore, refractories with > 30% SiO_2 serve as a constant source of oxygen in the presence of any significant RE residual with a result that the full potential of rare earths as deoxidizers can not be achieved.

If a more stable refractory such as magnesia is used, much lower oxygen contents can be achieved.

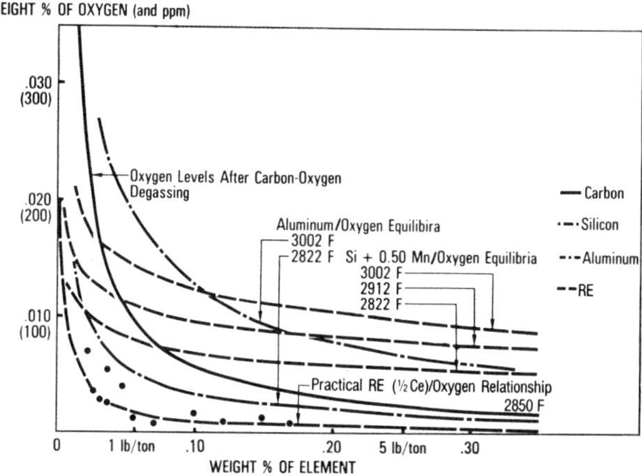

Fig. 9 - Information on deoxidation with elements commonly used in steelmaking

Fig. 10 shows the RE-oxygen relationship when melting on magnesia.[5] The oxygen content associated with a 0.10% addition of rare earths (0.05% retained RE) can be deduced from Fig. 10 to be approxiamately 4 ppm, close to the limit of accuracy for oxygen determination by vacuum fusion. A deoxidation constant, K' = % $(RE)^2$ x % O^3 = 1.6 x 10^{-13} is obtained when melting on magnesia refractories. (This K' was computed in the same manner as Leary's).

In practice, this means that with .05% rare earths in the steel, the oxygen content when melting on ordinary refractories is 20 ppm whereas melting on magnesia the oxygen content would be 4 ppm. Therefore, rare earths are five times more effective deoxidizers at the 0.05% RE level in special refractories than they are in ordinary refractories which are high in silica.

Desulfurization

Thermodynamic calculations have indicated that sulfur precipitates from steel either as oxysulfides or as one of the sulfide forms. Langenberg and Chipman have determined that for the reaction:

$$(Ce)_{1w/o} + (S)_{1w/o} \rightleftharpoons CeS_{(s)}$$

$$\%Ce \times \%S = K' = 1.5 \times 10^{-3}$$

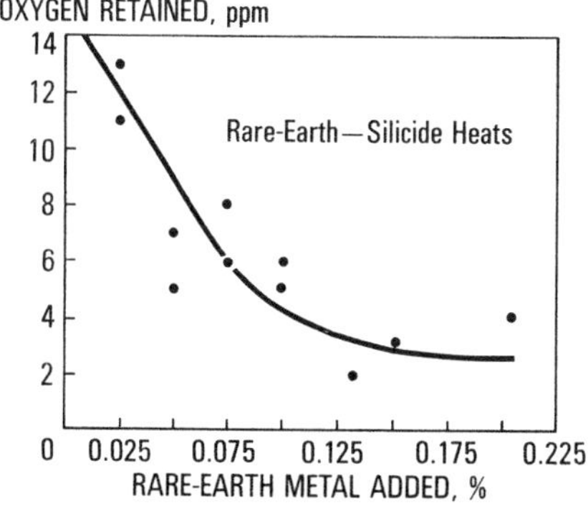

Fig. 10 - Deoxidation curve for RE in vacuum induction melting in MgO crucibles

at 1600°C. The value of K' decreases as the temperature decreases. (Lines representing these K' values at lower temperatures are included in Fig. 11.) Langenberg and Chipman's work was based on the use of pure cerium to determine the solubility product, but since the standard free energies of formation of all of the rare earth sulfides are quite comparable, this solubility product can be used to predict the precipitation of the rare earth sulfides when commercial rare earth metals and alloys containing 50% cerium, 30% lanthanum, 15% neodymium, and 5% praseodymium are used.

For example, using the above assumption, when a steel contains 0.025% sulfur prior to the addition of RE's, a rare earth addition in excess of 0.036% at a temperature of 2822°F will result in the precipitation or flotation of (RE) S from the steel.

With the large amount (0.20 to 0.30%) of RE added in tonnage steelmaking to achieve 0.05% residual (due to the reaction of rare earths with oxygen), it is not difficult to visualize that at

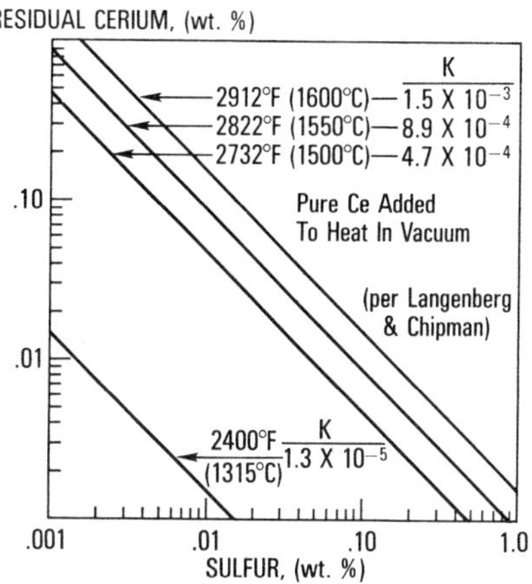

Fig. 11 - Solubility data for cerium sulfide over a range of temperatures

some time during the process of solution that a rare earth content in excess of 0.036% would exist in the steel causing the rare earth sulfides to precipitate from the melt. Rare earth additions to steel made with a single ladle practice (4-6lbs/ per ton) regularly reduce 0.025% sulfur to about half that value or 0.012%.

A second mechanism for sulfur removal proposed by Chipman[6] is summarized as follows: "A more complete and more rapid desulfurization depends upon three factors a) the basicity of the slag, b) the effectiveness of slag metal contact, and c) the oxygen potential." Laboratory results[5] have certainly confirmed that this method of desulfurization is capable of reducing the sulfur contents of steel to very low levels in a very short time when melting on magnesia refractories. Fig. 12 shows the amount of desulfurization achieved with various additions of rare earths, when melting

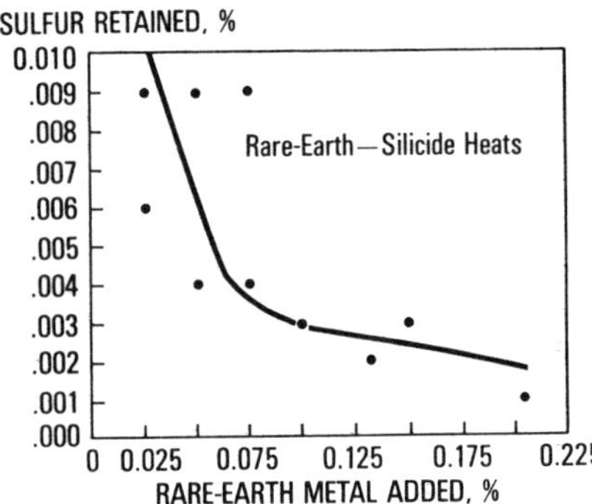

Fig. 12 - Desulfurization with RE's in vacuum induction melting in MgO crucibles with high lime slags

on magnesia, if an addition of 0.20% CaO (based on charge weight) is made prior to the addition of the rare earths. The low levels of sulfur can be obtained by allowing the reaction to proceed for only five minutes. Since this melting was done at a temperature of 2850°F, it is unlikely that the relationship proposed by Langenberg is the one responsible for the dramatic sulfur drop because insufficient rare earths had been added to be in equilibrium with the very low sulfur contents achieved. Therefore, the Chipman relationship is the more likely one to have been responsible for these dramatic sulfur removals. The function of the rare earths in this type of reaction is to produce extremely low oxygen contents, which in combination with lime and the stirring of the induction furnace result in dramatic sulfur reductions.

Inclusions found in steel

Rare earths are normally added to steel after an appropriate aluminum addition is made to lower oxygen and to improve the recovery of rare earths. Since aluminum is only 1/5 the atomic weight of the rare earths, it is most economical to get the lowest oxygen content with aluminum prior to the addition of rare earths. (With aluminum additions, it is possible to get oxygen contents of less than 100 ppm.)

Further confirmation of the ability of the thermodynamic calculations to predict the behavior of rare earths in steelmaking is shown by microprobe analyses of the inclusions found in rare earth containing steels.[7] Fig. 13 shows an inclusion found in a steel containing .020% rare earths, .025% Al, .0015% S. The dark gray core has been shown by the microprobe to be a oxysulfide and this is surrounded by rare earth sulfides of the $(RE)_xS_y$ type. The little bit of alumina around the outside may have resulted from polishing. The vast majority of oxide-sulfide containing inclusions in this heat were of this type.

A small but significant portion of the inclusions found in this heat are complex RE oxide-alumina combinations of two general types.[7] They are either (RE) Al O_3 or (RE) $Al_{11}O_{18}$. They are in many cases associated with MnS or may occur as small isolated inclusions. The two types are shown in Fig. 14.

The inclusion precipitation diagram indicated that the first inclusions to form in the presence of cerium, oxygen, and sulfur would be an oxide. No attempt was made to use thermodynamic calculations to predict a possible interaction of RE and Al to form a complex oxide inclusion when RE's are added to a steel previously deoxidized with Al. However,

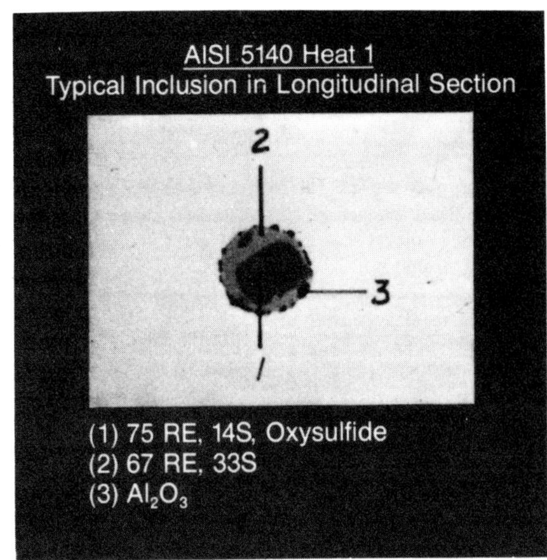

Fig. 13 - Typical RE oxysulfide $(RE)_2O_2S$ -RE sulfide $(RE)_xS_y$ inclusion found in a steel containing 0.020% RE and 0.025% aluminum (500X)

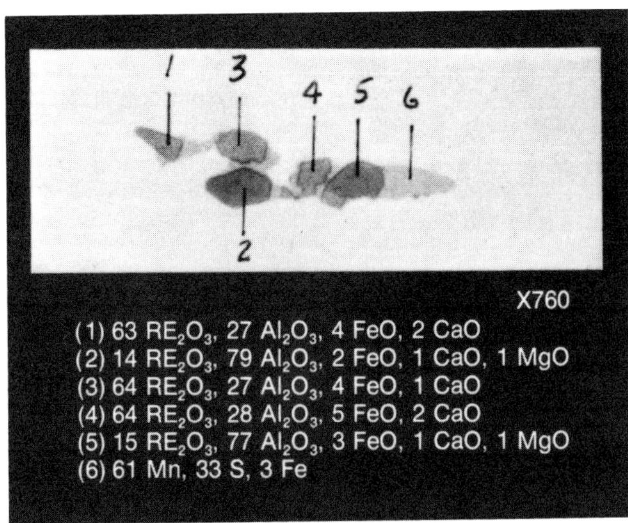

Fig. 14 - Typical (RE) Al_xO_y inclusions in a manganese sulfide matrix found in a steel containing 0.020% RE and 0.025% aluminum (760X)

the presence of a few complex RE-Al oxides rather than RE_2O_3 oxides is not surprising. The central position of the oxysulfide portion of the inclusion shown in Fig. 13 would indicate that, in fact, this portion of the inclusion was the first to precipitate. Finally, the calculations show that the next most likely inclusion to form would be some form of a rare earth sulfide. This again appears to be the case with the oxysulfide surrounded by some $(RE)_xS_y$ sulfide.

Fig. 15 shows an inclusion found in steel with approximately .08% rare earths and .010% sulfur. Again, we see the central core containing the oxysulfide inclusions, obviously the first to precipitate in this inclusion surrounded by the $(RE)_xS_y$ inclusions. Finally, as indicated, when the rare earth content increases the likelihood of formation of the compounds with lower standard free energies of formation would occur. In this case, the (RE) S sulfide which has a more positive standard free energy of formation than the $(RE)_2S_3$ or $(RE)_3S_4$ has precipitated around the edge of the inclusion. The (RE) S type sulfide is readily identified by its gold color.

Furthermore, Luyckx's[8] contention that if the RE/S ratio exceeds 3.0 then no MnS inclusions remain in steel, has been confirmed.

In all of the microprobe studies of RE-Al deoxidized steels that have been made, with which this writer is aware of, no $(RE)_2O_3$ oxides have been found.

Concentration of lanthanum in the RE_2O_2S inclusion

Thermodynamic calculations indicate that a

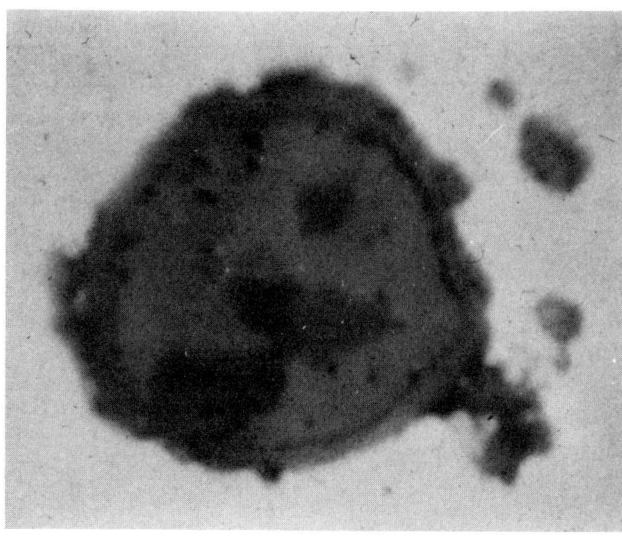

Fig. 15 - Typical inclusion in steel containing 0.080% RE. Outer rim (RE)S-inner core RE oxysulfides $(RE)_2O_2S$ in a matrix of $(RE)_xS_y$

concentration of lanthanum in the RE_2O_2S sulfides is quite likely. Table IV[9] shows the concentration of the individual rare earths in mischmetal and some of the inclusions examined under the microprobe. The microprobe analysis of $(RE)_2O_2S$ confirms a concentration of lanthanum in the inclusions.

La content of (RE)S inclusions

(RE) S type inclusions are easily distinquishable by their golden color[3] when examined microscopically using direct illumination. Limited data indicates lower lanthanum concentration in this type of inclusion confirming the thermodynamic predictions.

Absence of rare earth carbides

When rare earths are added in sufficient quantity to achieve complete globularization of the inclusions, the yield strength and tensile strength of these steels is unaffected. This indicates that rare earths have not formed carbides that would interfere with the strengthening mechanisms.[8] This collateral information confirms the predictions made on the basis of the thermodynamic data. Furthermore, no RE carbides have ever been detected in any of the microprobe analyses of RE treated steel.

Absence of rare earth nitrides

The addition of RE's to steels that use nitrogen as a major component of the strengthening mechanism has not interfered with that strengthening mechanism.[8] Again, the thermodynamic data indicated that RE nitrides were unlikely to form and steel works data confirms this conclusion. No RE nitrides have been found in any microprobe analyses that have come to the writer's attention.

The use of rare earths to counteract the deleterious effect of lead

N. Breyer[10] has shown the detrimental effect of lead on the ductility of 4130 at elevated temperatures. Two heats with similar compositions, except that an addition of .50% rare earths was made to one heat during induction melting, confirmed rare earths are able to overcome the deleterious effect of lead, Fig. 16. This supports deductions based on inspection of phase diagrams, that rare earths will form very high melting point compounds with lead, which would obviate the difficulty Breyer encountered in steels containing trace amounts of lead without RE's.

CONCLUSIONS

Thermodynamic data have indicated that when RE's are added to steel that:

1) Rare earths are extremely strong deoxidizers
2) Rare earths are able to desulfurize steels
3) The inclusions formed as a result of RE deoxidation should be first-oxides, these should be followed by the precipitation of some form of RE oxysulfides and these in turn by the precipitation of rare earth sulfides

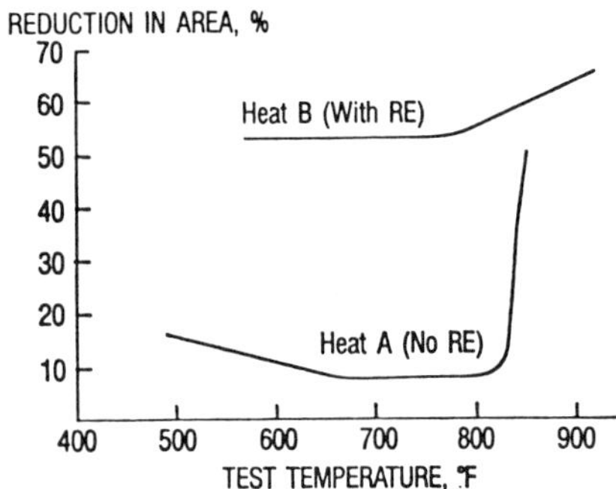

Fig. 16 - Effect of RE on the hot ductility of a medium carbon, low alloy steel containing lead

4) When the RE content is high enough, no MnS inclusions should remain in the steel
5) There should be a concentration of lanthanum in the oxysulfides
6) The (RE) S type inclusions should form at higher RE levels
7) No RE carbides should be formed
8) No RE nitrides should be formed
9) There should be a concentration of cerium in (RE) S type inclusions
10) Phase diagrams indicate the possibility of forming very high temperature RE-Pb compounds which would offset the detrimental effects of lead.

The thermodynamic analysis of the system Fe-Ce-S-O would not be expected to predict the formation of the complex RE oxide-alumina inclusions found in RE-Al containing steels. The analysis did predict that an oxide could form with certain combinations of Ce, O and S. Microprobe data does not indicate whether the RE oxide-alumina inclusions form prior to the formation of the RE oxysulfide-RE sulfides. Otherwise, the other nine predictions made have been shown to be true.

Thermodynamics fails to accurately predict the weight percentage of cerium, lanthanum, etc. necessary to accomplish all of these results because the activity coefficients and the interaction parameters are not available.

ACKNOWLEDGMENTS

The appendix with detailed mathematical derivations is available and can be obtained by writing to the authors. It was compiled by D. A. R. Kay and A. Vahed. Their work has been funded by grants from the American Iron and Steel Institute, the National Research Council of Canada, and Molybdenum Corporation of America.

REFERENCES

1. K. A. Gschneidner, N. Kippenhan, O. D. McMasters; "Thermochemistry of the Rare Earths," Rare Earth Information Center, Ames, Iowa 50010, Report #IS-RIC-6.

2. K. A. Gschneidner, N. Kippenhan; "Thermochemistry of the Rare Earth Carbides, Nitrides, and Sulphides for Steelmaking," Rare Earth Information Center, Ames, Iowa 50010, Report #IS-RIC-5.

3. J. A. Gibson and G. S. Harvey; "Properties of the Rare Earth Metals and Compounds," Battelle Memorial Institute, Technical Report AFML-TR-65-430.

4. R. J. Leary et al; U. S. Department of Interior, Bureau of Mines, R.I. 7091, March 1968, p. 28.

5. H. A. Tucker, R. T. Coulehan and W. G. Wilson; "Rare Earth Silicide Additions to an Alloy Steel to Increase Toughness and Ductility," U. S. Department of the Interior, Bureau of Mines, RI 7153, June, 1968.

6. J. Chipman; "Chemical Behavior of Sulfur in Iron and Steelmaking," Metals Progress, December, 1952, pp. 97-107.

7. W. G. Wilson and R. G. Wells; "Identifying Inclusions in Rare Earth Treated Steels," Metals Progress, Vol. 104, No. 7, December, 1973, pp. 75-77.

8. R. G. Wells; Private Communication, Crucible Materials Research Center, Colt Industries, Pittsburgh, Pennsylvania.

9. L. Luyckx et al.; "Sulfide Shape Control in High-Strength Low-Alloy Steel," Metallurgical Transactions, Vol. 1, 1970, pp. 3341-3350.

10. N. Breyer et al.; "Elimination of Lead Embrittlement in Steel," U. S. Patent 3,726,669, April 10, 1973.

ADDITIONAL READINGS

D. C. Hilty and J. W. Farrell; Private Communication, Union Carbide, Ferroalloy Division, Niagara Falls, New York.

E. D. Eastman et al.: (J) American Chem. Soc., 72, 2248, 1950.

L. N. Permyakov, Yu. V. Kryakovskiy, A. F. Vishkarev and V. I. Yavoyskiy: "Russian Metallurgy and Mining," 4, 37, 1964.

F. H. Spedding, A. H. Daane, eds.: The Rare Earths, John Wiley and Sons, Inc., New York, 1961.

W. Haughton, Ed.: Constitution of Alloys Bibliography, London, Institute for Metals, 1966.

J. Chipman, J. C. Fulton, N. Gokcen, G. R. Caskey: Acta Metallurgica, 2, 439, 1954.

C. W. Sherman, H. I. Elvander and J. Chipman: Trans. AIME, 188, 334-340, 1950.

Yu. V. Kryakovskii; I. Nikolaev; V. I. Yavoiskii; Chemical Abstracts, 70, 1969, 220, 54t.

F. C. Langenberg and J. Chipman: Trans, Met. Soc. AIME, June, 1958, p. 291.

CONTROL OF INCLUSIONS IN

HIGH-STRENGTH, LOW-ALLOY STEELS

A. McLean
Associate Professor
Department of Metallurgy & Materials Science
University of Toronto
Toronto, Ontario

D. A. R. Kay
Professor
Department of Metallurgy & Materials Science
McMaster University
Hamilton, Ontario

This paper originally appeared in Micro Alloying 75 Proceedings, an international symposium on high-strength, low-alloy steels, Page 215

Abstract: When microalloying elements are added to high-strength, low-alloy (HSLA) steels, non-metallic reaction products such as carbides, nitrides, oxides, sulfides and oxysulfides can be formed. The level and type of microalloying additions control the volume fraction, morphology, composition, and physical properties of these reaction products, which in turn influence the mechanical properties and corrosion behavior of the finished product.

This paper discusses the principles that control the formation of silicates, aluminates, sulfides, and other non-metallics. The effects of re-oxidation on the composition and morphology of the inclusion phases are related to the method of adding the microalloying elements. Also covered are practical methods of controlling reoxidation during ingot and continuous casting and the beneficial effects of this control on product performance.

INTRODUCTION

The manufacture of high-strength, low-alloy (HSLA) steels today depends on the controlled rolling of economical plain-carbon steels to which microalloying elements have been added by conventional steelmaking practices.

Microalloying elements play a critical role in controlling the various time-and temperature-dependent precipitation reactions which take place during the production of HSLA steels. The deoxidation and desulfurization of liquid steel by microalloying elements, together with re-oxidation, control the volume fraction and morphology of the oxide and sulfide inclusions in the finished product. These inclusions, in turn, control the impact-shelf energy of the HSLA steel.

INFLUENCE OF INCLUSIONS

The effects of inclusions on the mechanical properties of HSLA steels, primarily toughness, is largely due to their shape, size, and distribution in the final wrought product. In turn this is controlled by their relative deformability at working temperatures. Silicates and manganese sulfides are plastic at working temperatures, unlike aluminates and most of the oxysulfides. Control of inclusion plasticity, and thus the inclusion shape, can be obtained by small additions of elements such as the rare earths, titanium, zirconium, or calcium. The use of aluminum to produce a semi-killed silicate-free steel can also provide adequate control of the inclusion shape in some cases.

The general effects of second-phase particles on the total ductility and Charpy V-notch shelf energy of aluminum-killed steels have been investigated by Gladman et al. and are illustrated in Figure 1. The detrimental effect of unmodified sulfide inclusions is clearly shown. The volume fraction and morphology of carbides are also important, as crack initiation is probably caused by decohesion between the matrix and sulfide inclusions at low strains and the cracking of carbide particles at high strains.

ROLE OF MICROALLOYS

The yield strength of microalloyed steels is increased by both grain refinement and dispersion strengthening. The ductile-brittle transition temperature is decreased by grain refinement but increased by dispersion strengthening. There is therefore a balance between grain refinement and dispersion strengthening for a required toughness. Precipitation of microalloy carbides and nitrides can control grain growth during hot working and increase the yield strength through precipitation hardening.

Grain refinement, essential for a low ductile-to-brittle transition temperature, is achieved primarily by controlled rolling, which is designed to give a fine ferrite-grain size. Here microalloy additions play an important role in inhibiting the recrystallization and grain growth of austenite during hot working. This function can be critical in the production of plate, with the inherent slow-cooling rates and the relatively long time between reversing-mill passes enhancing recrystallization and grain growth.

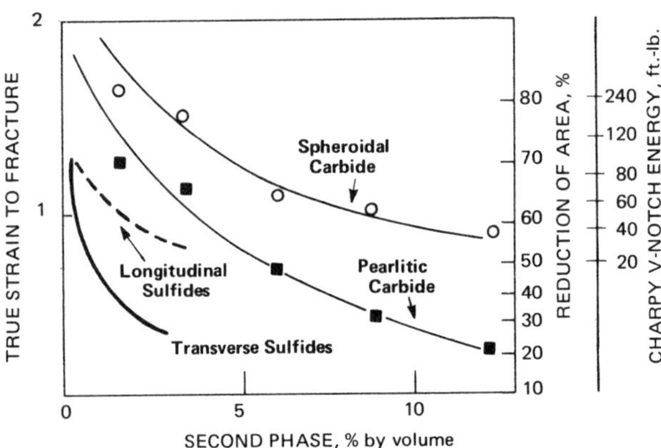

FIGURE 1 – The effect of second-phase particles on total ductility and Charpy-shelf energy.[16]

In producing microalloyed steel, oxygen and sulfur should be removed from steel by the formation and flotation of oxides and sulfides, which are stable at steelmaking temperatures of about 1600°C. The microalloying elements required to be in solution in austenite should not form stable compounds at temperatures greater than 1200°C. The compounds, which are required for precipitation hardening, should be in solution at 1200°C. and precipitate at lower temperatures.

It is therefore essential that the right amount of microalloy be available at the correct temperature and for the correct time during steelmaking, solution treatment, controlled rolling, cooling, and surface finishing if this step involves elevated temperatures. This requires greater control than that achieved in conventional plain-carbon steel production. One of the anomalies of conventional steelmaking is that after the relatively controlled oxidation used in the steelmaking process itself, microalloying additions are made in a ladle lined with highly reducible alumino-silicate refractories and the steel is teemed in air, resulting in reoxidation and nitrogen pick-up.

This paper will look at the thermodynamic stability of compounds in terms of the function of specific microalloy additions and steelmaking variables.

SOLUBILITY PRODUCTS

A measure of the thermodynamic stability of a compound is the solubility product of its constituent elements in steel. In the case of alumina formation during deoxidation at 1600°C., for example, the deoxidation reaction is:

$$2[Al]_{1\ w/o} + 3[O]_{1\ w/o} = Al_2O_{3(s)} \qquad (1)$$

where [ ] = a component in solution in steel and
1 w/o = the standard state at 1 weight per cent.

The standard free-energy change for this reaction:

$$\Delta G^o_{(cal)} = -293960 + 97.47T \qquad (2)$$

obtained from data in the literature[1-3] gives a solubility product in terms of Henrian activities (h_i) at 1600°C. of

$$(h^2_{Al})(h^3_O) = 3.1 \times 10^{-13} \qquad (3)$$

Although a precise knowledge of solution thermodynamics is needed to determine solubility products in terms of weight per cent concentrations, the activity solubility product may be used to give the relative stabilities of compounds. In dilute solutions, there is little error involved in equating h_i with the weight per cent of the ith element.

The authors have therefore limited their consideration to activity solubility products in order to outline the roles of specific microalloying elements in HSLA steelmaking. The standard free energies of formation of some important compounds of microalloying elements have been derived from literature data and used to calculate activity solubility products for oxides, sulfides, nitrides and carbides. Except where specifically stated, all activities are Henrian.

OXIDES AT 1600°C.

The activity solubility products for oxides at 1600°C. are presented in Figure 2, where the activity of the microalloying-element (h_M) is shown as a function of the activity of oxygen (h_O).

In killed steels, the oxides of vanadium and columbium, required for inhibiting austenite recrystallization and grain growth as well as precipitation hardening, are not stable at 1600°C. and have correspondingly high solubility products, even higher than that of silica. These microalloying elements are unlikely to be reoxidized by silica-containing refractories. Steels containing them can be produced in the semi-killed condition with a correspondingly high yield, although silicate inclusions adversely affect impact properties.

Protecting Manganese and Other Elements

Manganese, added to HSLA steels in amounts greater than 1% for its effect on austenite-ferrite transformation characteristics, should be protected by other elements in solution which form more stable oxides if the formation of manganese silicates is to be avoided. The presence of manganese silicates indicates excessive reoxidation and a complete loss of steelmaking control.

Boron should have the protection of aluminum deoxidation, and the proximity of the Ti_3O_5 and Al_2O_3 lines indicates that titanium also requires the protection of full-aluminum deoxidation. Insufficient deoxidation or reoxidation can result in the loss of the titanium required for solid-state precipitation reactions and control of sulfide-inclusion morphology.

Zirconium forms a very stable high-temperature oxide. Figure 2 indicates that zirconia should form even in aluminum-killed steels at 1600°C. However, the temperature dependence of the respective standard free energies of formation[4] is such that alumina becomes more stable at

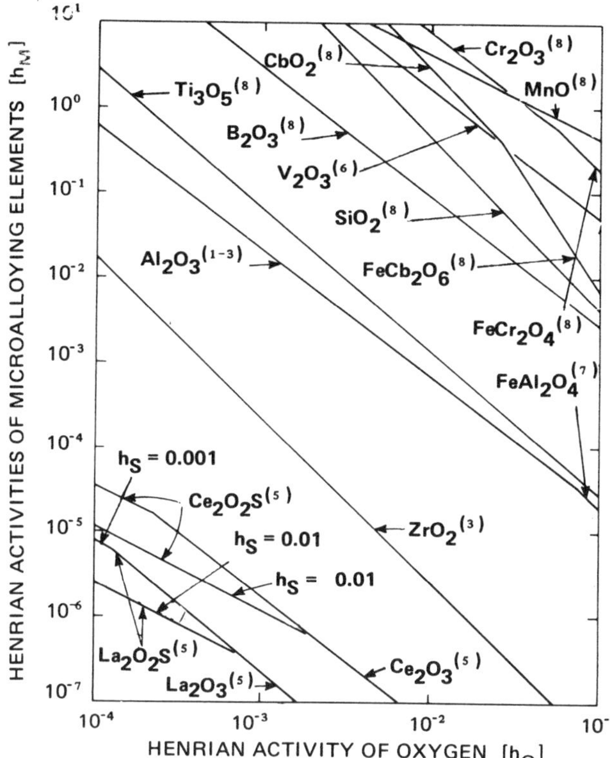

FIGURE 2 – The Henrian activities of microalloying elements versus the Henrian activity of oxygen in equilibrium with selected oxides in steel at 1600°C. The Ce_2O_2S and La_2O_2S values are given for two different Henrian activities of sulfur. Superscripts identify literature references.

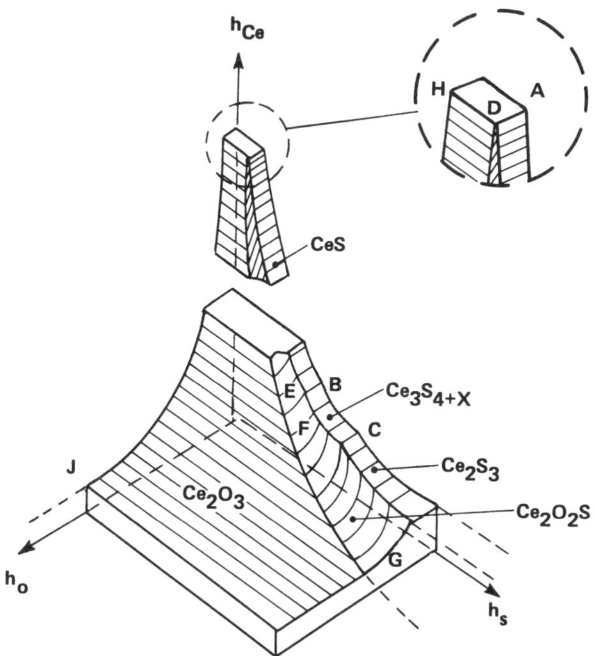

FIGURE 3 – The Fe-Ce-O-S precipitation diagram.[5]

temperatures below 1000°C. In practice, it has been found that about 0.1% aluminum is required to suppress zirconia formation.[9]

Rare-Earth Deoxidation

The rare-earth elements are among the most powerful deoxidizing elements used in steelmaking. In steels containing very low sulfur, rare-earth oxides should be the stable-deoxidation product. However, in commercial steels with relatively high sulfur-oxygen ratios, the oxysulfides are stable and substantial rare-earth deoxidation of aluminum-killed steels takes place. Therefore, the rare-earth elements are most susceptible to reoxidation during teeming and by interaction with alumino-silicate ladle refractories.

A precipitation diagram for the Fe-Ce-O-S system has been developed at steelmaking temperatures[5] and is reproduced in Figure 3. A projection of the equilibrium surfaces is given in Figure 4 which also gives the activities of oxygen and sulfur at points where the three solid phases are in equilibrium with liquid steel. The precipitation sequence which occurs when cerium is added to steel may be determined from this figure.

The compositions of supersaturated solutions of rare-earth, oxygen, and sulfur in steel are represented by points above the equilibrium surface of the precipitation diagram. Since pure rare-earth oxides are usually absent from rare-earth treated aluminum-killed steels, the projection of a typical supersaturated composition may be represented by the point J in Figure 4, which is vertically above the $(RE)_2O_2S$ equilibrium surface. Since the dissolved oxygen-to-sulfur ratio is less than the 2:1 stoichiometric ratio in $(RE)_2O_2S$ and since two atoms of oxygen to one of sulfur are consumed during oxysulfide precipitation, the steel composition must follow the precipitation path projection JK. At the point K, $(RE)_3S_{4+x}$ co-precipitates with the oxysulfide, the precipitation path following the line KE′ on the iso-stability surface. It is thermodynamically impossible for the steel composition to enter and remain in the sulfide-phase fields.

At E′ the precipitation path crosses over onto the $(RE)_2O_2S$ – (RE)S iso-stability surface, as (RE)S coprecipitates with oxysulfide. Precipitation ends at the point L where the precipitation path hits the equilibrium surface. The shape of the oxysulfide-sulfide iso-stability surfaces are such that the amount of oxysulfide coprecipitating with sulfide is very small. In practice oxysulfide inclusions will be coated first with $(RE)_3S_{4+x}$ and then (RE)S.

The precipitation path MNO in Figure 4 represents different initial compositions of oxygen and sulfur, showing a case where $(RE)_2O_2S$ precipitation is followed immediately by (RE)S precipitation. In either case, oxysulfide formation is unavoidable in the rare-earth treatment of aluminum-killed steels and must precede rare-earth sulfide formation. The presence of alumina in rare-earth inclusions indicates the loss of rare-earth deoxidation control.

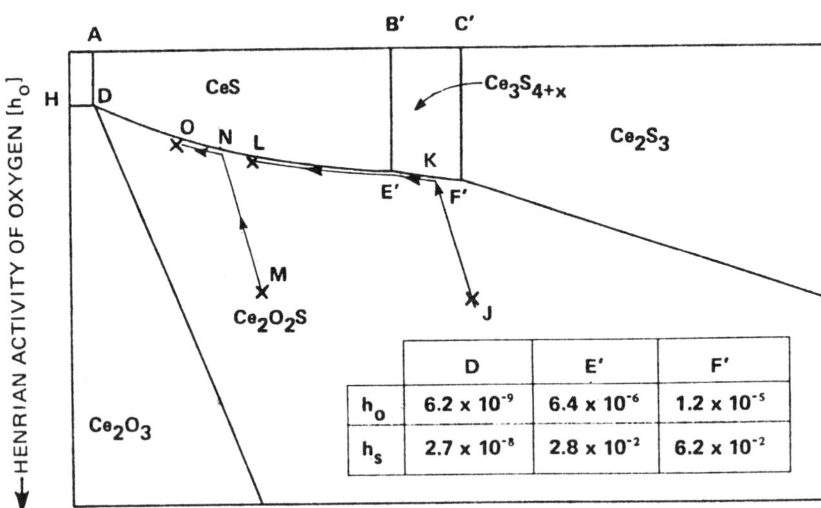

	D	E'	F'
h_O	6.2×10^{-9}	6.4×10^{-6}	1.2×10^{-5}
h_s	2.7×10^{-8}	2.8×10^{-2}	6.2×10^{-2}

FIGURE 4 – Precipitation paths for supersaturated Fe-Ce-O-S solutions at about 1600°C.

Effect of Calcium and Magnesium Deoxidation

Deoxidation with calcium or magnesium vapor at 1 atmosphere pressure gives equilibrium activities of 3×10^{-9} and 6×10^{-7} respectively for soluble oxygen.[3,4] Here the deoxidation efficiency is greatly decreased by the very low solubility of these elements in steel, so that they provide no protection against subsequent reoxidation. Treatment of aluminum-killed steel with calcium or magnesium can control oxide-inclusion morphology through the formation of aluminate compounds.

In general, adequate aluminum deoxidation is essential to avoid using unnecessarily large amounts of expensive sulfide shape-control agents for deoxidation. If the aluminum addition is too large, however, aluminum nitride rather than vanadium or columbium nitride could be precipitated during subsequent processing.

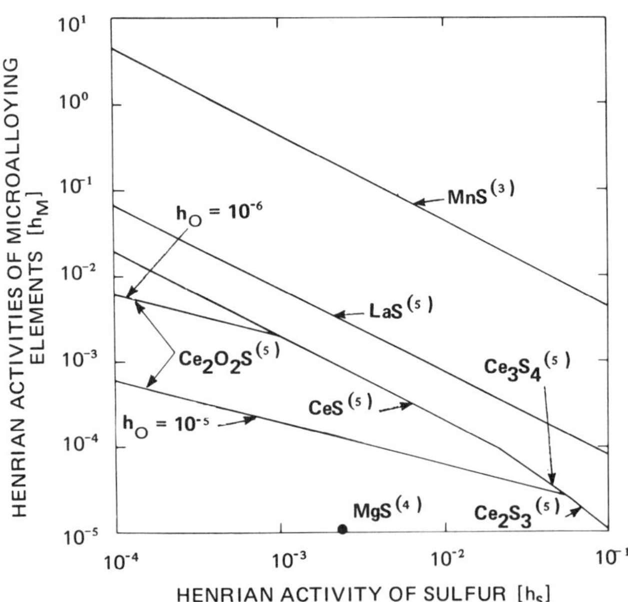

FIGURE 5 – The Henrian activities of microalloying elements versus the Henrian activity of sulfur in equilibrium with selected sulfides in steel at 1600°C. The Ce_2O_2S values are given for two different Henrian activities of oxygen.

SULFIDES AT 1600°C.

The activity solubility products for sulfides at 1600°C. are presented in Figure 5, where the activity of the microalloying element (h_M) is shown as a function of the activity of sulfur (h_S). Since there is very little thermo-mechanical data available on the sulfides of microalloying elements, Figure 5 essentially shows the difference between the original sulfide-inclusion modifier, manganese, and the rare-earths.

CeS is substantially more stable than lanthanum sulfide, and both Ce_3S_4 and CeS should be stable at the sulfur levels usually found in HSLA steels. Ce_2S_3 should not be found in HSLA steels but has been identified in cerium-treated, high-sulfur cast irons.[15] At high-oxygen contents, the rare-earth oxysulfides are the most stable. Because of the relatively high oxygen-sulfur ratio existing after aluminum deoxidation, steels are first deoxidized by rare-earth oxysulfide formation and then the rare-earth sulfides become stable. The data in Figure 5 indicate that very low soluble-oxygen contents are required for rare-earth sulfide formation.

Available low-temperature thermochemical data for titanium sulfide[4] indicate that its stability is very similar to that of manganese sulfide. Therefore, titanium should modify Type III sulfide inclusions as a partial substitute for manganese. Since the titanium content in HSLA steels is very much less than the manganese content, the substitution might be expected to take place during the last part of the solidification process or even in the solid state. The behavior of titanium therefore differs from that of the rare earths, which are very strong sulfide formers. The presence of manganese in a rare-earth sulfide indicates a low rare-earth residual, possibly through reoxidation, and a loss of rare-earth control of sulfide morphology.

Results from zirconium-treated steels[17] suggest that zirconium is a slightly stronger sulfide former than titanium. Again, modification is by partial substitution for manganese, but the stability of the modified sulfide could be high enough for its precipitation as a solid in the liquid steel. The precipitation of primary zirconium sulfide, in addition to the compound sulfide, is reported in steels containing 0.1% zirconium.[18]

Desulfurization with $Ca_{(g)}$ and $Mg_{(g)}$ at one atmosphere pressure gives equilibrium residual sulfur activities of 4.6×10^{-7} and 2.4×10^{-3}, respectively. The low-residual solubility of calcium and magnesium in liquid iron is not as critical here as in deoxidation, since resulfurization normally does not occur. However, consideration of reactions of the type:

$$CaO_{(s)} + [S]_{1\ w/o} = CaS_{(s)} + [O]_{1\ w/o} \quad (4)$$

and

$$MgO_{(s)} + [S]_{1\ w/o} = MgS_{(s)} + [O]_{1\ w/o} \quad (5)$$

indicate that $CaS_{(s)}$ and $MgS_{(s)}$ may be oxidized to their respective solid oxides at h_O/h_S ratios greater than 2.7×10^{-2} and 2.5×10^{-4} respectively, with sulfur reverting into solution. In this respect, $MgS_{(s)}$ is more susceptible to reoxidation than $CaS_{(s)}$. Together with the rare earths, calcium and magnesium have the ability to modify both oxide and sulfide inclusions.

NITRIDES AT 1600°C.

The activity solubility products for nitrides at 1600°C. are presented in Figure 6, where the activity of the microalloying element (h_M) is shown as a function of the activity of nitrogen (h_N).

The feature of this diagram is the high stability of the rare-earth nitrides. Thus, residual rare earths in solution should effectively remove nitrogen required for solid-state reactions from solution, even at 1600°C. In practice, however, reoxidation processes probably reduce the soluble rare-earth contents to such a low level that the nitrides are not precipitated.

The nitrides of vanadium, columbium, and aluminum cannot precipitate at 1600°C., but the nitrides of zirconium and titanium can appear as large cuboids precipitated in the melt. These large inclusions are not effective in grain refinement or precipitation hardening.

Treatment of liquid steel with $Ca_{(g)}$ and $Mg_{(g)}$ at one atmosphere pressure gives nitrogen activities, in equilibrium with the corresponding nitrides, of 0.049 and 0.62 respectively. Such a treatment should not remove nitrogen from solution.

CARBIDES AND NITRIDES IN AUSTENITE

Microalloying element carbides are unstable in steel at 1600°C. However, at solution-annealing temperatures and below, the stability and solubility of both the carbides and nitrides are critical to grain-growth inhibition and precipitation hardening. Here the precipitated fraction forming on cooling from the austenitizing temperature is critically dependent upon the alloy composition as well as the solubility, density, and stoichiometry of the precipitating phases.

Columbium and Titanium Carbide

The effect of these variables for columbium and titanium carbide is shown in Figure 7, which is based on the work of Gladman et al.[19] The figure shows the solubility curve for both carbides at an austenitizing temperature of 1100 to 1300°C., together with the volume fraction of the precipitated carbide as a function of carbon content.

For a steel containing 0.1% columbium, the volume fraction of precipitating carbide increases with carbon content up to the stoichiometric level (Area A). At higher carbon contents, the volume fraction of carbide is controlled by the columbium content and remains constant. At carbon contents between the stoichiometric line and the solubility curve, the maximum amount of fine carbide is precipitated (Area B). If the carbon content is such that the solubility limit is exceeded at 1200°C., then the volume fraction of coarse carbide (Area D) increases at the expense of that of the fine carbide (Area C). In this case, a higher austenitizing temperature, with consequent enhanced grain growth, is required to increase the volume fraction of fine precipitate.

In the case of titanium carbide, the lower stoichiometric ratio and inclusion density give a higher volume fraction of precipitated carbide. The higher solubility

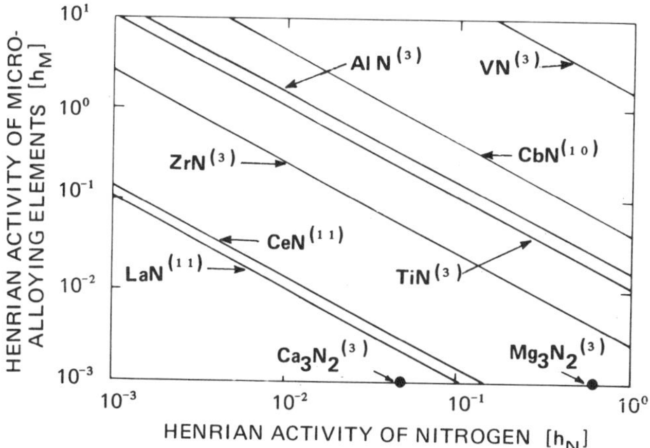

FIGURE 6 – The Henrian activities of microalloying elements versus the Henrian activity of nitrogen in equilibrium with selected nitrides in steel at 1600°C.

product for titanium carbide given by Gladman et al.[19] extends the carbon range over which the maximum volume fraction of fine carbide can be precipitated. Alternatively, a lower austenitizing temperature may be used. Stoichiometric alloys give the maximum fraction of fine precipitates when the alloy composition exceeds the austenitizing solubility limit.

The effect on grain refinement of a decreased fraction of fine precipitate is illustrated in practice[16] when the solubility limit of aluminum nitride at the austenitizing temperature is exceeded at high-aluminum contents. The fraction of fine precipitate decreases with increasing aluminum content giving a resultant increase in the final ferrite-grain size.

Free Energies of Formation

Standard free energies of formation of carbides and nitrides, obtained from the literature, have been combined with the free energies of solution of graphite and nitrogen in austenite, respectively, which have been derived from solubility data.

$$C_{(gr)} = [C]_{1\ w/o} \quad (6)$$

$$\Delta\bar{G}_{(cal)} = +7704 - 6.86T \quad (7)$$

$$\tfrac{1}{2}N_{2(g)} = [N]_{1\ w/o} \quad (8)$$

$$\Delta\bar{G}_{(cal)} = -2060 + 8.94T \quad (9)$$

In the case of vanadium carbide, for example:

$$V_{(s)} + C_{(gr)} = VC_{(s)} \quad (10)$$

$$\Delta_{1273} = -18000\ cal \quad (11)$$

and combination with equation (6) gives

$$V_{(s)} + [C]_{1\ w/o} = VC_{(s)} \quad (12)$$

$$K = a_{VC}/(a_V \cdot h_C) \quad (13)$$

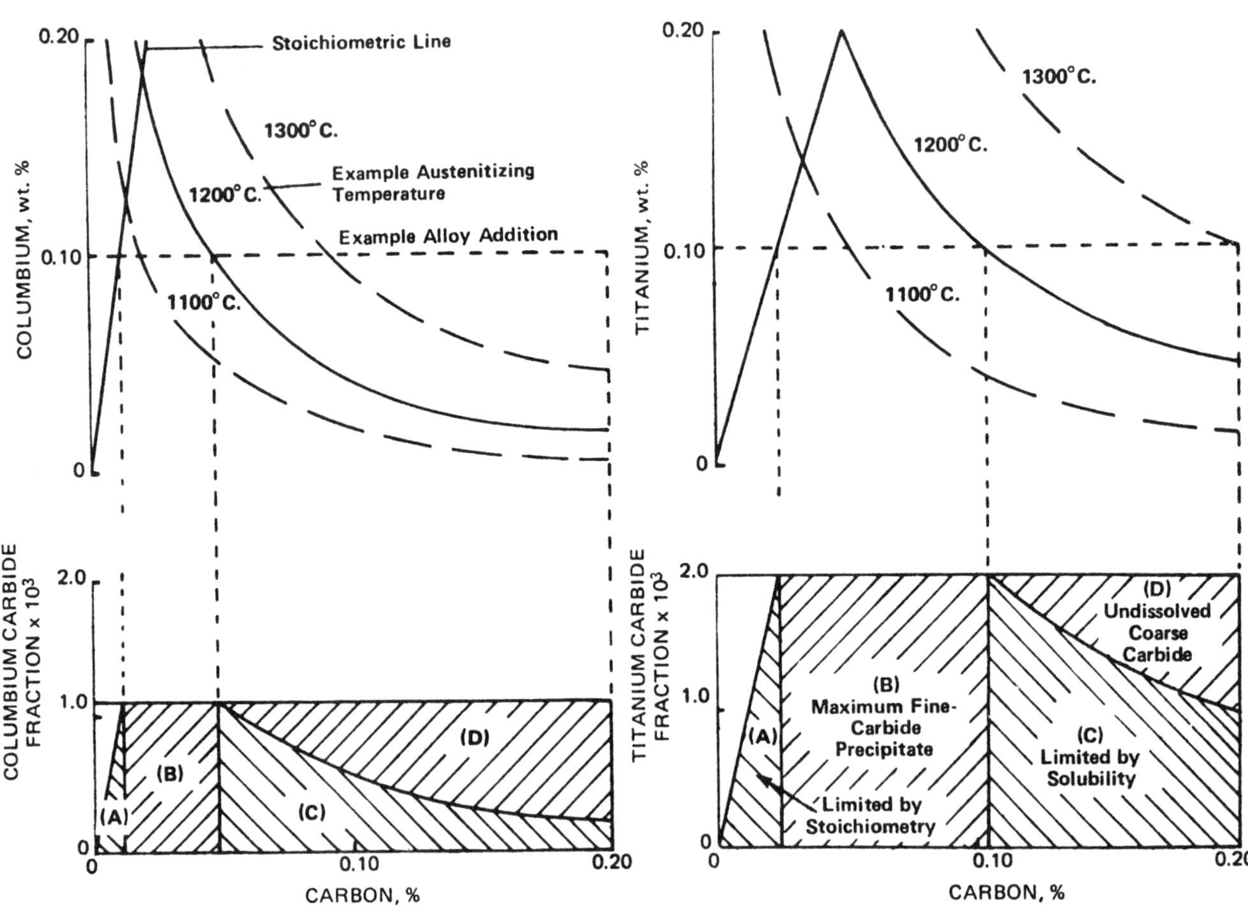

FIGURE 7 – The effect of alloy composition, solution temperature, and precipitate stoichiometry on the fine-carbide fraction produced on cooling from 1200°C. for a microalloying addition of 0.10%. In the case of titanium carbide, the total titanium content must be corrected for the formation of coarse, insoluble titanium nitride, which does not participate in precipitation strengthening.[19]

The solubility product of vanadium carbide is therefore expressed in terms of the Henrian activity of carbon and the Raoultian activity of the microalloying element, which takes the pure element as the standard state. These solubility products for some microalloying-element carbides and nitrides at 1100°C. are presented in Figures 8 and 9. The interpretation of these figures is more qualitative than those of the previous figures, because of the use of a Raoultian activity. In addition, the stoichiometry of some of the microalloy carbides and nitrides is not the simple one assumed here, and indeed many are non-stoichiometric. Microalloy carbides and nitrides are now recognized to be oxy-carbides, oxy-nitrides or oxy-carbonitrides where oxygen can stabilize one structure at the expense of another.[20]

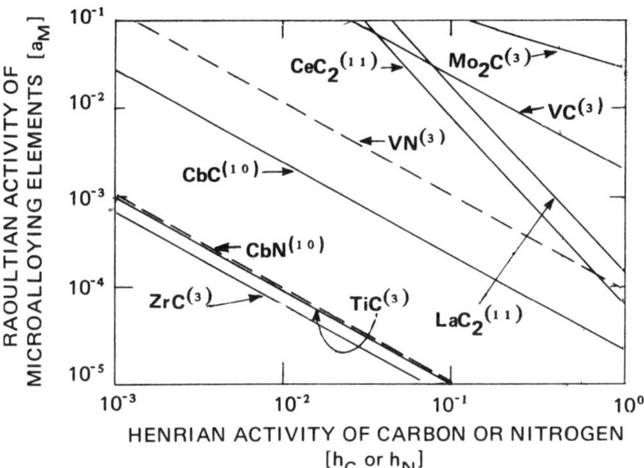

FIGURE 8 – The Raoultian activities of microalloying elements versus the Henrian activities of carbon and nitrogen in equilibrium with selected carbides and nitrides in steel at 1100°C.

Microalloy Nitrides and Carbonitrides

Figures 8 and 9 show that microalloy nitrides are generally more stable than the corresponding carbides. Carbonitrides will have intermediate stabilities and compositions dependent on the soluble carbon-nitrogen ratio. Columbium carbonitride, for example, is predominantly carbide in a steel of low-nitrogen content containing 0.10% carbon. A nitrogen-rich carbonitride, however, forms in the presence of nitrogen.

Molybdenum carbide has a high-solubility product consistent with the use of molybdenum in solution in austenite to inhibit recrystallization and grain growth. Chromium carbides (not shown) have an even higher solubility product. Vanadium carbide and nitride are more stable, but they still have a high solubility in austenite, with vanadium carbide precipitating at high carbon-nitrogen ratios and vanadium nitride at low ratios. The high solubility of vanadium carbide in austenite permits the development of high strengths through precipitation hardening in normalized carbon-manganese-vanadium steels. Vanadium carbonitrides are reported to precipitate in ferrite.[21]

Large contents of soluble residual rare earths should interfere with molybdenum carbide, chromium carbide, and vanadium-carbonitride precipitation reactions through the precipitation of rare-earth carbonitrides (see Figure 9 for cerium and lanthanum nitrides). However, in practice, the soluble rare-earth residual is probably very low because of reoxidation losses.

Columbium carbonitrides have intermediate stabilities and can still dissolve in austenite at 1100°C. The role of columbium is complex. In solution in austenite, it inhibits recrystallization and grain growth with the coarser precipitates inhibiting grain-boundary movement. At lower temperatures, the fine precipitates precipitation-harden the ferrite.

Aluminum nitride is traditionally accepted as an austenite grain refiner. Its solubility in austenite at 1100°C. is too low for it to participate in precipitation-hardening reactions. At high-aluminum contents, the presence of coarse-nitride particles even decreases its grain-refining potential.[19]

Varying Role of Titanium

The behavior of titanium carbonitride varies with the carbon-nitrogen ratio. At low ratios, it is more stable than aluminum nitride, is virtually insoluble in austenite, and commonly appears as large cuboids which are generally

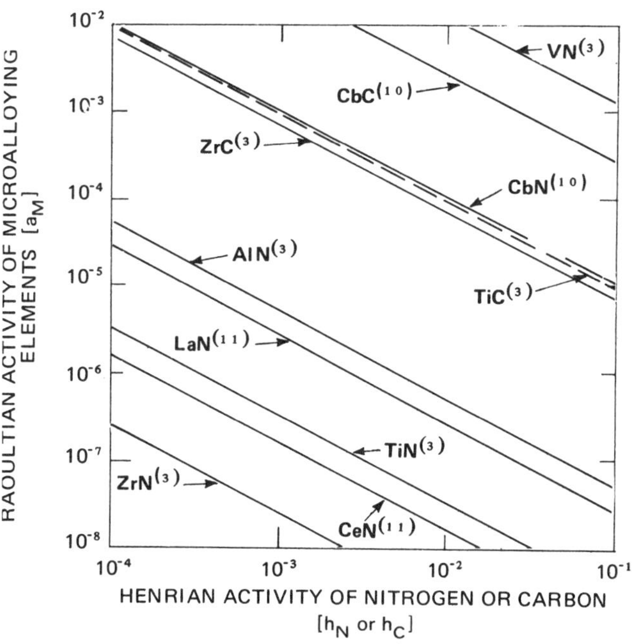

FIGURE 9 – The Raoultian activities of microalloying elements versus the Henrian activities of nitrogen and carbon in equilibrium with selected nitrides and carbides in steel at 1100°C.

incapable of contributing to grain refinement. At high ratios, it has a stability similar to that of columbium nitride and can contribute to dispersion hardening. At intermediate ratios, it could contribute to grain refinement.

It should be noted that the role of titanium in microalloying steels is complex and difficult to control. Titanium forms a stable oxide, modifies sulfide inclusions, and forms a carbonitride which can contribute to precipitation hardening and grain refinement or effectively remove nitrogen from solution, depending on the carbon-nitrogen ratio. In contrast, a microalloying element like columbium does not form a stable oxide or modify sulfide inclusions, and its role in grain refinement and dispersion hardening is more readily controlled.

Zirconium nitride is the most stable microalloying element nitride and its solubility product is correspondingly low at 1100°C. For this reason, zirconium is avoided when precipitation hardening by carbonitride formation is required. At high carbon-nitrogen ratios, zirconium carbonitride should contribute to grain refining and dispersion-hardening precipitation reactions (Figure 9).

Strong nitride formers in general have a beneficial effect on toughness through the removal or 'nitrogen fixing' of nitrogen in solution.

Precipitation Hardening

The degree of precipitation hardening which can be obtained by a precipitate is qualitatively related to its stability. More vanadium than columbium is required, for example, to obtain the same degree of precipitation hardening at low carbon-nitrogen ratios. Precipitates like zirconium carbonitride will not even go into solution to any great extent during a solution anneal. Aging characteristics must also depend on precipitate stability. The tendency to over-age, for example, is greater for a precipitate with a high-solubility product.

The simple trends given above have been based on the thermodynamic stabilities of only some of the many possible microalloying compounds. In the case of zirconium, for example, the formation of the carbosulfide, $Zr_4C_2S_2$, is critical at high-zirconium contents[21] and the formation of an oxysulfide cannot be overlooked.[22] The kinetics of precipitation reactions in the solid state are just as important as the thermodynamics, particularly when the effect of a precipitate is critically dependent on its size.

These factors emphasize that the control of microalloying elements in HSLA steels may be difficult and should be implemented at the steelmaking stage.

SOLUBILITY PRODUCTS AT 1100°C.

The solubilities of some microalloying-element carbides and nitrides in austenite have been determined experimentally as a function of temperature[10,12,13,14] and are given in Table I. Solubilities at 1100°C. have been calculated from the data in Table I and are shown in simplified form in Figure 10, together with the approximate solubility of titanium nitride.

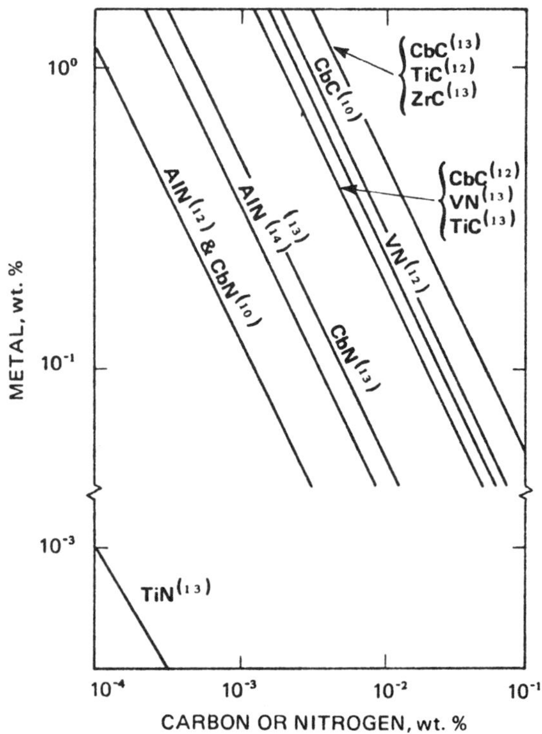

FIGURE 10 – Solubility products of selected carbides and nitrides at 1100°C.

TABLE I – SOLUBILITY PRODUCTS IN AUSTENITE (WEIGHT PERCENT)

Compound	Solubility
CbC	$\log_{10}[Cb][C] = -6770/T + 2.26$ (12) $= -7900/T + 3.42$ (13) $= -10960/T + 5.43$ (10)
TiC	$\log_{10}[Ti][C] = -7000/T + 2.75$ (12) $= -10475/T + 5.33$ (13)
VC	$\log_{10}[V][C] = -9500/T + 6.72$ (13)
ZrC	$\log_{10}[Zr][C] = -8464/T + 4.26$ (13)
AlN	$\log_{10}[Al][N] = -6770/T + 1.03$ (12) $= -7400/T + 1.95$ (14) $= -7184/T + 1.79$ (13)
CbN	$\log_{10}[Cb][N] = -8500/T + 2.80$ (13) $= -12230/T + 4.96$ (10)
VN	$\log_{10}[V][N] = -8330/T + 3.46$ (12) $= -8700/T + 3.63$ (13)

NOTE: Superscripts identify literature references

As predicted in Figures 8 and 9, nitrides are more stable than the corresponding carbides. With the exception of titanium nitride, the weight percent solubilities are closer in value than the predicted solubilities. In addition, according to the work of Narita,[13] titanium carbide is more stable than columbium carbide while the reverse is true according to the work of Irvine et al.[12]

The solubilities given in Figure 10 emphasize the experimental difficulties involved in determining solubility products in the solid state, particularly their dependence on alloy composition. Narita,[13] for example, determined solubility products in ternary Fe-M-N and Fe-M-C alloys, although the effect of oxygen cannot be discounted. Irvine et al.[12] determined solubilities in carbon-manganese steels after a solution treatment of normal duration. As a result, the solubilities may not necessarily represent equilibrium conditions. In this respect, equilibrium solubilities may be of little quantitative use in practice, and more experimental work is required to determine the relative importance of different microalloying elements in solid-state precipitation reactions.

In production, the basis of control must begin with the precise control of the steel composition itself at the steelmaking stage. The effect of conventional steelmaking variables on HSLA steel production is considered in the remainder of this paper.

LADLE STEELMAKING

In the production of HSLA steels, microalloying additions are made to the ladle. Therefore, the ladle cannot be regarded as simply a method of transporting molten steel from the furnace to the casting station. Actually, after preliminary refining in the furnace, steel of the required composition is finally "made" in the ladle, except for some possible additions to the mold.

While slag/metal/refractory interactions have always occurred in the ladle, such reactions become critical when they involve elements which are present only in parts per million. Since these elements determine whether or not the steel will possess the required mechanical properties, particular attention must be given to these reactions.

For this reason, the ladle should be treated as a steelmaking vessel. The large sums of money that have been expended on improving the lining life of basic-oxygen furnaces from 100 to 200 heats per lining to over a thousand have paid off. The same philosophy should be applied to the ladle vessel, but unfortunately this is seldom the case.

Interaction with Ladle Glaze

Steelmaking ladles generally contain silica and alumino-silicate phases, which are unstable in contact with molten steel containing strong deoxidizing elements such as titanium, aluminum, zirconium, and rare earths. In addition, the reaction between the slag – which is often basic in character to enhance sulfur removal – and the acidic refractories will result in the formation of a ladle glaze which may be rich in iron oxide, silica, and sulfur from the slag. Refractory/slag interaction with the formation of a glaze, and subsequent glaze/melt interaction during the next heat in the ladle, will cause a loss of the deoxidizing elements from the steel with a resulting increase in the oxygen content of the melt during the time it is held in the ladle.

In a similar way, when the sulfur dissolved in the steel has been lowered by the addition of strong desulfurizers, subsequent sulfur pick-up may occur from the ladle glaze which was formed during the previous heat. With sulfur present in the glaze at a higher activity than sulfur dissolved in the steel, there exists a driving force for sulfur transfer to the metal. This is a repetitive process from heat to heat. As the steel from one heat leaves the ladle, a new "sulfur-rich" glaze is formed by the reaction between the ladle slag and refractory. When the next heat of steel is desulfurized, there is again the opportunity for sulfur pick-up from the glaze formed during the previous heat.

Sulfur Reversion

The ladle glaze therefore acts as a reservoir for both oxygen and sulfur. These elements, present in the slag and subsequently stored in the glaze during one heat, are effectively pumped back into the steel during the next heat, after the addition of deoxidizing/desulfurizing elements. The increase in the oxygen content of the steel can have a secondary, but still important, effect on the composition and the morphology of the sulfides suspended in the melt. Since the strong sulfide formers are even stronger deoxidizers, any increase in the oxygen content of the melt implies that the sulfides, which were originally formed from strongly deoxidized steel, become unstable and undergo transformation to oxysulfides and/or oxides, with the sulfur reverting to the melt. For example:

$$2RES + 2[O] \rightarrow RE_2O_2S + [S]$$

$$RE_2O_2S + [O] \rightarrow RE_2O_3 + [S]$$

If reversion reactions of this type are excessive, iron-manganese sulfides may eventually form with consequent loss of shape control.[23,24] In conventional steelmaking, one would not expect to remove sulfur in an acid-lined furnace. It is therefore not surprising that low and irregular recoveries, with erratic product performance, are observed when microalloy additions of deoxidizing/desulfurizing elements are made to acid-lined ladles. Improvements are to be anticipated when the working lining of a ladle consists of a stable, basic oxide such as magnesia, rather than alumino-silicate. Ladle spraying with a magnesia or burned-dolomite base mix should also improve recoveries and performance.

POURING SYSTEMS

Behavior of Smooth and Rough Streams

In the casting of steels containing microalloying elements, particular care is required to minimize oxygen and

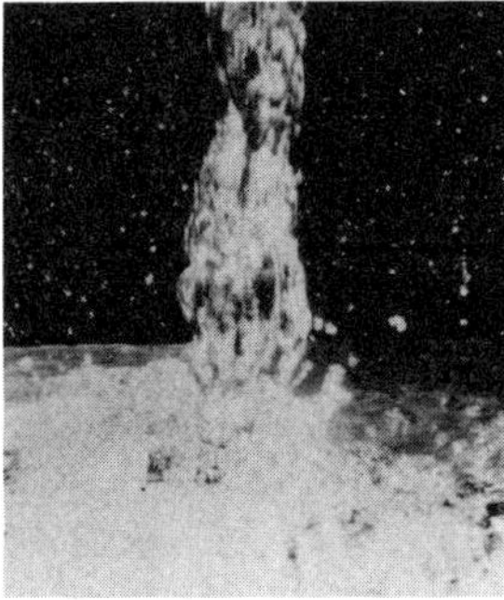

FIGURE 11 – Water-model study of smooth and rough streams showing the effects of gas entrainment on conditions in the mold pool.[25]

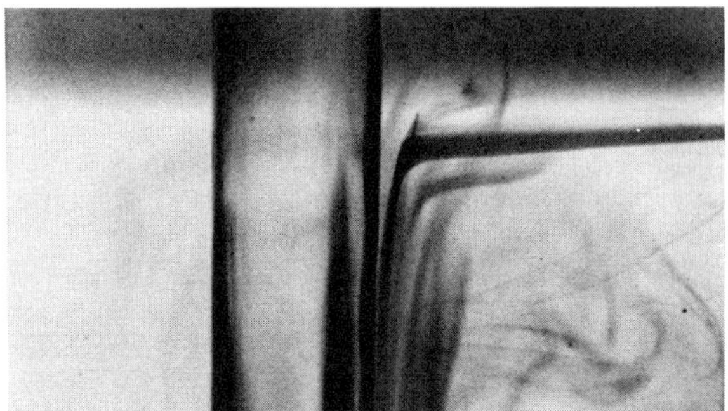

FIGURE 12 – Smoke injection onto a smooth water stream, illustrating how air is drawn into a mold by boundary-layer attachment.

nitrogen pick-up during pouring. Gas pick-up may occur as follows:

- By direct contact between the falling stream and the atmosphere;
- By reaction between the steel and bubbles of air carried down into the mold pool with the falling stream;
- By reaction at the pool surface with the atmosphere in the mold.

Water-model studies of casting streams and the effects of gas entrainment have been conducted by Maddever et al.[25,26] A comparison between smooth and rough flaring streams is shown in Figure 11.

With smooth streams, the gas that is dragged downwards by boundary-layer attachment (Figure 12) is peeled off at the pool surface with very little entering the mold pool. Under these conditions, gas pick-up will occur only by direct contact with the falling stream and at the pool surface. In the first case, the time of contact is only a fraction of a second. The surface-area/volume ratio of metal exposed to the atmosphere is small and thus gas pick-up should be minor. Similarly at the pool/gas interface, conditions are calm and relatively quiescent so that gas pick-up should be slight.

Exactly the opposite takes place with a rough, broken

a. Smooth Stream

b. Rough Stream

c. Close-Up of Rough Stream

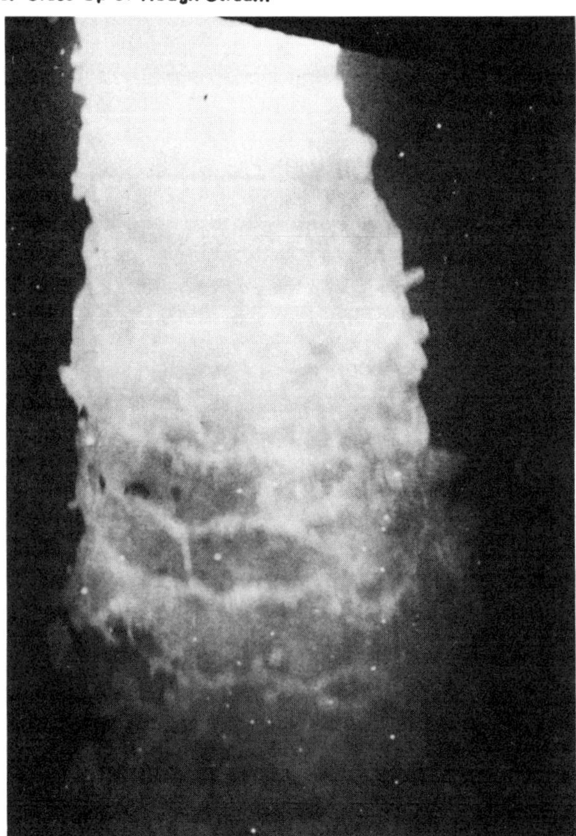

FIGURE 13 – Examples of smooth and rough steel streams.

FIGURE 14 – Smooth-stream penetration in the mold pool.[26]

or flaring stream. Copious quantities of gas are entrained by the falling fluid and carried into the bulk of the mold pool. The swarms of bubbles rising to the pool surface, together with a "wild" stream, create extremely turbulent conditions at the surface. In this case, conditions for gas pick-up are greatly enhanced. The surface-area/volume ratios of metal exposed to the atmosphere during the fall, within the pool, and at the pool surface are all drastically increased and severe contamination of steel can occur. Examples of smooth and rough steel streams are shown in Figure 13.

FIGURE 15 – Rough-stream penetration in the mold pool.[26]

Mixing Patterns in Mold

By using polyethylene beads as tracers, it has been possible to record photographically the mixing patterns in the mold associated with smooth and rough streams. In the former case (Figure 14), the particulate matter remains on the surface except when it drifts directly under the incoming stream, where it is driven deep into the mold pool. In practice, this could lead to the entrapment of mold scum within the internal structure of the ingot, billet, or slab. On the other hand, the metal surface next to the mold walls should be relatively clean and free from splash defects.

With a rough stream (Figure 15), the non-metallic material is forced outwards away from the incoming stream and towards the mold walls. Here it is carried downwards along the freezing interface, leading to the formation of massive surface or sub-surface defects.[25]

Reoxidation

Farrell et al.[27] have discussed the formation of inclusions by reoxidation during casting. They have shown that in aluminum-killed steel containing manganese and silicon, the first reoxidation products formed consist of large alumina galaxies. With continuing reoxidation, the newly formed inclusions contain decreasing amounts of aluminum and increasing amounts of manganese and silicon. The presence of silicate inclusions in aluminum-killed steel is strong evidence that reoxidation has occurred. The same conclusion applies to steels deoxidized with aluminum followed by additions of calcium, zirconium, or rare earths.

In the author's laboratories, the types of inclusions formed by reoxidation have been investigated by levitating droplets of steel in an atmosphere of controlled oxygen potential and recording the process of oxide

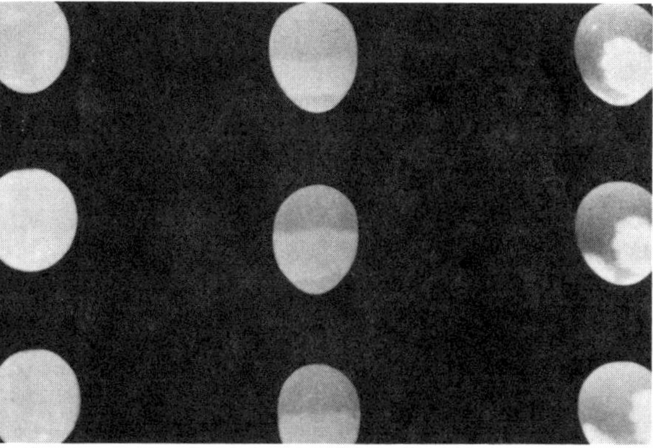

FIGURE 16 – Formation of reoxidation products on a molten-steel droplet.

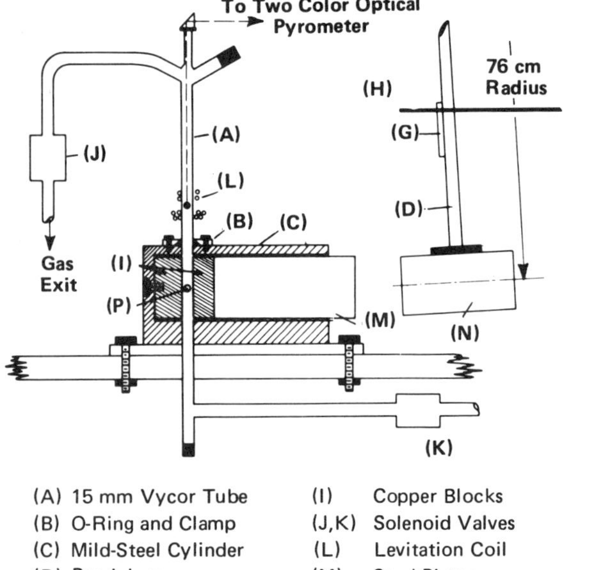

(A)	15 mm Vycor Tube	(I)	Copper Blocks
(B)	O-Ring and Clamp	(J,K)	Solenoid Valves
(C)	Mild-Steel Cylinder	(L)	Levitation Coil
(D)	Pendulum	(M)	Steel Piston
(G)	Sliding Brass Contact	(N)	12 lb. Lead Hammer
(H)	Brass Slide-Rod	(P)	Gas Inlet

FIGURE 17 – Droplet-quenching system used to study reoxidation of steel droplets.

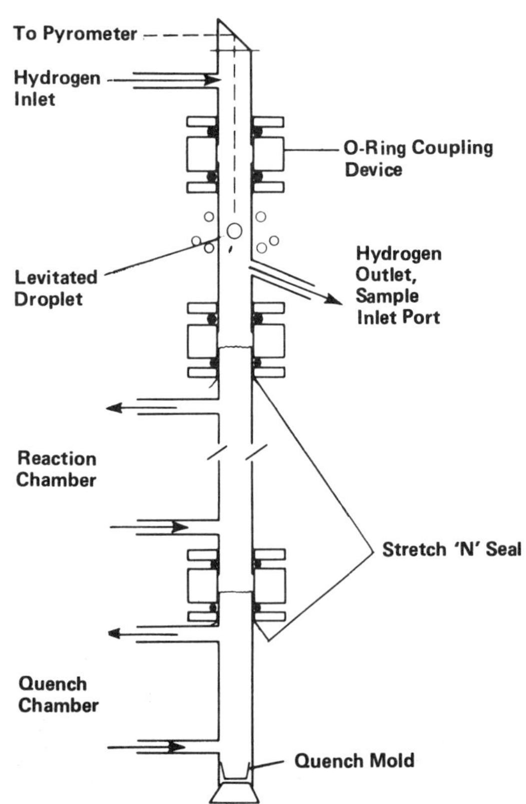

FIGURE 18 – Equipment for studying the rate of oxygen and nitrogen pick-up by falling steel droplets.

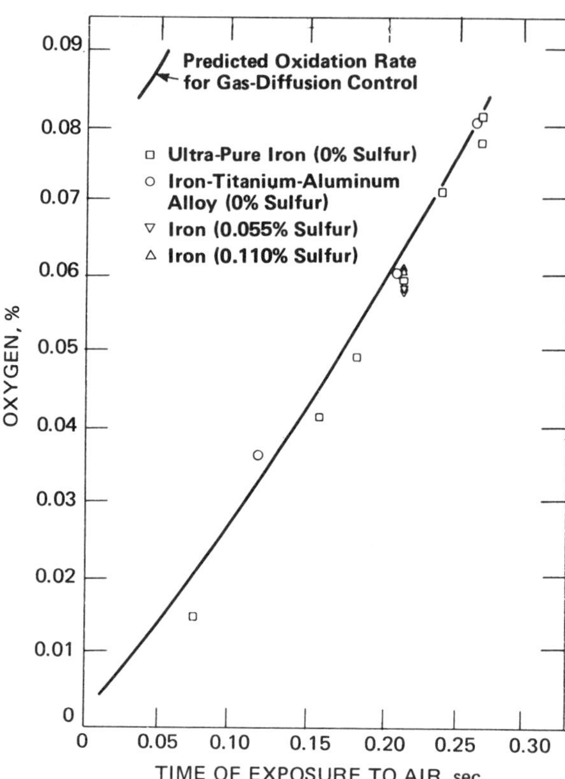

FIGURE 19 – Oxygen pick-up by falling droplets. Droplets were mold-quenched under nitrogen.[28]

formation with high-speed photography (Figure 16). At any point during the reaction, the power to the levitation coil can be cut off and the falling droplet quenched between two copper blocks with the piston and anvil device (Figure 17).

In other laboratory experiments on reoxidation, Greenberg[28] has measured the rate of oxygen pick-up by iron droplets falling through air (Figure 18). Steel droplets were melted within the levitation coil, allowed to fall for various times through air in the reaction chamber, solidified in a copper mold located within the quench chamber, and then analyzed for oxygen.

It was found that the reoxidation rate was unaffected by changes in the concentration of sulfur, titanium or aluminum (Figure 19). The measured rates were in good agreement with predicted rates for a reaction controlled by oxygen diffusion in the gas phase. This implies that the rate of reoxidation of steel exposed to the atmosphere is rapid and unhindered by the presence of sulfur or strong deoxidizers such as titanium or aluminum. The extent of reoxidation is minimized by decreasing the contact time between the steel and the atmosphere, i.e. in the fall height, and by adjusting conditions in the ladle and/or tundish to provide smooth pencil-like streams rather than coarse flaring streams with associated tearing, splashing, air entrainment and droplet formation.

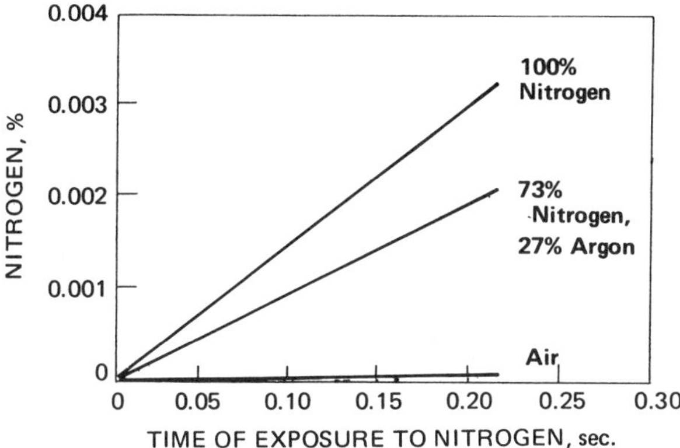

FIGURE 20 – Nitrogen pick-up by falling droplets. The initial droplet temperature was 1600°C.[29]

Nitrogen Absorption

In similar experiments, Greenberg[29] measured the rate of nitrogen pick-up by droplets falling through air, nitrogen, and argon-nitrogen mixtures (Figure 20). The pick-up rates are very much slower than those observed for oxygen. Further, the rate of nitrogen pick-up from air is very much less than that from pure nitrogen or nitrogen-argon mixtures of about the same nitrogen concentration as air. This implies that the amount of nitrogen picked up by direct contact with a pouring stream is small due to the lack of time available for reaction.

However, with rough or spraying streams, there is further opportunity for reaction due to the following considerations:

- The increase in the surface-area/volume ratio of metal exposed to the atmosphere;
- Air carried down into the mold pool forming swarms of gas bubbles, thus improving conditions for gas-metal reactions;
- Turbulent conditions generated at the mold-pool surface continuously exposing fresh metal for nitrogen pick-up.

During this investigation it was also found that the rate of nitrogen pick-up greatly increased as the dissolved sulfur level in the steel decreased below 0.02% (Figure 21). This implies that nitrogen absorption could be excessive in steels which had been strongly desulfurized by calcium, zirconium, or rare earths.

Koros reported plant data for the pick-up of nitrogen during tapping and teeming of rimmed and aluminum-killed steel.[30] In rimmed steel, there was no observable change in the nitrogen content of the steel between the furnace and the mold. On the other hand, the nitrogen content of aluminum-killed steel increased by 15 to 20 ppm during tapping and a further 15 to 20 ppm during teeming. These results were attributed to the surface-active effects of oxygen in rimmed steel decreasing the absorption rate of nitrogen. This effect is similar to that observed by Greenberg on the influence of surface-active sulfur on nitrogen absorption by falling steel droplets.

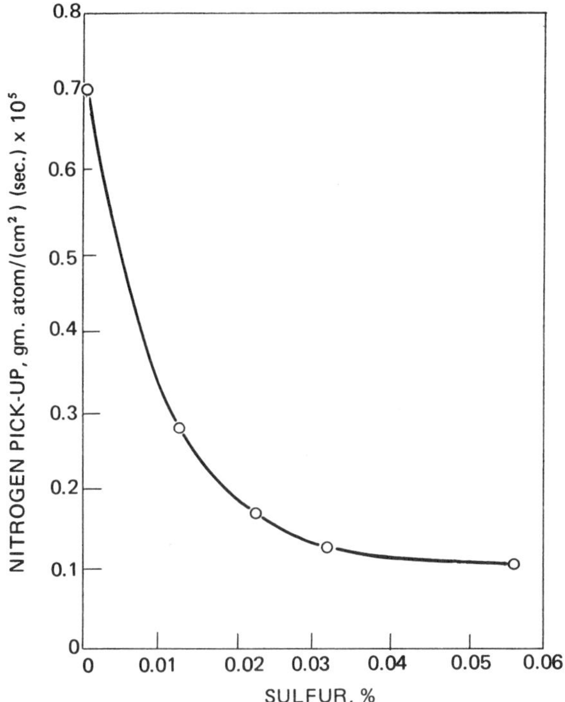

FIGURE 21 – The effect of sulfur on the rate of nitrogen pick-up by droplets falling through pure nitrogen at 1600°C.[29]

a. Pouring Stream without and with Castellated Nozzle

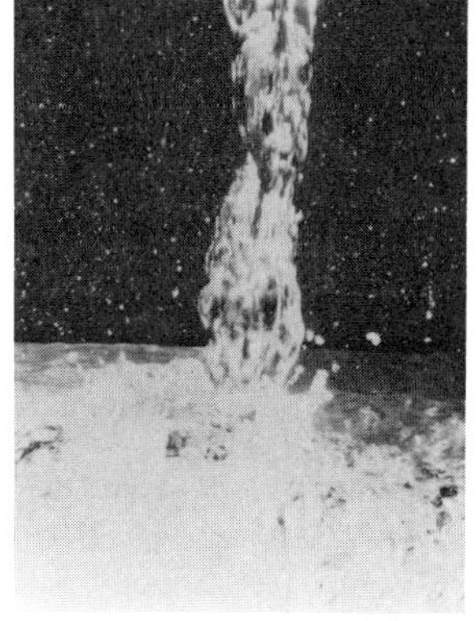

b. Castellated Nozzle before and after Use

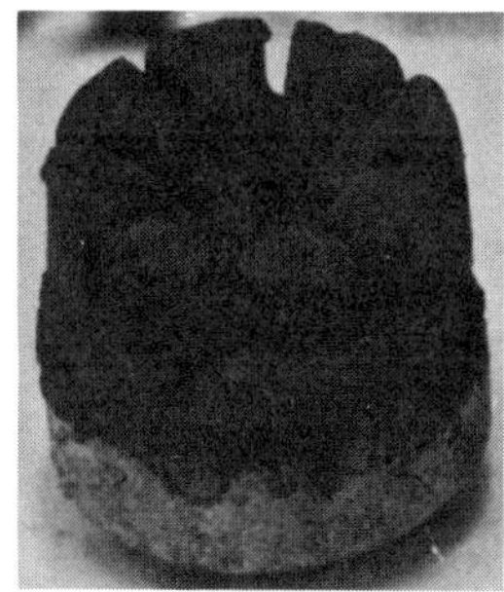

FIGURE 22 – Reduction in air entrainment of a pouring stream by use of a castellated nozzle.[25]

a. Stopper-Rod Control

b. Slide-Gate Valve

FIGURE 23 – Reduction in turbulence when a slide gate is used to meter the flow rather than a stopper rod.

a. Regular Stream

b. Use of Multi-Port Nozzle

FIGURE 24 – Improvement in tundish-stream conditions when a multi-port nozzle is used on the ladle to destroy turbulence within the tundish.[25]

Thus, for steels which have been strongly deoxidized and also strongly desulfurized by the addition of microalloying elements, particular care is required during pouring of the steel if the nitrogen content is to be kept under control.

Techniques that Reduce Gas Pick-up

Improvements in steel cleanliness, mechanical properties, and heat-to-heat consistency will be observed when pouring conditions are adjusted to minimize oxygen and nitrogen pick-up. Practical methods of achieving these objectives include proper design of pouring systems, i.e. ladles, tundishes, pouring boxes, the use of inert-gas shrouds, castellated nozzles which impart a controlled degree of turbulence to a pouring stream (Figure 22), multi-port nozzles, slide-gate valves and synthetic slags. Figure 23 shows how turbulence, air entrainment and hence, gas pick-up, are reduced when a slide-gate valve, rather than a stopper rod, is used to meter the flow from a ladle.[26] Figure 24 shows the dramatic improvement obtained in a tundish stream when a multi-port nozzle was used from ladle to tundish to destroy the turbulence created by the incoming ladle s ream in the tundish.

ALLOY-ADDITION PRACTICE

Alloy-addition methods have a profound effect on alloy recoveries and hence on the properties of the final product. Luyckx and Jackman[31] have reviewed various procedures which have been employed for the addition of rare-earth materials. Bingel and Scott[32] have discussed results obtained with several mold and ladle practices for both misch metal and rare-earth silicides.

Ladle and Mold Additions

An advantage of ladle additions is the ease of operation. However, unless precautions are taken, recoveries can be very low with incomplete control of inclusion morphology. Rare-earth additions to the ladle appear to be satisfactory when the alloys are added as rings, bars, or in canisters and plunged into fully-killed steel under a basic slag. Melt agitation by induction stirring, gas injection, or the use of magnesium-containing additives[31] helps to remove reaction products and reduce the sulfur level, particularly if the ladle is lined with a stable, basic refractory to minimize melt reoxidation.

While higher alloy recoveries and more consistent shape control are generally associated with mold additions, there is a greater tendency for the formation of large inclusion clusters and hence poorer cleanliness. Variability in stream conditions will certainly affect alloy recovery, cleanliness, and product performance, even when all of the additions have been made to the ladle. For example, unsatisfactory ingots may be obtained at the beginning of a cast due to a wide flaring stream caused by a partially opened nozzle with the stopper rod disturbing the flow.

Mold and stream shrouding, bottom pouring, the use of rare-earth-filled steel tubes suspended in the mold, plunging after teeming is complete, and the use of wire feeding have all contributed to improvements in the reliability of alloy recovery, inclusion control, and service performance of the final product.

Optimizing Calcium Additions

In using highly reactive alloying materials, the extent of success is mainly determined by the method adopted for their addition. This was well illustrated at this symposium by Pircher and Klapdar.[33] They described an operating procedure for injecting calcium-silicon and calcium carbide into aluminum-killed steel in the ladle using an argon-carrier gas and a refractory-coated lance which is immersed about 2.7 m below the metal surface.

In this way, calcium losses due to air oxidation are completely avoided. In addition, by using ladles with a dolomite lining rather than fireclay, oxygen pick-up by refractory-melt interaction is largely eliminated and consequently the desulfurization efficiency of the calcium is increased. With this treatment, the oxygen and sulfur content of the steel is reduced, the number of inclusions is decreased, and elongated sulfides and oxide stringers are replaced by globular calcium aluminates containing sulfur which maintain their shape during hot-rolling. A further advantage is that the resulting inclusions are randomly distributed throughout the final product.

SUMMARY

Inclusion morphology can be controlled by alloy additions. The elements used and the means of addition have profound influences on the final results. Aluminum-killed steels generally have good oxide morphology, but manganese-sulfide inclusions are generally present as stringers found during hot rolling. The addition of strong sulfide formers, such as zirconium or the rare earths, normally leads to less plastic sulfides.

Advantages of Semi-Killed Steels

Because of the higher oxygen content, semi-killed steels generally form oxysulfides, which are much less plastic than the iron-manganese sulfides. In this case, additions of sulfide-controlling elements may not be necessary. If silicon is used for deoxidation, the formation of silicates constitutes a problem. However, with aluminum control of the oxygen level and in the absence of silicon, there is the possibility of producing silicate-free steel, in which the oxide inclusions will consist mainly of alumina and iron-manganese aluminates. All of these oxides are solid in the molten steel and non-plastic during hot rolling.

With aluminum semi-killed steel therefore, both sulfide and oxide inclusions are in a form which favors uniformity of properties in both the longitudinal and transverse directions. In such steels, precautions must be taken to prevent silicon pick-up from furnace slag carried into the ladle and from both refractories and alloy additions such as ferromanganese. If this is not done, manganese aluminosilicates are likely to be formed as a stable phase, leading to oxide stringers during hot rolling and anisotropic properties.

Aluminum semi-killed steels have other advantages over aluminum-killed steels containing sulfide-shape control additives. They give higher yields during ingot casting and improved yield by continuous casting, since problems of nozzle blockage caused by large residual concentrations of aluminum or other strong deoxidizers are avoided.

Overcoming Sulfide Problems

Sulfide inclusions are the most serious problem in HSLA steels. With the use of higher sulfur coals for iron production, increasing attention is being given to external desulfurization of hot metal to produce low-sulfur steels. From a chemical viewpoint, it is far easier to reduce the sulfur content of hot metal than to reduce the sulfur content of steel, since both carbon and silicon strongly increase the activity coefficient of sulfur and thus enhance its removal. As a result, consideration is being given today to the relative merits and cost benefits associated with the production of extra-low sulfur steel versus the use of sulfide shape-control additives. At the present time, a combination of both desulfurization at the higher sulfur levels and shape control at the lower levels seems to yield the best economics.

Rare earths (such as silicide or mischmetal), zirconium, titanium, calcium, magnesium, or alloy mixtures of these materials are the most common shape-control agents. All of these additives have a high affinity for oxygen and can only be used effectively in killed steels. For the same reason, problems are encountered by the interaction between steels containing these elements and aluminosilicate ladle refractories. This is particularly so if attempts are made to desulfurize in the ladle, possibly with the addition of basic slag formers, rather than simply altering the sulfide morphology. Under these circumstances the ladle must be regarded as a reaction vessel and treated accordingly.

High Oxygen in Low-Carbon HSLA Steels

Since microalloyed steels are characterized by a low-carbon content, the oxygen content prior to deoxidation will be high. While this chemistry leads to a low volume content of carbides, which is desirable, there is the inherent danger that the volume content of oxides will be increased.

In silicon-killed steels, silicate stringers represent a major problem. In aluminum-killed steels, particularly those which have undergone reoxidation, the formation of alumina clusters can be dangerous. From this standpoint, there is an advantage if a higher carbon content can be tolerated, since the oxygen content is then lower and the volume fraction of oxides will be decreased.

On the other hand, similar benefits will be obtained by carbon removal under vacuum or reduced CO pressures – for example, by the use of argon-oxygen injection. Theoretically, if the partial pressure of CO is reduced by a factor of 10, then for any particular carbon level, the equilibrium oxygen content is also reduced by the same factor. This reduction will decrease the volume fraction of oxides.

Interaction of Many Factors

In this paper the thermodynamic factors which govern the formation of oxides, sulfides, nitrides and carbides when microalloying elements are added to steel, have been expressed diagrammatically in terms of the relative stability of the solubility products. These diagrams help to understand the different types of behavior associated with the microalloying additions.

In the manufacture of HSLA steels, control of the oxygen and nitrogen residuals is of practical importance, not only from the standpoint of controlled precipitation of oxides and nitrides, but also with respect to desulfurization and control of sulfide morphology. Consideration has therefore been given to the effects of stream turbulence, air entrainment, and sulfur content of the steel on the pick-up of oxygen and nitrogen from the time the steel leaves the furnace until it solidifies. The influence of refractory/glaze/melt interactions on melt reoxidation and subsequent loss of sulfide-shape control has been described and the importance of considering the ladle as a steelmaking vessel has been emphasized.

Finally, the implications of various alloy-addition practices have been outlined in relation to alloy recovery, inclusion control, heat-to-heat consistency, and final-product quality.

ACKNOWLEDGEMENTS

The authors should like to acknowledge the financial support of the American Iron and Steel Institute and Molycorp, Inc. They would also like to express their appreciation to Mr. J. Hiam, Dominion Foundries and Steel Ltd., for many helpful discussions.

REFERENCES

1. D. Ghosh and D.A.R. Kay: The Electromechanical Society, Inc., *International Symposium on Metal-Slag-Gas Reactions and Processes,* 1975, proceedings volume, p. 267-283.
2. R.J. Freuhan: *Metallurgical Transactions,* 1970, vol. 1, p. 3403-10.
3. J.F. Elliott, M. Gleiser, and V. Ramakrishna: *Thermochemistry for Steelmaking,* Addison-Wesley Publishing Co., Reading, Massachusetts, 1963, vol. II.
4. J.F. Elliott and M. Gleiser: *Thermochemistry for Steelmaking,* Addison-Wesley Publishing Co., Reading, Massachusetts, 1960, vol. 1.
5. W.G. Wilson, D.A.R. Kay and A. Vahed: *Journal of Metals,* May 1974, p. 14-23.
6. A. Kontopoulos: Ph.D. Thesis, McMaster University, 1971.
7. J.C. Chan, C.B. Alcock and K.T. Jacob: *Canadian Metallurgical Quarterly,* 1973, vol. 12, p. 439-443.
8. C. Bodsworth and H.B. Bell: *Physical Chemistry of Iron and Steel Manufacture,* Longman Group Limited, London, 1972.,
9. B. Pollard: *Metals Technology,* July 1974, p. 343-7.
10. R.C. Hudd, A. Jones and M.N. Kale: *Journal of Iron and Steel Institute,* 1971, vol. 209, p. 121-25.
11. K.A. Gshneidner and N. Kippenham: "Thermochemistry of the Rare-Earth Carbides, Nitrides and Sulfides for Steelmaking," Rare-Earth Information Center (Iowa State University, Ames) report IS-RIC-5, 1971.
12. K.J. Irvine, F.B. Pickering and T. Gladman: *Journal of Iron and Steel Institute,* 1967, vol. 205, p. 161-82.
13. K. Narita: *Transactions, Iron and Steel Institute of Japan,* 1975, vol. 15, p. 145-52.
14. L.S. Darken et al., *Journal of Metals,* 1951, vol. 3, p. 1174.
15. Yu. V. Kryakovskii, I. Nikolaev and V.I. Yavoisku: (U.S.S.R.) Sb., Mosk. Inst. Stali Spltivov, 1968, no. 48, p. 135-42. Chemical Abstract, 1969, vol. 70, 220546.
16. T. Gladman, B. Holmes and I.D. McIvor: "Effects of Second Phase Particles on Strength, Toughness and Ductility", BSC/ISI Conference, The Effects of Second Phase Particles on the Mechanical Properties of Steel, Scarborough (U.K.), 1971.
17. D. Brown: British Steel Corporation Group Research Report, SNW(C) F 77/8.
18. M. Espenhahn, H.-E. Buhler and H. Litterscheidt: *Thyssenforschung,* 1969, vol. 1, p. 136-43.
19. T. Gladman, D. Dulieu and I.D. McIvor: "Structure – Property Relationships in High-Strength Microalloyed Steels": *Microalloying 75 Proceedings.*
20. D.H. Jack and K.H. Jack: *Materials Science and Engineering,* 1973, vol. 11, p. 1-27.
21. J.M. Arrowsmith: British Steel Corporation Group Research Report, SNW(C)/E 7/21.
22. K.A. Gshneidner: *International Symposium on Sulfide Inclusions in Steel,* American Society for Metals, 1974, p. 159-177.
23. W-K. Lu and A. McLean: *Ironmaking and Steelmaking,* 1974, vol. 1, no. 4, p. 228-233.
24. D.A.R. Kay, W-K. Lu and A. McLean: *International Symposium on Sulfide Inclusions in Steel,* American Society for Metals, 1974, proceedings volume, p. 23-43.
25. W. Maddever, A. McLean, J.S. Lucket and G. Forward: *Canadian Metallurgical Quarterly,* 1973, vol. 12, no. 1, p. 79-88.
26. W. Maddever: M.A. Sc. Thesis, University of Toronto, 1974.
27. J.W. Farrell, P.J. Bilek and D.C. Hilty: *Proceedings, Electric Furnace Conference,* AIME, 1970, vol. 28, p. 64-86.
28. L.A. Greenberg and A. McLean: *Transactions, Iron and Steel Institute of Japan,* 1974, vol. 14, p. 395-403.
29. L.A. Greenberg and A. McLean: (to be published).
30. P.J. Koros, S.S. Lewis, J.G. Shawgo and J. Silver: *Proceedings, National Open Hearth and Basic Oxygen Steel Conference,* AIME, 1970, vol. 53, p. 97-107.
31. L. Luyckx and J.R. Jackman: *Proceedings, Electric Furnace Conference,* AIME, 1973, vol. 31, p. 175-181.
32. C.J. Bingel and L.V. Scott: Ibid., p. 171-174..
33. H. Pircher and W. Klapdar: "Controlling Inclusions in Steel by Injecting Calcium into the Ladle," *Microalloying 75 Proceedings.*

Biographies – A. McLean studied metallurgy at the Royal College of Science and Technology in Glasgow and obtained his Ph.D. from the University of Glasgow in 1963. Following five years at McMaster University in Hamilton, Ontario, he joined the Graham Research Laboratories of Jones and Laughlin Steel Corporation in Pittsburgh, supervising the deoxidation and casting group. In 1970, he moved to the University of Toronto as Associate Professor.

D.A.R. Kay also studied at the Royal College of Science and Technology and received his degrees from the University of Glasgow – a B.Sc. in 1956 and a Ph.D. in 1960. He began his career with the United Kingdom Atomic Energy Authority and then joined the University of Sheffield in 1963 as Lecturer in Metallurgy. He came to McMaster University in 1969, where he specializes in the application of metallurgical physical chemistry to iron and steelmaking.

QUESTIONS FOR AUTHORS

1. RARE-EARTH TREATMENT OF TITANIUM STEELS

Question: Is rare-earth treatment of titanium steels feasible in view of the relative stabilities of the titanium and cerium sulfides and nitrides, particularly in those steels where the atomic percentage of titanium greatly exceeds that of cerium?

Answer: It is possible that rare-earth treatment of titanium steels could be beneficial. In an aluminum-deoxidation practice, good deoxidation and reoxidation control is essential for the protection of soluble titanium.

It can be calculated from Figure 2 of our paper that 0.05% soluble aluminum is required to protect about 0.2% titanium in solution at 1600°C. If the soluble aluminum is 0.02%, then only about 0.06% titanium can be retained in solution. Deoxidation control with aluminum alone is therefore difficult.

Adequate rare-earth treatment should preclude both Al_2O_3 and Ti_3O_5 formation, control sulfide morphology, eliminate lamellar-eutectic carbosulfides, and retain titanium in solution. The role of titanium is then confined to the solid state and to carbonitride precipitation required for grain refinement and precipitation strengthening.

Although cerium nitride is more stable than titanium nitride at 1100°C. as shown in Figure 10, the soluble rare-earth content in most commercial rare-earth-treated steels is probably very low. In addition, 10 ppm of soluble cerium can combine with only 1 ppm of soluble nitrogen to form the nitride, and the amount of nitrogen "fixing" by cerium is likely to be small.

Deoxidation with calcium or magnesium would also keep titanium in solution, but due to the low residual solubility of these elements in molten steel, there would be essentially no protection against reoxidation.

2. REOXIDATION OF RARE EARTHS

Question: In rare-earth-treated steels, have you found inclusions which were unequivocally caused by reoxidation from air during teeming? What would be their location in the ingot? From a reoxidation viewpoint, does it make any difference whether rare earths are added to the ladle or to the mold?

Answer: Unless precautions are taken, opportunities for reoxidation are greater when rare earths are added to the ladle than in the mold, since oxygen pick-up may occur from an oxidizing slag carried over from the furnace, from ladle glaze or refractories, and from a chemical reaction with the atmosphere during pouring. Much of this reoxidation is avoided by mold additions, although other problems may arise, such as materials handling, variable recovery between ingots, and final-steel cleanliness.

If rare earths are added to the ladle, we doubt if it is possible to specifically pinpoint air as the particular cause of reoxidation, since the oxygen may be derived from several other sources. During teeming, a substantial amount of air is carried into the mold pool so that oxygen pick-up from bubbles rising up through the steel is rapid and essentially complete. Under these circumstances, if rare earths are added to the mold having incomplete sulfide-shape control, one would expect inclusions containing rare-earth oxides, alumina, and in extreme cases even manganese silicates. Complex inclusions of this type indicate transient reactions within localized zones. Such reactions form solid or liquid products which are often out of equilibrium with each other and with the bulk composition of the steel.

The location of such reaction products within the ingot will be strongly influenced by the type of flow patterns generated by the incoming ladle stream, the character of which is affected by such factors as the relative position of the stopper rod with respect to the nozzle opening, and the head of steel in the ladle.

3. OPTIMUM METHOD OF ADDING RARE EARTHS

Question: What do you recommend to the steelmaker as an optimum method of adding rare earths?

Answer: We would recommend adding rare earths to an aluminum-killed steel in a basic-lined ladle after tapping is complete and the steel is protected from reoxidation by a basic, reducing slag. The rare earths could be added with a plunging device similar to that used for desulfurizing hot metal with magnesium-impregnated coke. As an alternative, the material could be added with a gas-injection technique similar to that described at this meeting by Pircher and Klapdar for calcium and magnesium additions. Argon injection through a porous plug would enhance separation of the sulfides into the slag or ladle lining, thus reducing the total sulfur content of the steel. To retain shape control of the sulfides remaining in the steel, atmospheric reoxidation during casting should be minimized.

PHYSICAL CHEMISTRY
OF EXTERNAL DESULFURIZATION
OF MOLTEN IRON AND STEEL

Robert D. Pehlke

Dept. of Materials and Metallurgical Engineering
The University of Michigan
Ann Arbor, Michigan 48109

This paper originally appeared in the Steelmaking Proceedings, Vol. 61, pp. 55, 1978.

ABSTRACT

A wide variety of external desulfurization processes for molten iron and steel are thermochemically analyzed. The thermodynamics of the desulfurization reactions and the resultant phase equilibria including sulfides, oxysulfides and oxides for various processes are reviewed. The potential capabilities for desulfurization are evaluated in each case. Desulfurization of iron and steel using magnesium, calcium, rare earths, lime, calcium carbide and other compounds are discussed in terms of the fundamentals of processing techniques and their integration into an iron and/or steelmaking operation.

INTRODUCTION

Control of sulfur can be accomplished by external desulfurization of the molten iron to be supplied to the steelmaking vessel or by external desulfurization of the refined steel prior to casting. Several desulfurizing systems or agents can be used including magnesium, calcium, rare earths, lime, calcium carbide and several other compounds. This work develops a direct comparison of the desulfurization reactions in iron and in steel in terms of the relevant thermodynamics and phase equilibria. The comparative feasibility of the desulfurization systems is utilized to define the limits of potential application in production of low sulfur steel.

THE SULFUR PROBLEM

The presence of sulfur in steel is considered to be undesirable at all concentration levels, except for high machinability grades. Every effort is made to achieve low sulfur content which in turn reduces rejections in the mill, and also will meet property requirements for finished products.

Low sulfur steel, according to current steelmaking standards, can be considered to be that with less than 0.02 wt% S, with concentrations down to 0.005 wt% being specified as maximum for certain grades. A 0.015 wt% sulfur maximum can be produced in the oxygen steelmaking processes with hot metal desulfurization and careful selection of scrap, and even in the electric furnace with scrap control. However, sulfur content of the fuel in the open hearth furnace, even if the hot metal were desulfurized and scrap controls were invoked, would require external desulfurization of the steel because of the high sulfur input from the fuel. At a sulfur specification of 0.010 wt% maximum, the grade can be produced in the oxygen steelmaking processes with hot metal desulfurized below 0.01 wt%, or by external desulfurization of the steel. At 0.005 wt% maximum sulfur, steel desulfurization is required.(1)

The external desulfurization systems which are commercially utilized will be examined in terms of their thermochemistry to define limitations and potential applications. The kinetics of desulfurization and the specific equipment and procedures utilized to implement desulfurization will not be considered.

THERMODYNAMIC APPROACH

The solution behavior of sulfur in liquid iron has been studied extensively and, at concentrations below about 1/2 wt%, sulfur exhibits Henrian behavior, i.e. the activity or tendency to react is directly proportional to concentration. However, the presence of other alloying elements in appreciable concentrations can influence the activity of sulfur. In particular, carbon dramatically increases the activity of sulfur. Silicon also increases the activity, whereas manganese slightly decreases the activity of sulfur. Consequently, in the case of hot metal with its high carbon and silicon contents, the activity of sulfur at a given sulfur concentration is substantially higher than it would be in most steels where the concentration of carbon and silicon are relatively low.

As a basis for comparing the potential effectiveness of various desulfurization reactions, the activity of sulfur in hot metal is taken as being 10 times its concentration, whereas for steel the activity is taken as being equal to its concentration, i.e. the activity coefficient, f_S, is unity.

$$\text{Hot Metal} \quad a_S = f_S[\%\underline{S}] \sim 10[\%\underline{S}] \qquad [1]$$

$$\text{Steel} \quad a_S = f_S[\%\underline{S}] \sim [\%\underline{S}] \qquad [2]$$

The activity of the desulfurizing agent and of the reaction product, a sulfide or oxysulfide, usually can be assumed to be unity. In the case of magnesium, and of calcium in steel, the reactants are vaporized and the activities are taken as unity for the vapor at approximately one atmosphere. For calcium in hot metal, the limited solubility usually results in the calcium being dispersed in the liquid pig iron as droplets of liquid at unit activity. For rare earths, various concentration, i.e. activity, levels are considered. The reaction products, sulfides or oxysulfides, usually form as nearly pure compounds and their activity can be taken as unity.

Another factor which is important in considering the thermochemistry of desulfurization reactions is temperature. It will be assumed that the activity behavior of sulfur is independent of temperature and the solute interactions which for the most part have been measured at steelmaking temperatures, 1600°C (2912°F), can be applied at typical iron temperatures, 1370°C (2498°F). Evaluation of the equilibria involved for the desulfurization reactions will be calculated at these respective temperatures for desulfurization of steel and of iron.

ROLE OF OXYGEN

The role of oxygen in desulfurization should receive considerable emphasis because those elements which form stable sulfides also form stable oxides. Consequently, the utilization of reactive metallic elements for removal of sulfur requires that the oxygen content of the liquid metal be extremely low. This means that in the case of desulfurization of steels that the steels be fully killed. In addition, in the case of certain elements, namely rare earths, zirconium and possibly titanium, the formation of an oxysulfide as a stable separate phase must be considered. The formation of the oxysulfide can be important in regard to the form of residual nonmetallic inclusions and can also affect the stoichiometry of the desulfurization reaction.

DESULFURIZATION SYSTEMS

Magnesium

The desulfurization systems considered and the comparative results are summarized below. Magnesium desulfurization can be selected as a detailed illustrative example where the desulfurization reaction is

$$Mg(g) + \underline{S} = MgS(s) \qquad [3]$$

The equilibrium constant has been calculated at 1370 and 1600°C for hot metal and steel, respectively.

$$K = \frac{a_{MgS}}{P_{Mg} a_S} \qquad \text{Hot metal } K = 18{,}500 \quad \text{Steel } K = 412 \qquad [4]$$

The thermochemical data have been taken primarily from references 2 and 3. Noting that magnesium will exist as a gas at these temperatures and assuming that the desulfurization product is pure magnesium sulfide, the equilibrium residual sulfur contents can be calculated.

$$K = \frac{a_{MgS}}{P_{Mg} a_S} \qquad a_{MgS} = 1 \quad P_{Mg} \sim 1 \text{ atm.} \qquad [5]$$

Hot Metal $[\%S] \underset{\sim}{\sim} 5.4 \times 10^{-6}$

Steel $[\%\underline{S}] \underset{\sim}{\sim} 0.0024$

Calcium

Calcium metal as a desulfurization agent can be shown to have a potential effectiveness both for hot metal and for steel. It should be noted that calcium boils at a temperature intermediate between those typical of liquid pig iron and of steel and this difference in reacting phase must be accounted for in the thermochemical analysis.

$$Ca(\ell) + \underline{S} = CaS(s) \qquad [6]$$

Hot Metal $K = 7.34 \times 10^{8}$

$[\%S] \underset{\sim}{\sim} 1.4 \times 10^{-10}$

$$Ca(g) + \underline{S} = CaS(s) \qquad [7]$$

Steel $K = 9.3 \times 10^{6}$

$[\%\underline{S}] \underset{\sim}{\sim} 10^{-7}$

Rare Earths

Rare earths which are typically categorized by the thermochemical behavior of cerium and lanthanum, but also include several other metals, are used in steel for purposes of desulfurization and sulfide shape control. The theoretical aspects of these equilibria have been developed by Turkdogan (4) wherein the solubility products have been shown to be low for the oxides, sulfides and oxysulfides.

The thermodynamics and phase equilibria of rare earths in steelmaking have been treated in detail by Wilson, Vahed and Kay (5,6) and by McLean and Lu.(7) Using the data of Vahed and Kay,(6) reported for 1900°K, the equilibrium constants for the formation of cerium and lanthanum (the principal constituents of rare earth additions to steel) oxides, sulfides and oxysulfides are as follows:

Table I. Equilibrium Constants at 1900°K for Some Cerium and Lanthanum Compounds

Reaction	$K_{1900°K}$
$2\,\underline{Ce} + 3\,\underline{O} = Ce_2O_3$	3.3×10^{20}
$2\,\underline{La} + 3\,\underline{O} = La_2O_3$	1.2×10^{22}
$\underline{Ce} + \underline{S} = CeS$	2.1×10^{5}
$3\,\underline{Ce} + 4\,\underline{S} = Ce_3S_4$	1.4×10^{18}
$2\,\underline{Ce} + 3\,\underline{S} = Ce_2S_3$	2.5×10^{12}
$\underline{La} + \underline{S} = LaS$	1.0×10^{5}
$2\,\underline{Ce} + 2\,\underline{O} + \underline{S} = Ce_2O_2S$	7.7×10^{19}
$2\,\underline{La} + 2\,\underline{O} + \underline{S} = La_2O_2S$	1.4×10^{21}

The large values for these equilibrium constants indicate a very high effectiveness for the desulfurization of steel by rare earths.

The rare earths are very effective deoxidizers, the equilibrium oxygen content, for example, being calculated as less than 1 ppm in the presence of 10 to 100 ppm cerium for the equilibrium with Ce_2O_3. (6) Also, a residual cerium concentration of 100 ppm could reduce the sulfur to about 10 ppm. In fact, however, Ce_2O_3 is normally not found in aluminum-killed steels and the precipitation sequence is $(RE)_2O_2S$ followed by the rare earth sulfides. The oxysulfide equilibrium for cerium predicts a very low sulfur content.

$$2\,\underline{Ce} + 2\,\underline{O} + \underline{S} = Ce_2O_2S \qquad [8]$$

$$K = 7.7 \times 10^{19} = \frac{1}{[\%Ce]^2[\%O]^2[\%S]}$$

For [% Ce] = 0.01 and [% O] = 0.001;

$$[\%\ S] = 1.3 \times 10^{-10}$$

The rare earth desulfurization would be very effective as well if used for treatment of hot metal.

Titanum and Zirconium

Titanium and zirconium which are effective deoxidizers, and can be used for removal of sulfur from molten pig iron and/or steel, have not been used commercially to any great extent because of cost, and also because of the hot shortness which zirconium may cause or the low temperature brittleness which has been found in steels containing titanium and/or zirconium. It should be noted that the thermochemical data for these systems is limited.

Calcium Carbide

Calcium carbide is an effective desulfurization agent. As shown below at hot metal temperatures and under conditions where the carbon activity is essentially that of pure graphite the carbide does not tend to decompose in hot metal as it would in steel. Nevertheless, calcium carbide is highly effective as a desulfurizer of hot metal, provided adequate surface area is available for reaction.

$$CaC_2(\beta) + \underline{S} = CaS(s) + 2C(s) \qquad [9]$$

Hot Metal $K = 4 \times 10^5$

$$a_S \sim 10\%\ \underline{S} \qquad a_{CaS} = 1$$

$$a_{CaC_2} = 1 \qquad a_C = 1$$

$$[\%S] \sim 2.5 \times 10^{-7}$$

A similar effectiveness can be shown for calcium carbide if used with steel, although the carburizing potential of the material restricts this application.

The decomposition of calcium carbide at typical molten pig iron temperatures can be described in terms of the equilibrium constant as shown in Eq. 10.

$$CaC_2(\beta) = Ca(\ell) + 2C(s) \qquad [10]$$

At Hot Metal Temperature

$$K = \frac{a_{Ca}a_C^2}{a_{CaC_2}} = 5.42 \times 10^{-4}$$

The fact that the equilibrium constant is substantially less than unity, under circumstances where the decomposition of the carbide at unit activity would produce calcium liquid at unit activity in the presence of carbon saturated iron, shows that calcium carbide would not tend to decompose in this situation. Consequently, the utilization of calcium carbide as a desulfurizer of hot metal requires that maximum surface area of the solid carbide be provided for reaction with the molten iron. The thermodynamics are highly favorable for desulfurization which can be enhanced substantially through the provision of maximum reaction surface area.

Lime

Lime is an effective desulfurizer and is utilized in steelmaking slag to provide a high basicity and the reaction mechanism for removal of sulfur from steel in the furnace. The injection or contacting of lime with liquid iron or with steel can provide a means for sulfur removal. The desulfurization reaction for metal-slag systems is normally written as

$$\underline{S}(Metal) + O^{=}(Slag) = \underline{O}(Metal) + S^{=}(Slag) \qquad [11]$$

In the case of hot metal, where carbon and silicon are present, the desulfurization reaction is markedly enhanced because of the oxygen released with carbon or silicon as indicated in equations 12 and 13.

$$\underline{S} + CaO(s) + \underline{C} = CaS(s) + CO(g) \qquad [12]$$

$$\underline{S} + 2CaO(s) + 1/2\underline{Si} = CaS(s) + 1/2Ca_2SiO_4 \qquad [13]$$

For desulfurization of hot metal by injection of lime, the equilibrium relationship which depends upon the activity of calcium sulfide and lime is as follows:

$$K = \frac{a_{CaS}a_O}{a_{CaO}a_S} = 0.012 \qquad [14]$$

$$a_S = 8.3[a_O]\frac{a_{CaS}}{a_{CaO}}$$

for Hot Metal $a_S \sim 10[\%S]$

$$a_O < 10^{-4}$$

$$[\%\underline{S}] < 8.3 \times 10^{-4}\frac{a_{CaS}}{a_{CaO}}$$

Table II. Summary of Results of Thermochemical Analysis of Desulfurization Systems

Reactant	Desulfurizing Reaction	Approximate weight percent Equilibrium Residual Sulfur in Hot Metal	Steel
Magnesium	$Mg(g) + \underline{S} = MgS(s)$	5.4×10^{-6}	0.0024
Calcium	$Ca(\ell) + \underline{S} = CaS(s)$	1.4×10^{-10}	-
	$Ca(g) + \underline{S} = CaS(s)$	-	1.0×10^{-7}
Rare Earths* (e.g. Cerium)	$2\underline{Ce} + 2\underline{O} + \underline{S} = Ce_2O_2S$	-	$\frac{1.3 \times 10^{-20}}{[\%Ce]^2[\%O]^2}$
Calcium Carbide	$CaC_2(\beta) + \underline{S} = CaS(s) + 2C(s)$	2.5×10^{-7}	-
Lime	$CaO(s) + \underline{S} = CaS(s) + \underline{O}$	$8.3\ [a_O]$	$40[\%\ O]$

* 1900°K

Thus the desulfurization of hot metal using lime is a favorable process.

For steel,

$$CaO(s) + \underline{S} = CaS(s) + \underline{O} \quad [15]$$

$$K = \frac{a_{CaS}a_O}{a_{CaO}a_S} = 2.5 \times 10^{-3}$$

$$\text{Steel } [\%S] \simeq 40[\%O]\ \left(\frac{a_{CaS}}{a_{CaO}}\right)$$

Thus, desulfurization of steel with lime, even for killed grades, is a marginal process.

The desulfurization systems considered and the comparative results are summarized in Table II.

SUMMARY OF RESULTS

Various desulfurization systems have been examined from a thermochemical viewpoint showing that several have the potential to reduce sulfur concentrations in liquid iron and steel to extremely low levels. These effective systems include the rare earths, calcium, and in the case of molten pig iron also magnesium. Magnesium treatment of killed steel can also remove sulfur but to a limited extent. Lime injection is marginal even for killed steels, but can be effective in pig iron. Calcium carbide does not tend to decompose in molten pig iron, and consequently will be dependent upon a high surface contact area for effectiveness. Titanium and zirconium also have potential for use as desulfurizers. The strategy developed to implement a desulfurization system will be highly dependent upon oxygen control in the molten iron or steel.

The practical application and implementation of these systems in commercial practice will depend upon the specific physical or mechanical systems utilized to carry out the reactions and the related kinetics for desulfurization. This form of treatment to produce the high quality products required at present is becoming of increasing necessity. The understanding of the thermochemistry of sulfur control can provide the opportunity to produce a high quality steel product with an optimum processing strategy.

REFERENCES

1. Orton, J.P., Koros, P.J., and Bosley, J.J., "New Developments in Low Sulfur Steel - Needs and Production," AISI 85th General Meeting, New York, May 1977.
2. Chipman, J. and Elliott, J.F., "Physical Chemistry of Liquid Steel," Chapter 16, Electric Furnace Steelmaking, AIME, New York, 1963.
3. Elliott, J.F. and Gleiser, M., Thermochemistry for Steelmaking, Vol. I, Addison-Wesley, Reading, Massachusetts, 1960.
4. Turkdogan, E.T., "Theroretical Aspects of Sulfide Formation in Steel," Sulfide Inclusions in Steel, Symposium, Nov. 1974, American Society for Metals, Metals Park, Ohio, 1975, pp. 1-22.
5. Wilson, W.G., Kay, D.A.R., and Vahed, A., "The Use of Thermodynamics and Phase Equilibria to Predict the Behavior of the Rare Earth Elements in Steel," Journal of Metals, Vol. 26, No. 5, May 1974, pp. 14-23.
6. Vahed, A. and Kay, D.A.R., "Thermodynamics of Rare Earths in Steelmaking," Metallurgical Transactions, Vol. 7B, 1976, pp. 375-383.
7. McLean, A. and Lu, W-K., "The Thermodynamic Behaviour of Rare Earth Elements in Molten Steel," Metals and Materials, October 1974, pp. 452-457.